GUIDE TO THE SUN

GUIDE TO
the Sun

Kenneth J. H. Phillips
Rutherford Appleton Laboratory, Oxfordshire, UK

CAMBRIDGE
UNIVERSITY PRESS

Published by the Press Syndicate of the University of Cambridge
The Pitt Building, Trumpington Street, Cambridge CB2 1RP
40 West 20th Street, New York, NY 10011–4211, USA
10 Stamford Road, Oakleigh, Melbourne 3166, Australia

First published 1992
First paperback edition 1995

Printed in Great Britain at the University Press, Cambridge

A catalogue record for this book is available from the British Library

Library of Congress cataloguing in publication data
   Phillips, Kenneth J. H.
      Guide to the sun / Kenneth J. H. Phillips.
         p.   cm.
      Includes bibliographical references and index.
      ISBN 0–521–39483–X (hardback)
      1. Sun.   I. Title.
      QB521.P45   1992
   532.7–dc20   91–35910 CIP

ISBN 0 521 39483 X hardback
ISBN 0 521 39788 X paperback

Front cover illustration: see figure 3.8 (Courtesy F. Diego & D. Walker)
Back cover illustrations: (centre) see figure 5.11 (Courtesy L. Golub, IBM Research
                          and Smithsonian Astrophysical Observatory)
                          (bottom) see figure 2.8 (Courtesy J. W. Leibacher)

Contents

Preface

A book about the sun could be written from several widely different points of view. One could, for example, describe its worship as a god over the ages, or its indispensability for all life-forms on the earth, or the conversion of its energy for purposes useful to everyday life.

In these pages, I have described the sun largely from the point of view of astronomy and physics, giving an account of our present knowledge gained from observations and theory (Chapters 2–8). I have included more important recent developments in solar science, in particular the exciting findings from spacecraft and modern earth-based instrumentation, which include the probing of the sun's interior by helioseismology and some understanding of the sudden releases of energy in the solar atmosphere known as flares. Our understanding of the sun, imperfect though it may still be, has not been arrived at without much careful observation and thought over a long period of time, and I have therefore thought it fitting to place solar astronomy in an historical context by outlining the chief developments from the earliest recorded times (Chapter 1). The subject of solar energy is a very current one, and I have felt that the book would be incomplete without a chapter (Chapter 9) dealing with how solar energy is or can be harnessed. Finally, I have described the rather specialized way in which the sun is observed, using telescopes and auxiliary equipment and spacecraft instruments. Chapter 10 dealing with this aspect also has a section on suitable methods for observing the sun for amateur astronomers.

The range of subject matter, then, is very broad, and therefore to keep the book to a moderate size I have had to leave out very detailed considerations. It has always been my aim, though, to make clear the essence of any particular subject in a form that is intelligible to anyone with an elementary knowledge of science, in particular physics or astronomy. The level aimed at is broadly the same as that of popular science magazines such as *Scientific American*, *New*

Scientist, *Sky and Telescope* or *Astronomy Now*. In particular, the use of mathematics has been kept to the barest minimum; I have instead relied on illustrations to make clear what might be new concepts. The long history of solar science has meant that it has acquired a jargon that can be puzzling to the uninitiated. I have attempted to do without this jargon as much as possible, but have included a Glossary defining those terms whose use was practically indispensable in the text. There is a bibliography for each chapter which should assist in directing further reading.

In summary, the book is intended for a broad readership, ranging from those with science interests at high-school level to post-graduates and professionals who want to enlarge their interests.

Warning

It is extremely easy to do irreparable damage to one's eyesight by looking at the sun and more especially to try to observe it through even a small telescope or pair of binoculars. The only safe way for observing sunspots with a telescope is by projecting an image of the sun on to a card; the method is described in full in Chapter 10 (§10.4).

Units and notation

To describe the sun as an astronomical object inevitably brings in distances, masses, time-scales, etc., so vast that they transcend ordinary human experience. To specify these requires units and a way of writing numbers that may not be completely familiar to some readers, and the following describes the scheme I have followed in this book.

I have used the so-called International System of Units (Système International, SI) for nearly all quantities. Imperial or British-style units such as miles and pounds, as well as those in the c.g.s. (centimetre, gram, second) system (still extensively used in the astronomical community) are therefore avoided as far as possible. The principal units of measurement and their standard abbreviations are specified in Table 1, with their equivalents in the c.g.s. and Imperial systems.

Other conventions are also followed here. Large, astronomical numbers (or those that are correspondingly small, i.e. a small fraction of one) are expressed using powers of ten: thus one million (1 000 000) is written 10^6, or a number like 5 000 000 as 5×10^6. Small numbers like 0.000 005 are similarly written 5×10^{-6}. Prefixes to SI units are also used to express multiples of a thousand: thus one kilometre is 1000 metres. The commonest such prefixes are: giga- (G: a thousand million or 10^9); mega- (M: a million or 10^6 times); kilo- (k: 10^3); milli- (m: 10^{-3}); micro- (μ: 10^{-6}); nano (n: 10^{-9}).

Angles are measured in the usual way (1 right angle$=90°$; $1°=60'$,

Table 1. *Principal SI units and their conversion to other units*

Quantity	SI unit	Conversion to corresponding	
		c.g.s unit	Imperial unit
Length	metre (m)	100 centimetres (cm)	39.37 inches (in)
Mass	kilogram (kg)	1000 grams (g)	2.205 pounds (lb)
Time	second (s)		
Temperature	kelvin (K)		1 K=1 deg C (Celsius)=1.8 deg F (Fahrenheit)
Velocity	metre per second (m/s)	100 cm/s	39.37 in/s
Acceleration	m/s^2	$100\,cm/s^2$	
Force (mass× acceleration)	newton (N) $(m\,kg/s^2)$	100 000 dynes	0.22 lb weight
Pressure (force/area)	pascal (Pa) (N/m^2)	$10\,dyne/cm^2$	
Energy or work	joule (J) (N m)	10 000 000 erg	
Frequency	hertz (Hz) and (cycles/second)		
Power	watt (W) (J/s)		
Magnetic flux density	tesla (T)	10 000 gauss (G)	

$1'=60''$); a second of arc is written as 1 arc second (or 1 arc sec). Note that 1 radian$=57.30°$ or 206 265 arc secs.

A few non-SI units are used. These include some used in astronomy but not generally in other sciences, in particular some distance units: the astronomical unit (AU) is the mean distance of the sun from the earth, equal to 149.6×10^6 km (93.0×10^6 miles); the light-year is the distance a ray of light travels in one year, often used to express stellar distances, equal to 9.46×10^{12} km; and the parsec, similarly used, is the distance at which an astronomical unit would subtend an angle of 1 arc sec, equal to 3.09×10^{13} km, or 3.26 light-years. Photon energies are sometimes expressed in electron volts (eV): $1\,eV=1.6\times10^{-19}$ J.

Two mathematical constants often used in the text are $\pi=3.1416$ and $e=2.7183$, the base of natural logarithms. Appendix 1 gives values of some other astronomical and physical constants referred to, while statistics of the sun are listed in Chapter 8 (§8.1).

Acknowledgements

I thank all those who have allowed me permission to use illustrations and other published material, and particularly the following people for their help and advice for providing illustrations:

Staff at the Rutherford Appleton Laboratory, in particular Dr Paul Williams (Director), Professor John Harries (Head of Space Science Dept) and Professor Michael Cruise (Head of Astrophysics Division), for permission to use RAL material; the staff of the RAL Library; Mr Reginald Jones; and Drs Gordon Bromage, Trevor Edwards, Leslie Gray, David Hall, David Infield and David Stickland.

Professor Cornelius de Jager, Dr Brian Dennis, Dr Joseph Gurman, Dr Hugh Hudson, Dr Arthur Hundhausen, Dr Einar Tandberg-Hanssen, Dr Thomas Vestrand and Dr Richard Willson and others involved with instruments on the *Solar Maximum Mission* spacecraft.

Dr Christopher Walker, British Museum, London.

Dr Jon Darius, Science Museum, London.

Mr Peter Hingley, Royal Astronomical Society, London.

Drs William Stuart and Edward Harris, British Geological Survey, Edinburgh.

Drs Francisco Diego and David Walker, University College London.

Dr Alan Fitzsimmons, Queen's University, Belfast.

Mrs Janet Sandland, Royal Observatory, Edinburgh.

Professor George Isaak, Birmingham University.

Dr Kevin K. C. Yau, University of Durham.

Dr David Clark, Science and Engineering Research Council, Swindon.

Dr Munetoshi Tokumaru, CRL, Ibaraki, Japan.

Professor Yoshi Ogawara, Institute of Space and Astronautical Science, Tokyo, Japan.

Dr Loren Acton, Lockheed Palo Alto Research Laboratories, California, USA.

Professor Dennis Pardee, Oriental Institute, University of Chicago, USA.

Dr John Leibacher (Director), Dr Jack L. Harvey and Ms Ann Barringer, National Solar Observatory/Kitt Peak, Tucson, Arizona, USA.

Mr Lou B. Gilliam and Mr D. Todd Brown, National Solar Observatory/Sacramento Peak, Sunspot, New Mexico, USA.

Drs Neil Sheeley, Kenneth Widing and Richard Tousey, US Naval Research Laboratory, Washington, DC, USA.

Ms Julie Clausen, Sandia National Laboratories, US Department of Energy, Albuquerque, New Mexico, USA.

Drs Steven S. Vogt and A. Hatzes, Lick Observatory, California, USA.

Professor Ronald J. Angione, Astronomy Dept, San Diego State University, California, USA.

Dr Patrick S. McIntosh, Space Environment Laboratory, National Oceanic and Atmospheric Administration, Boulder, Colorado, USA.

Dr Richard Canfield, Institute for Astronomy, University of Hawaii, USA.

Dr Harold Zirin, Big Bear Observatory, Pasadena, California, USA.

Dr Leon Golub, Smithsonian Astrophysical Observatory, Cambridge, Massachusetts, USA.

Dr John McKinnon, National Geophysical Data Center, Boulder, Colorado, USA.

Dr David Speich, Space Environment Laboratory, National Oceanic and Atmospheric Administration, Boulder, Colorado, USA.

Dr Minze Stuiver, University of Washington, Seattle, USA.

Kenneth J. H. Phillips

Thatcham, Berkshire, England

Chapter 1

The history of solar observation: from sun worship to the space age

1.1 Early ideas about the sun

The sun must have been a source of wonder for the earliest people who watched its constant rising and setting and passage across the sky, and very likely it was worshipped as a god by most ancient communities. Many pre-historic monuments show evidence from their alignments that the sun was an important object of study. One of the most famous is Stonehenge, in Wilt-shire, England, built between 2800 and 1500 BC. From the centre of the henge a large stone known as the Heelstone, its top level with the horizon, marks the point of sunrise at the summer solstice (21 June).

Sun worship was widespread among ancient eastern and Mediterranean peoples. Thus, the sun for the ancient Egyptians was the god Re, a fiery disc conveyed in a boat across the sky during the day and along a great river surrounding the earth during the night to reach the east by next morning, contending on the way with obstacles; during the day, a serpent would sometimes attack the sun for a time, causing a solar eclipse. For the ancient Babylonians, the sun was a living being which moved against the solid vault of the sky, emerging each morning through a door in the east and disappearing in the west in the evening. Figure 1.1 shows a stone tablet (*ca.* 900 BC) depicting the sun-god Shamash in his temple with a solar disc in front of him, with the Babylonian king being led into his presence. The author of the first chapter of Genesis tells how the sun was made by God, who placed it in the vault of heaven 'to separate light from darkness'. The ancient Israelites are warned that, on entering Canaan, they should not be tempted to worship the gods of the Canaanite peoples, including the sun and other heavenly bodies. This seems to account for the surprisingly small number of astronomical references in the Old Testament. In the religion of the ancient Hindus, the sun-god was one of the principal triad of deities, called by them Surya in their

Vedas or Books of Divine Knowledge. Early Greek lore, as expressed for example by Homer (*ca.* 700 BC), was similar to the Egyptians', with the sun-god Helios rising each morning from the lake of the sun, a gulf in the huge river Oceanos surrounding the earth (Fig. 1.2). From about the fifth century BC, the Greeks associated the sun with their god Phoebus Apollo.

The earliest significant attempts to understand the sun and indeed the universe, including the earth, were made by the Greeks from the sixth century BC onwards. Despite the prevalence of religious beliefs concerning the sun among the ancient Greeks, several philosophers began to speculate freely about its physical nature. Among the earliest ideas was that of Anaximenes (sixth century BC), who thought the sun a flat body supported in the sky by air. Xenophanes (sixth century BC) considered the sun and stars to be fiery clouds, lit up by their own motion in the cosmos; the sun was supposed to be renewed each morning, the stars each evening. The fall of a meteorite in daylight led the Athenian Anaxagoras in the mid-fifth century BC to conclude that the sun was composed of a mass of red-hot iron with a size no larger than the Peloponnessus, roughly 160 km.

Anaxagoras' value for the sun's diameter implied a comparatively small distance to the sun: about 30 000 km, compared with the modern value of

Fig. 1.1 Stone tablet, *ca.* 520 BC, showing the Babylonian sun-god Shamash in a shrine with a solar disc in front of him. From Sippar, Iraq. (Reproduced by courtesy of the Trustees of the British Museum)

Fig. 1.2 Athenian wine bowl, *ca.* 430 BC, depicting a dawn scene with the sun-god in a chariot pulled by winged horses out of the ocean, with stars shown as boys diving and disappearing into the water. (Reproduced by courtesy of the Trustees of the British Museum)

149 600 000 km. An attempt to make an actual measurement of the solar distance, or at least to find the ratio of the moon's distance to that of the sun's, was made by the astronomer Aristarchus of Samos (*ca.* first half of third century BC). The method consisted of measuring the angle between the moon, an observer on earth and the sun at the time when the moon appeared precisely half-illuminated – an angle that is almost indistinguishable from a right angle. Although his method was in principle perfectly valid, the uncertainty of timing the instant of the moon's being half-illuminated led to a considerable underestimation of the sun's distance, just 19 times the lunar distance (the modern value is about 400 times the lunar distance). Unfortunately for the progress of astronomy, a solar distance similar to Aristarchus' was adopted by Claudius Ptolemy of Alexandria in his astronomical compendium known as the *Almagest* (AD 140), and was accepted for the next 1500 years!

A measurement of the solar and of the lunar distance separately, involving solar and lunar eclipses, was attempted by Hipparchus (*ca.* 140 BC). He gave a surprisingly accurate result for the distance to the moon: 59.1 times the mean radius of the earth (the modern value is 60.3), but the solar distance was too great for him to determine. He used Aristarchus' value that the sun was 19 times more distant than the moon to deduce that the sun's diameter was seven times that of the earth.

Though much ancient Greek astronomy was purely speculative, there began in time to be an accurate understanding of some aspects of the cosmos, for example the shape of the earth, and of some astronomical phenomena such as eclipses. The idea of a spherical earth, replacing the earlier one of a flat earth, had been put forward by Parmenides and the school of Pythagoras as early as *ca.* 500 BC, and generally held by the time of Plato and Aristotle in the fourth century BC. Empedocles (*ca.* 450 BC) knew that a solar eclipse was caused by the moon passing over the sun, and Anaxagoras correctly explained solar and lunar eclipses and the phases of the moon. Helicon, a follower of the more well-known astronomer Eudoxus, even predicted a solar eclipse in 361 BC, illustrating that he knew about the complex nature of the moon's path in the sky. Herakleides of Pontus (*ca.* fourth century BC) put forward the advanced notion that, despite the appearance of the heavens rotating about the earth, it was in fact the spherical earth that rotated on its axis.

For most Greek astronomers, as indeed for astronomers in the west and east right up to the sixteenth century, the earth, rotating or not, was at the centre of the universe, with the sun, moon, planets (five were visible to the unaided eye: Mercury, Venus, Mars, Jupiter and Saturn) and stars all revolving about the earth. Opinions differed somewhat as to the relative disposition of the planets – whether Venus and Mercury, for example, were closer to or farther from the earth than the sun – but it was generally taken that the moon was the closest body, followed by the sun (or Venus and Mercury); the slower-moving planets Jupiter and Saturn were more distant, and the stars most distant of all the visible heavenly bodies. The irregularities of planetary motions which had been noted were sufficiently explained by the complex system of epicycles, expounded by Ptolemy in the *Almagest*. In this, each planet moved round the earth not simply in a circular orbit but rather performing a yearly motion in a small circle – the epicycle – the centre of which moved round the earth.

There were some dissenters among the Greeks who did not hold the prevailing opinion of the geocentric universe, among whom was Aristarchus. He appears to have suggested, some 1700 years before the idea was again seriously raised, that it was the earth that revolved about the sun, not the sun about the earth. The Pythagorean school (*ca.* 500 BC) also had a scheme in which the earth and other planets, including the sun, revolved about a 'central fire'; this was invisible to inhabitants of the Mediterranean area as their part of the earth was supposed to be always directed away from it. Despite these unorthodox ideas, the world of astronomy, both in the west and the near-east (where a considerable interest had been kindled by the Arabs in the seventh century AD), was content right up to the Middle Ages to place the earth at the centre of the universe and the heavens revolving round it. A good illustration of this is to be found in the early fourteenth-century *Divine Comedy* of Dante, particularly the third book (*Paradise*), in which the cosmography of Ptolemy's *Almagest* is faithfully adhered to.

The Ptolemaic system was finally challenged by Nicolaus Copernicus, the Polish astronomer, who in his book *De Revolutionibus Orbium Coelestius* (1543) proposed the sun as the centre of the planetary system. The earth with its companion the moon was held to be in an orbit about the sun between the orbits of Venus and Mars, with Mercury closest to the sun, Jupiter and Saturn the most distant. This is the picture of the solar system we are familiar with today, though Uranus, Neptune and Pluto have since been discovered as part of the sun's family, together with the asteroids and comets. Copernicus envisaged a 'sphere of stars' that was much more distant from the sun than the earth since they showed no motion as the earth went round the sun. He retained circular orbits for the planets with the Ptolemaic epicycles in his new scheme to explain residual irregularities in planetary movements.

The precise nature of the planetary orbits was discovered by Johann Kepler, using the extensive and accurate observations of Mars made by the Danish astronomer Tycho Brahe with whom he had worked. Kepler summarized his findings in his three laws of planetary motion (1609). One of these describes the form of a planetary orbit as a conic section known as an ellipse, produced when a cone is intersected by a plane making a smaller angle to the base than the cone sides. Thus the Ptolemaic epicycles were ultimately buried.

According to Kepler, the planets were impelled to move round the sun by a force which was tangential to the planet's orbit rather than directed at the sun, and which had its origin in the rotation of the sun on its axis. In addition, there was supposed to be alternately an attraction and repulsion from the sun as the planet's north and south magnetic poles were directed towards the sun: hence arose the deviation of a planet's orbit from a circle to an ellipse, with the planet sometimes nearer to, sometimes farther from, the sun. Kepler's rather complicated explanation for how the sun kept its attendant planets in their orbits was superseded half a century later by the work of Sir Isaac Newton, who found that Kepler's laws had the consequence that the force controlling each planet's motion round the sun was directed towards the sun, falling off as the inverse square of the distance to the sun. This force of gravitation was universal in character, since it was the same as that which causes objects near the surface of the earth to fall. The mechanics of the solar system could thus be understood in terms of something with an everyday familiarity.

The gross underestimates that Aristarchus and Hipparchus had made of the sun's distance and diameter were still used by astronomers up till the time of Kepler. Though the absolute scale of the solar system was poorly known, the relative distances of all the planets could be determined with great accuracy: Kepler's third law stated how the orbital period of a planet (which was easily measured) was related to its mean distance from the sun, so that the distances of planets from the sun could be given in terms of the earth's distance. A means of determining the earth's distance was provided by the very rare occasions when the planet Venus passed over the sun's disc – a 'transit' of Venus. A young Lancashire curate and amateur astronomer, Jeremiah Horrocks, was the first to see such an event, in 1639, which he had predicted. The next transits of Venus were due to occur in 1761 and 1769. In 1716, Edmund Halley, famous for the comet that bears his name, suggested that Venus would be sufficiently close to the earth during transit that observers at different locations would see Venus take different tracks across the sun's disc. Comparison of the observations would give the distance of Venus to the earth and, since the relative distances of the two planets from the sun were known, the distance of the earth to the sun could be found. The 1761 and 1769 transits occurred after Halley's death, but his suggestion was put into effect, especially for the 1769 occasion for which Captain James Cook was commissioned to make his first voyage. From his observations in New Zealand and those made in Europe, a value was eventually (1824) obtained for the earth's distance to the sun – 153 000 000 km, or about 2% more than the present estimate.

1.2   Sunspot observations

Aristotle had maintained that the sun was perfect and blemish-free, but solar observers long before the invention of the telescope around 1608 had discerned small dark features within the face of the sun. Many of their records give what are now considered accurate accounts of genuine 'naked-eye' sunspots on the solar disc, some fancifully described as objects familiar to the culture of the time, though some may be merely the sighting of, for example, birds or small terrestrial clouds. The earliest reasonably certain reference to a sunspot seems to have been due to Aristotle's own pupil Theophrastus of Athens in the mid-fourth century BC, to whom is also credited a sighting of the aurora.

There has been much interest in large numbers of oriental records of what appear to be naked-eye sunspots. A 1988 catalogue (by K. K. C. Yau and F. R. Stephenson) lists 157 such records before the invention of the telescope. They are mostly from China but also from Korea, Vietnam and Japan, and occur in dynastic histories, chronicles and local gazettes. Some are from very early on, but the earliest definite sunspot sighting is Chinese, dated 165 BC. The descriptions in these accounts suggest that the sun when observed was dimmed by fog or dust in the atmosphere. There are long periods when there were no sunspot records, but it is uncertain whether these were when sunspots were really scarce or were simply times when there was a lack of scientific interest due, for example, to political unrest.

There are records from other areas of the world also – for example, some Russian chronicles record sunspot sightings when the sun was dimmed by a forest fire in 1371 – but those of western Europe up to the seventeenth century are generally lacking, owing apparently to the prevailing Aristotelian view, vigorously upheld by the Church, of the immaculate sun. One exception was the observation of a spot by Kepler in 1607, which he attributed at the time to Mercury crossing the solar disc though later he realized it must have been a sunspot.

The invention of the telescope offered a completely new means of exploring the sky. Galileo is the most celebrated of the early astronomers to use the telescope. He was already well known for his experiments on the pendulum and his theory of motion, in which he disproved Aristotle's assertion that different weights fall at different speeds. When a professor at the University of Padua, he made his own telescope with which he observed the moon, planets, stars and the sun. He found that the solar surface was marked by dark spots which appeared and disappeared, with lifetimes that were variable but on average a few days (Fig. 1.3). The more persistent were seen to cross the sun's disc from east to west in about two weeks, which he explained by the rotation of the sun on an axis. Although he first saw sunspots in 1610, Galileo evidently did not feel confident enough to announce his discovery for another two years; he then wrote a discourse on their nature to his former pupil the Grand Duke of Florence in which he says: 'Having made repeated observations I am at last convinced that the spots are objects close to the surface of the solar globe, where they are continually being produced and then dissolved, some quickly and some slowly; also that they are carried round the sun by its rotation, which is completed in a period of about one lunar month.'

At around the time of this discourse, Galileo became more interested in other astronomical studies and made no further systematic solar observations. A more extensive series of sunspot observations was made by his contemporary Christoph Scheiner, a Jesuit priest from Swabia in south Germany. His observations cover the period 1611–27, and were published in an impressively long volume entitled *Rosa Ursina sive Sol* in 1630. This was dedicated to the Duke of Orsini in Rome, whose badge was the rose and the bear mentioned in the Latin title of the book. The illustrations in this book show Scheiner and his assistant recording sunspots from an image projected by a telescope on to a screen (Figs. 1.4, 1.5). This is still a common way to observe the sun with a small telescope and a safe one, unlike direct viewing which Galileo had apparently tried at great risk to his eyesight.

Scheiner and Galileo had both noted that the paths of spots as the sun rotated were in general not straight lines, but rather somewhat curved, which

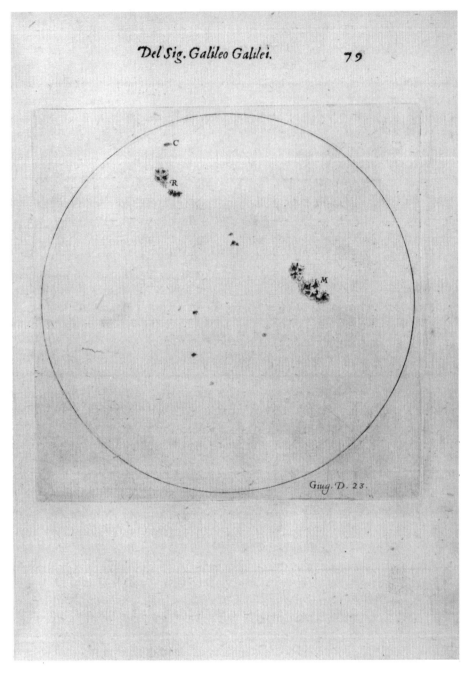

Fig. 1.3 One of Galileo's sunspot drawings from the early seventeenth century,
from his manuscript *Delle Macchie Solari*. (Courtesy Royal Astronomical
Society)

ROSA VRSINA

SIVE

SOL

EX ADMIRANDO FACVLARVM
& Macularum fuarum Phœnomeno VARIVS,

NECNON

Circa centrum fuum & axem fixum ab occafu in ortum annua,
circaq.alium axem mobilem ab ortu in occafum conuerfione
quafi menftrua, fuper polos proprios, Libris quatuor
MOBILIS oftenfus,

A

CHRISTOPHORO SCHEINER
GERMANO SVEVO, E SOCIETATE IESV.

AD PAVLVM IORDANVM II.
VRSINVM BRACCIANI DVCEM.

BRACCIANI,
Apud Andream Phæum Typographum Ducalem.

Impreſſio cœpta Anno 1626. finita vero 1630. Id.Iunij. *Cum licentia Superiorum.*

Fig. 1.4 Title page of Christoph Scheiner's volume on observing sunspots *Rosa Ursina sive Sol* (1630). (Courtesy Royal Astronomical Society)

Fig. 1.5 A page from Scheiner's *Rosa Ursina* showing the projection method used
by him for recording sunspots. (Courtesy Royal Astronomical Society)

they both correctly explained by the fact that the solar equator is inclined by a
small angle to the plane of the earth's orbit round the sun. For one half of the
year, the solar north pole is tilted towards the earth, and for the other half the
south pole. They also both showed that sunspots were not distributed over all
the sun but were confined to a band extending to latitudes roughly 30° north
and south. When near the edge (or *limb*) of the sun, the spots were often
associated with small bright patches which became known as faculae (Latin
for 'little torches'). The two astronomers at first differed in their opinions of
the nature of the spots. Scheiner's observations that the sun was not after all
spotless was unpalatable to his fellow Jesuits, one of whom refused to believe
that there were spots since he had not found mention of them in Aristotle's
works. Scheiner himself at first thought that the spots were not strictly solar in
origin, but attributable to small planets revolving near the solar surface. This
idea was advanced in documents, written under a false name, to a friend of
Galileo's. Galileo replied asserting his own view that sunspots were on the
solar surface, this being proved by the fact that their shapes changed when
near the east or west limbs, becoming elongated or foreshortened. Galileo
also stated that sunspots only appeared dark by contrast with the brilliant solar
surface, but were in fact as bright as the brightest areas on the moon. His
drawings, like Scheiner's, showed that larger sunspots had a darker inner
region, the *umbra*, and a lighter outer part, the *penumbra* (this can be seen in
Fig. 1.3). The intensity of the umbra is in fact as much as a quarter of that of
the unspotted sun. It was in Galileo's reply to Scheiner that he first mentioned his support of the Copernican system of the sun-centred universe in
place of the Ptolemaic, an opinion which led to a collison with the authorities

of the Roman Church and to his trial before the Inquisition at Rome in 1633 and finally to his being forced to recant.

For some 230 years succeeding the first telescopic sightings of sunspots by Galileo, Scheiner and other observers, very little was added to the knowledge of their nature. Alexander Wilson, a professor at Glasgow University, had deduced in 1769 that spots were funnel-shaped depressions in the solar surface from the fact that the umbrae of spots were often displaced within the spot when near the limb. He deduced that the umbra was actually the solid surface of the sun proper (a similar idea had been advanced as far back as the fifteenth century by Nicolaus de Cusa). Sir William Herschel, writing in 1794 about his own observations of umbral displacements, went further in stating that it was entirely possible that living beings inhabited the solid interior of the sun, adapted to the extreme circumstances there! Such was Sir William's high repute that this idea remained current for a large part of the nineteenth century.

Some notable advances were then made in the early nineteenth century, when many solar observers were active. Heinrich Schwabe, an apothecary in Dessau, started recording sunspots in 1826, and by 1843 had concluded that their numbers had a periodicity of about ten years. His findings were published in a German astronomical journal but they attracted little attention until mentioned in a publication by the explorer Alexander von Humboldt a few years later. Schwabe's discovery of what has become known as the 'eleven-year sunspot cycle' is perhaps the most well-known fact about solar activity, and its cause is still being widely discussed. An important concept that Schwabe had introduced was that of sunspot *groups*, for spots are not randomly scattered over the solar surface but rather are confined to groups which, when simple, may consist of only a pair of spots or, when complex, may contain very many spots.

Rudolf Wolf, an astronomer at the University of Berne and later at Zurich Observatory, became greatly interested in Schwabe's discovery and in 1848 set out to investigate the periodicity firstly by standardizing observations that were then being made of sunspot numbers and secondly by an historical reconstruction of such numbers back to the earliest telescopic times in the seventeenth century. He himself observed the sun as frequently as possible for some years using a small telescope, and sought collaborators to do the same to fill in gaps due to poor weather conditions at his own location. Wolf defined the number of spots (known as the *relative sunspot number*) both in terms of the number of sunspot groups and the total number of spots, with a correcting factor representing the estimated efficiency of the observer and telescope used, so that relative sunspot numbers could be compared for any era regardless of the observer. The reconstruction of relative sunspot numbers from historical records had decreasing reliability the further back in time Wolf searched. Almost complete daily numbers could be obtained back to

1818, but only monthly means to 1749 and annual means to 1700. No more than the approximate years of maximum and minimum spot numbers could be determined from the seventeenth-century accounts. Wolf's data from about 1700 onwards clearly confirmed what Schwabe had concluded from his sample of numbers covering only 17 years: there was an almost regular cycle with peaks separated by a period averaging 11 years but which varied from 8 to 14 years. At the minimum of each cycle, spots would usually not disappear altogether, though there would often be several days when there were no spots on the visible hemisphere of the sun. Figure 6.5 (Chapter 6) shows a compilation of smoothed monthly sunspot numbers from 1749 to 1988.

Another remarkable solar observer at about this time was Richard Carrington, an amateur astronomer who had set up his private observatory in Redhill, Surrey, with an 11-cm refracting telescope. He made extensive observations between 1853 and 1861, making drawings of the sun and measuring sunspot latitudes and longitudes by a timing method. The precision of his measurements led to a much better determination of the direction of the sun's rotation axis against the fixed stars, which is still used today for calculating the orientation of the axis as seen from the earth. The notion of Carrington longitude of the sun (the solar longitude system has no convenient zero point as on the earth with the Greenwich meridian) also remains in use, the period of 25.38 days being adopted by Carrington as the length of time the sun rotates on its axis at the equator. This period is actually only an approximation, as Carrington noticed that sunspot rotation periods depended on latitude north or south: spots near the equator rotated in about 25 days, but those at higher latitudes up to about 27 days. Now, the earth revolves round the sun in the same direction as the sun rotates on its axis, and so as seen from the earth sunspots appear to take rather longer to complete a rotation: the period of 25.38 days, for instance, which is known as the sidereal period (i.e. with respect to the fixed stars) becomes 27.28 days with respect to the earth, the so-called synodic period.

Carrington started his observations when sunspot numbers were decreasing from a maximum in 1848. Near sunspot minimum, which occurred early in 1856, he noticed that most spots were confined to equatorial regions. Shortly after, two new zones of spot activity abruptly occurred at much higher latitudes north and south. All the spots in these new zones had latitudes above 20°, some as high as 40°. Spots continued to appear in the equatorial regions, but eventually petered out completely, while the spots in the new zones now appeared at slightly lower latitudes. This is another notable property of the eleven-year cycle: the migration of spots to lower latitudes as a cycle progresses. This discovery was extended by the German solar astronomer Gustav Spörer, who is credited by having his name given to the law expressing this latitude drift of sunspots. A striking illustration of Spörer's law was given by E. W. Maunder, superintendent of the solar department at the Royal

Observatory, Greenwich, who plotted out the distribution of spot latitudes over a complete solar rotation for the period 1877–1902. Figure 1.6 (upper part) is a more complete plot covering the entire period when sunspot measurements were made at the Royal Observatory, and shows the characteristic 'butterfly' shape that is obtained for each cycle.

To present-day astronomers, the eleven-year sunspot cycle is so familiar that it is difficult to conceive that, at some time in the past, it might not have been operating. Some 27 maxima have now been recorded between 1700 and 1989, with cycle lengths not very different from 11 years. The possible presence of an eleven-year cyclical pattern has been searched for in the pre-telescopic oriental sunspot record. Unfortunately, with less than 200 accounts of naked-eye spots stretching over almost 1800 years, it is not really possible to do this with any certainty. There is in any case some tendency for these sightings to be commonest on the first day of the lunar month, which may have been because special searches were then made for eclipses, which had great astrological significance. Also, a recent search for naked-eye spots shows them to be far more frequent than had been supposed, suggesting that the oriental observers were either less methodical in recording their sightings

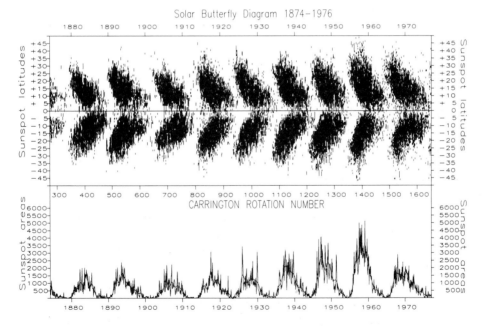

Fig. 1.6 A practically unbroken record of sunspot measurements was made at the Royal Observatory, Greenwich, for over a century, starting in 1874. These graphs summarize results for spot latitudes – the butterfly diagram, showing the migration of spots towards the equator over the course of an eleven-year cycle – and spot areas (unit=millionths of a solar hemisphere). (Courtesy Royal Greenwich Observatory and D. J. Stickland)

or that naked-eye spots were so common that only the largest spots were worthy of mention.

Not long after the invention of the telescope there was a period of over 50 years when sunspot records were surprisingly sparse. The Polish astronomer J. Hevelius had noted the presence of numerous spots in 1644, but very few records exist after this until the early eighteenth century. Although it cannot be certain that a continuous watch was being kept on the sun by the various observers active at the time, it would seem strange that little attention was being paid to the sun at a time when there was a wide interest in other astronomical phenomena and many discoveries were being made. One or two large sunspots were noted, such as one in 1671 seen by G. Cassini and one in 1684, seen by John Flamsteed, the first Astronomer Royal. But very few spots were seen again until 1715. Spörer used historical records to conclude that the comparative lack of spot observations between 1645 and 1715 was due to a real fall in spot activity, but his paper was not widely known about until Maunder drew attention to it in the 1920s. With further evidence to support this conclusion, the reality of the 'Maunder minimum' is now widely accepted. At any rate, nothing like it has occurred since, and sunspot numbers now faithfully obey the eleven-year periodicity.

1.3   The aurora

Solar activity comprising sunspots and other phenomena is strongly related to disturbances in the earth's magnetic field and gives rise to various effects in the earth's upper atmosphere. The *aurora*, better known as the northern or southern lights, is a visible manifestation of the atmospheric effects, appearing in the night sky as luminous arcs, rapidly changing rays and other structures, sometimes coloured red or green. Strong solar activity is often associated with spectacular auroral displays over wide areas of the earth, though more usually they are only seen in near-polar regions. Thus, auroral displays have the same eleven-year periodicity that sunspot numbers have.

Aurorae are such striking spectacles that it is not surprising that writers from very early times have commented on them. These historical accounts are of great interest since aurorae, being more easily seen than naked-eye sunspots, give a reliable indication of the level of solar activity at the time. There are, for example, many descriptions of aurorae in writings of ancient Greece and Rome, even though the frequency of aurorae from such low latitudes is generally no more than one display in several years. Thus, Plutarch mentions a record of an aurora seen from Greece in the fifth century BC, and Aristotle comments in his *Meteorologica* (330 BC) on 'condensations' and 'combustions' of the upper air, giving rise to the red coloration familiar in the brightest aurorae. There is almost certainly mention of aurorae in the writings of Livy and Pliny the Elder. Seneca gives an account of an occasion when soldiers were despatched to the Roman colony at Ostia, an auroral display being mistaken for a fire. As with naked-eye sunspots, oriental records, particularly from China but also Japan and Korea, include sightings of aurorae, a recent catalogue listing some 258 probable auroral references from 687 BC to AD 1600. There are also descriptions of aurorae in medieval and later European literature. They were often regarded as religious portents, but the accounts often accurately and unmistakedly describe the development of a large display. Gregory of Tours, writing in the late sixth century from Trier in southern France, relates this occasion: 'We beheld for two nights signs in the heavens, namely rays in the north so clear and splendid, that none such were ever seen before; on both sides were blood-red clouds ... In the middle of the heavens was a gleaming cloud to which these rays gathered themselves.' The red clouds evidently refer to the coloured arcs that are usually seen, while the 'gleaming cloud' is probably the phenomenon of converging rays known as an auroral corona. The Elizabethan historian John Stow tells of what seemed to be fire and smoke emitted from a black cloud in the north on a November night in 1574. The 'black cloud' in this and many other early references is the section of clear night sky between an auroral arc and the horizon which is dark by contrast against the auroral form.

There is a 1621 account by the French astronomer Pierre Gassendi, in which the name aurora borealis (for northern lights) is first used. After that, there are few references until a great display in March 1716. Reports were detailed in articles published by the Royal Society in London and the Académie des Sciènces in Paris. Edmund Halley collated the various descriptions for the Royal Society in an article entitled *An Account of the late surprizing Appearance of the Lights seen in the Air, with an attempt to explain the Principal Phænomena thereof.* The aurora started with a 'dusky cloud' over the northeast, tinged with reddish-yellow, out of which occurred moving rays, stretching up almost to the overhead point. There was eventually a convergence of the rays to form a corona, as in the account of Gregory of Tours. The great foresight of Halley is revealed by his suggestions that the height of auroral features could be estimated if their positions against the stars were noted by observers in different locations (this was eventually accomplished) and that the earth's magnetism was in some way responsible for the occurrence (as was later proved). Though Halley had been a keen observer for many years, he remarks that this was the first time he had seen this phenomenon; of all the various types of 'meteors', i.e. astronomical and meteorological sky phenomena, this was the only one he had not witnessed till that time, and almost despaired of doing so.

The small number of auroral sightings between the displays recorded by Gassendi and Halley is significant in view of the apparently reduced sunspot activity in these years. The reality of this reduced auroral frequency was investigated as early as 1733 by the French astronomer de Mairan, who cites a catalogue of aurorae by the Swede Anders Celsius (inventor of the centigrade thermometer) as evidence. It seems to confirm the reality of the Maunder minimum.

1.4   Solar spectroscopy

The story of solar spectroscopy begins with the experiments of Sir Isaac Newton in 1666, by which he found that white light was actually a combination of all the colours of the rainbow, red, orange, yellow, and so on to violet. He demonstrated this by allowing sunlight (an approximation to white light) to pass into a darkened room through a slot in a window shutter. A triangular glass prism was introduced into the light path, so that, instead of a white spot being produced on the wall opposite the window, there was a band – a *spectrum* – containing all the familiar rainbow colours. He found that the colours could be recombined into white light by introducing a second prism, upside-down with respect to the first and immediately behind it. W. H. Wollaston, in 1802, examined the details of the solar spectrum in a very similar set-up to Newton's but allowing sunlight in through a very narrow slit. Wollaston saw a number of dark lines crossing the spectrum – the lines being images at particular places in the spectrum of the slit through which sunlight entered. He thought that five of the most prominent lines constituted the boundaries of 'primary' colours, red, yellowish-green, blue and violet. Josef von Fraunhofer's experiments in 1814–15 were like Wollaston's, but his 'spectroscope' consisted of a prism mounted on a theodolite and a telescope to observe the resulting solar spectrum. With this arrangement, Fraunhofer saw hundreds of dark lines, some weak, others so strong that they appeared almost black. Using the theodolite, he was able to measure the positions of the lines and so construct a map, denoting the strongest by the letters A, B, C, etc. (Fig. 1.7). Some of these letters are still in use. Fraunhofer deduced from his experiments that the lines were some property of sunlight itself, i.e. not due to an illusion or an optical effect.

  The origin of the dark or 'Fraunhofer' lines was widely discussed for some time. It was noticed that certain lines were strengthened as the sun's altitude decreased in the sky, suggesting that the earth's atmosphere was responsible for these particular lines, since, when the sun is lower down, sunlight passes through a greater thickness of atmosphere. Laboratory investigations provided some clues. The spectra of incandescent gases were found to consist of intensely bright lines at certain spectral locations, while incandescent solids, such as glowing charcoal, gave a 'continuous' spectrum, i.e. like the sun's but without the dark lines. Two closely spaced bright lines located in the yellow part of the spectrum of vaporized sodium had been noted long before Wol-

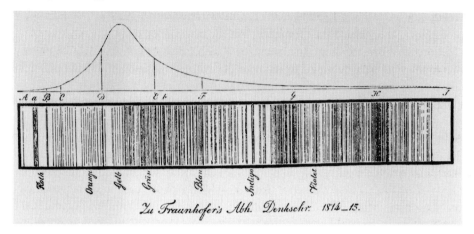

Fig. 1.7 Fraunhofer's map of the solar spectrum, showing locations of the chief absorption (Fraunhofer) lines in a schematic representation of spectral distribution of solar radiation. The letters A, B, C, etc., are retained in modern-day notation of Fraunhofer lines.

laston's experiment. They occurred frequently in laboratory spectra owing to contamination by common salt (sodium chloride), and were found to be in exact coincidence with two dark lines in the solar spectrum, labelled the D lines by Fraunhofer. Although it was not understood how the lines were formed – in particular why they should appear bright in laboratory spectra but dark in solar spectra – it was natural to infer that the sun itself contained vaporized sodium. Similarly, other metallic elements and hydrogen were identified from the unique pattern of bright lines in their laboratory spectra and the same pattern of dark lines in the solar spectrum.

In 1859, Gustav Kirchhoff found that the continuous spectrum of the light produced by a hot, incandescent solid was transformed into a solar-like spectrum, i.e. having dark lines, when the light passed through a gas that was cooler than the solid; moreover, the dark lines were in exactly the same spectral locations, or wavelengths, as the bright lines produced by the same gas when incandescent. L. Foucault had earlier noted this for the case of the sodium D lines. A gas emitting radiation at particular wavelengths when hot was thus deduced to be capable of absorbing at those same wavelengths when placed with a hotter source behind it. Kirchhoff generalized this to a law stating that the powers of emission and absorption of a body at some wavelength were the same for a given temperature. His conclusion was that the sun's surface – the *photosphere* – was an incandescent solid or liquid producing a continuous, line-free spectrum, while above it was an atmosphere. This atmosphere was supposed to be incandescent, producing a bright-line spectrum by itself, but seen against the photosphere the resulting spectrum was the observed continuous one crossed by the dark Fraunhofer (sometimes also called absorption) lines. The photosphere is not in fact solid or liquid as Kirchhoff had thought, but an opaque gas.

Shortly before the announcement of Kirchhoff's experiments, a discovery was made that was destined to have an enormous application to astrophysics. Christian Doppler, in Prague, found that the motion of a light source relative to an observer caused a change in the wavelength of the emitted light; the wavelength is increased as the light source recedes from the observer, and decreased as it approaches. The French physicist H. Fizeau pointed out that the Fraunhofer lines of the solar spectrum might be used to detect motion along the line of sight using this so-called Doppler–Fizeau (or more usually Doppler) effect. The rotation of the sun, which is from east to west, produces such motions: the Fraunhofer lines in the spectrum emitted by the east (approaching) limb are shifted towards the violet end of the spectrum, those in the spectrum emitted by the west (receding) limb are shifted towards the red. This was observed by H. C. Vogel.

The thousands of Fraunhofer lines in the solar spectrum that had been discovered since Fraunhofer's measurements were mapped by the Swedes K. Ångström and C. Thalén (1869) and H. A. Rowland (USA) in the 1890s, using a ruled grating to disperse the light rather than a glass prism. The measured positions were given in terms of wavelength, the unit of which was selected by Ångström to be one ten-millionth of a millimetre (1×10^{-10} m); this unit is still often used, and now has Ångström's name attached to it (symbolized Å, 1 Å $= 0.1$ nm). The wavelength range of the 'visible' spectrum, i.e. that accessible to the human eye, is about 420 nm (violet) to 700 nm (red). Thus, a strong dark line due to hydrogen in the red part of the spectrum, called line C by Fraunhofer but more familiar to solar astronomers as the Hα line, has a wavelength of 656.3 nm or 6563 Å, and the sodium D lines in the yellow have wavelengths of 589.0 and 589.6 nm. Progress was made in identifying the elements giving rise to the Fraunhofer lines: Rowland had found 36, including hydrogen, carbon and many metals, notably iron. The lines absorbed by the earth's atmosphere – the 'telluric' lines – were found to be due to diatomic oxygen (O_2) and water vapour molecules.

The identification of helium, the second lightest element (and second most abundant, after hydrogen), was a remarkable achievement for nineteenth-century solar scientists. A bright yellow line, very close to the sodium D lines at 589.0 and 589.6 nm, but slightly shorter in wavelength (587.6 nm), was observed in the spectrum of a prominence in 1868. Although at first mistaken for the D lines, the slight difference of wavelength was noted by two of the leading solar astronomers of the day, the Frenchman P. J. C. Janssen (who had seen the helium line during the total eclipse of that year) and the Englishman Sir Norman Lockyer. There was no known terrestrial material having a spectral line at this location, and so it was suggested that the solar line was due to a new element, named helium (after the Greek sun-god). It was finally

Fig. 1.8 Nineteenth-century shapers of solar astronomy I: (a) Norman Lockyer; (b) George Ellery Hale; (c) P. J. C. Janssen. (Courtesy Royal Astronomical Society)

discovered on the earth (by W. Ramsey in gas released from clevite rock) some 30 years later.

The explanation for the emission lines of particular elements was eventually given by the quantum theory of the atom in the 1920s and 1930s, advanced by Niels Bohr, Albert Einstein, Max Planck and others. The theory was to explain the structure of the atom in a radically different way from earlier, 'classical' ideas. The essence of the new theory was that an atom absorbed or emitted radiation in discrete amounts of energy – *quanta* – and that this energy was simply related to the frequency (or wavelength) of the radiation. Light was supposed to be made up of myriads of massless particles or *photons* each carrying a quantum of energy and moving at a speed of 300 000 km per second – the speed of light in a vacuum. Bohr proposed that an atom has particular states of energy, and that emission of radiation occurs when the atom releases energy by going from a higher-energy state to a lower-energy one. The atom is raised, or 'excited', to high-energy states by collison with another particle, e.g. a free electron, or absorption of a photon. Under normal circumstances, with no excitation processes occurring, the atom is said to be in its ground state. Thus, the emission-line spectrum of an element such as hydrogen arises from the excitation of hydrogen atoms to high-energy states followed by the release of photons with specific energies corresponding to specific wavelengths (e.g. that of the Hα line) as the atom is 'de-excited' to a lower-energy (e.g. the ground) state.

Ernest Rutherford's early experiments demonstrated that an atom consisted of a nucleus with much lighter electrons moving round it in orbits far larger than the physical extent of the nucleus, so that most of the atom is empty space. In the case of the hydrogen atom, the nucleus is a single proton, which under normal conditions is attended by a single electron. This electron moves round the nucleus in an orbit which may, like a planet's orbit round the sun, be circular or, more generally, elliptical. Excitation of the hydrogen atom to higher-energy states, as in the Bohr picture, occurs by the electron jumping from a close-in orbit to one further out; and conversely emission of a photon occurs by the orbiting electron jumping from an orbit with larger radius to one close in. Unlike planetary orbits, however, only those with particular radii and energy states occur, so that emitted photons are constrained to take certain values of energy which are proportional to the frequency of radiation or inversely proportional to the wavelength. The spectrum of hydrogen results from the emission of photons having particular frequencies or wavelengths that are emitted by transitions of electrons between particular orbits. Hydrogen, having a simple electron structure, has a simple spectrum, but more complex atoms give rise to more complex spectra. Relating the emission lines to the type of transitions involved was to become a subject of study for many years.

1.5   Solar eclipses and the solar atmosphere

As the moon goes round the earth, it may from time to time pass over the sun's disc. The apparent diameters of the sun and moon happen to be almost equal (both about half a degree) as seen from the earth. This curious coincidence gives rise to the outstanding spectacle of a total eclipse of the sun, when the sun's disc is just covered up by the moon, allowing an observer to view the outer, tenuous parts of the solar atmosphere which are normally invisible owing to sunlight scattered by the earth's atmosphere. As the solar and lunar apparent diameters are so nearly equal, the duration of totality in the very best eclipses does not exceed about seven minutes, and often it is much less. Sometimes, even though the moon might pass centrally over the sun, the moon appears smaller than the sun since, with the moon's orbit departing significantly from being a circle, it is occasionally further from the earth than usual. Under such conditions, the centrally eclipsed sun may take on the appearance of a ring, hence the name 'annular' for this type of eclipse. The small amount of direct sunlight left is enough to prevent the outer solar atmosphere from being seen.

Solar eclipses must have excited much attention in prehistoric times. The earliest existing records of them are from western Asia, particularly Babylonia, the most ancient one for which a date is certain being an Assyrian clay tablet recording a partial eclipse in 763 BC. Some believe that another tablet from the ancient Syrian city of Ugarit (Fig. 1.9) refers to an even earlier eclipse – in 1223 BC. There are also Chinese records dating back to about

Fig. 1.9 Obverse (right) and reverse of a clay tablet found in the ancient Syrian city of Ugarit, possibly recording a total solar eclipse on 5 March 1223 BC. If this is correct, it pre-dates known Babylonian eclipse records by several hundred years. (Photo courtesy D. Pardee, Mission de Ras Shamra (France), Oriental Institute (University of Chicago))

720 BC; those after 206 BC, the start of the Han dynasty, give detailed accounts of eclipses and indeed all astronomical phenomena, and continue till quite modern times. European records from ancient times are fairly sparse, though the Greek and Roman world were familiar with the Babylonian accounts mentioned in Ptolemy's *Almagest*. Monasteries were an important repository of eclipse records in medieval times. There are many eclipse accounts in Islamic writings, particularly around AD 800–1000. One of these, from Persia, refers to an annular eclipse in AD 873, from which it was wrongly deduced that the moon's apparent size was always less than the sun's. A method of predicting eclipses is known to have been used by ancient astronomers. This involves a period of time – just over 18 years – known as the Saros when the sun, moon, earth and the plane of the moon's orbit return to almost exactly the same relative configuration, so that eclipses are repeated after this time interval.

Any variation in the earth's rotation rate in the past two or three thousand years – e.g. a slowing down by tidal friction due to the moon and other solar-system bodies – can be investigated by referring to these ancient eclipse records. It has been found from Babylonian, Chinese and medieval eclipse records that the lengthening of the earth's day is less than what can be attributed to tidal friction, and it is inferred that there has been a slightly variable acceleration in the rotation rate.

The solar atmosphere is revealed in a spectacular way during a total eclipse. At the instant of totality, when the photosphere is just blotted out by the moon's disc, the red-coloured inner part of the sun's atmosphere, the *chromosphere*, appears for a moment, either as a crescent or with a 'diamond ring' appearance. Immediately after, the pearly white outer part of the solar atmosphere, the *corona*, becomes visible. It is generally irregular in shape, being composed of fine streamers that can often be traced out to many times the sun's diameter. Flame-like structures called *prominences* may also be present. During totality, the sky is dark enough to allow the brightest stars and planets to be visible, and there may be a hush among birds and animals confused by the sudden darkness (though sometimes the reverse: one observer remarked that a nearby calf that had been quietly grazing before the totality of an eclipse in 1879 'began to low most piteously' once totality began).

The solar corona may have been recognized as early as the time of Plutarch (first century AD) who remarked on the light surrounding the moon's disc during an eclipse in AD 83. There is a medieval account from the Italian astronomer Muratori who saw the corona during an eclipse in 1239, and reported a 'burning hole' in it which was probably a prominence. When Kepler saw a total eclipse in 1605, he attributed the corona to a lunar rather than solar atmosphere, a misconception widely upheld until the nineteenth century, as was that due to Wassenius in 1733 of solar prominences being 'clouds' in the lunar atmosphere. There are few references to the corona until the early eighteenth century, when observers began to use telescopes to study the eclipsed sun. The total eclipse visible in Europe in May 1715 was widely seen, and at the encouragement of Halley people in the British Isles took timings of the various stages of the eclipse, from which it was possible to deduce the sun's angular diameter very accurately. The corona was described by Professor R. Cotes in a letter to Sir Isaac Newton as being in the shape of a ring close to the sun, with four radiating streamers extending out forming a cross.

Francis Baily gave an account of a striking phenomenon during an annular eclipse in 1836. He noticed that, just as the annular phase was beginning, when the solar and lunar limbs were still in contact, a row of brilliant points

Fig. 1.10 Nineteenth-century shapers of solar astronomy II: (a) Francis Baily; (b) E. W. Maunder; (c) Warren de la Rue. (Courtesy Royal Astronomical Society)

suddenly formed along the moon's limb like 'the ignition of a fine train of gunpowder'. Baily's Beads, as they are now called, are rays of photospheric light shining along lunar crevices at the limb.

Baily's description of this eclipse helped to kindle great interest in succeeding total eclipses, and astronomers began to take their telescopes and other instruments on expeditions to remote parts of the world. They risked disappointment (as do those in our times) from unfavourable weather conditions, but in the early days of photography there was also the possibility of photographic equipment failing to operate during the few precious moments of totality. The enthusiasm of these observers is well illustrated by Janssen's spectacular attempt to see the 1870 total eclipse, visible in the Mediterranean. To escape the beleaguered city of Paris, surrounded by Bismarck's Prussian troops, he removed his equipment by balloon, arriving in northern Algeria, only to be clouded out at the time of totality!

The total eclipses of 1842, 1851 and 1860, all visible in Europe, were widely observed. During the 1851 eclipse, the first successful photograph was obtained, showing the inner corona and prominences (Fig. 1.11). The 1860 eclipse, visible in Spain, was observed by Father Secchi and Warren de la Rue; it was then that prominences were proved to be part of the solar, not lunar, atmosphere by the fact that the moon was observed to move relative to them. A total eclipse in 1868, visible in India, was the first to be observed spectroscopically, when Janssen noted the bright-line spectra of prominences, in particular the yellow helium line mentioned in §1.4. The coronal spectrum was then studied. It was found that the inner corona had a continuous spectrum with no absorption lines, but that the outer corona showed the familiar Fraunhofer spectrum of direct sunlight. But the most intriguing feature was a strong emission line in the green part of the spectrum, at a wavelength of 530.3 nm, discovered by C. A. Young and W. Harkness during an eclipse in 1869. The line was unknown in laboratory spectra, and a new element was suspected, named 'coronium'. Several more unidentifiable emission lines were noticed in subsequent years, in particular one in the red at 637.5 nm, and by the end of the nineteenth century, 14 were known. The coronium problem was not unique as there were unidentifiable lines in the spectra of nebulae, for which the element 'nebulium' was postulated. However, it became increasingly unlikely that such new elements could be admitted into the periodic table of elements, details about which were becoming better known. Eventually, clues to the problem began to emerge, nebulium in time being identified with oxygen with one and two electrons missing from its normal retinue (singly and doubly 'ionized'). In addition, it was found that some lines in a star that had gone through a nova-like outburst were due to six-times ionized iron, this being done by comparison with the emission-line spectrum of a high-temperature spark discharge produced in the laboratory by I. S. Bowen and B. Edlén in 1939. Eventually, W. Grotrian

Fig. 1.11 Daguerreotype of the total solar eclipse of 28 July 1851, taken by Berkowski with a small telescope at Königsberg. The inner part of the corona and some prominences are visible.

found by similar means that the coronal red line was due to iron no less than nine-times ionized; as the temperature of the spark discharge was about 500 000 K, it was inferred that the corona must have a temperature of this order. Edlén shortly after found the green line to be due to 13-times ionized iron.

During an eclipse in 1870, Young also observed for the first time the *flash* spectrum, visible at the instant the moon's limb is exactly in contact with the sun's east or west limb. It consists of numerous bright lines that, with some differences, notably in intensity, correspond to the dark Fraunhofer lines in the photospheric spectrum. This seemed to confirm Kirchhoff's idea that the solar atmosphere, which by itself had a bright-line spectrum, produced the Fraunhofer spectrum when seen against the bright continuous spectrum of the photosphere. Actually, Kirchhoff had supposed that the corona was producing the dark lines, but Young's observation showed that only a thin layer next to the photosphere – known for many years as the 'reversing layer' (because its spectrum is almost the reverse of the Fraunhofer spectrum) – was responsible. Young's observation was made with a conventional spectroscope, but he proposed that the flash spectrum might be better seen with a *slitless* spectroscope, the thin ring of the inner solar atmosphere acting as its own slit; the resulting spectrum would then appear as a series of crescents or rings at the positions of bright lines in a slit spectrum. All the spectroscopic observations described so far had been made by eye, though this was evidently unsatisfactory for attempting to record the details of the short-lived flash spectrum. Photography was applied in time (spectroscopes becoming *spectrographs*); the first successful attempt to photograph the flash spectrum was made with a slitless instrument during an eclipse in 1883.

The form of the corona had long been noticed to be very variable. At some eclipses the corona had lobe structures extending around the dark lunar disc with no preferred solar latitude; Janssen had likened the structures on one occasion to dahlia petals. At other times, most notably in 1878, the corona took the form of long streamers extending from opposite sides of the sun and approximately aligned with the solar equator. Sometimes delicate plumes at the solar poles were noted, reminding some observers of the pattern iron filings make in the field of a bar magnet; this gave rise to the speculation that the sun had a general magnetic field. It was realized that coronal form was related to the stage in the sunspot cycle, the irregular lobe-form occurring near or at sunspot maximum, the form with symmetrical streamers and polar plumes (as with the 1878 eclipse) at sunspot minimum.

Although the corona could still only be observed during total eclipses, the spectroscopic observations of Janssen and Lockyer in 1868 provided a means of seeing prominences at any time the sky was clear. Secchi and L. Respighi at Rome were pioneers in such observations, and with others were responsible for classifying prominences and finding how they were distributed with solar latitude. Broadly, they found there were two types of prominence: 'eruptive' (quickly changing) and 'nebulous' (better known today as quiescent), the latter mainly occurring at high latitudes. Many of the rapidly changing prominences were associated with sunspot groups on the limb of the sun. Their velocities could attain 200 km/s, as was confirmed by their Doppler-shifted spectra.

The chromosphere itself began to be observed at times outside of eclipses in the late nineteenth century. Observers found it to consist of a narrow region extending to a few thousand kilometres above the photosphere. Seen especially in the bright Hα line in the red part of the spectrum, the chromosphere's outer edge was seen to have a jagged edge, with tiny pointed structures (now known as spicules) directed outwards but not always exactly radially. Secchi's description of this as a 'burning prairie' is very apt, the spicules generally lasting for only a few minutes.

Attempts to observe the corona without having to wait for a total eclipse remained unsuccessful until the twentieth century. The problem was the extremely low surface brightness (about a millionth of the photosphere) which was overwhelmed by the photospheric light scattered by the earth's atmosphere. In 1930, Bérnard Lyot was finally successful in building an instrument – the 'coronagraph' – that was able to record the delicate coronal emission (see 10.1). It consisted of a lens focusing an image of the sun on to a circular 'occulting' disc, slightly larger than the photosphere to blot out all its light, beyond which there was a further lens and diaphragm to eliminate diffracted and reflected light produced by the first lens. An image of the coronal emission beyond the occulting disc was then formed by a further lens. All possible precautions were taken to eliminate stray light. The instrument was installed at the Pic du Midi Observatory in the Pyrenees, to give the clearest possible skies. Lyot obtained the first non-eclipse coronal photograph in 1930. Later developments by Lyot included the observation of the corona through interference filters which narrowed the spectral range to include only the green and red coronal spectral lines.

1.6   Application of photography

Photography was, so to speak, invented 'gradually', as the search for light-sensitive chemicals was a long one, made by many investigators. However, the first incontestably successful results were produced by the Frenchmen L.J.M. Daguerre and J.N. Niepce in the 1820s, their discovery being announced in 1839. They used a plate with a thin film of silver sensitized with iodine on which a focused image was allowed to fall. The plate was afterwards exposed to mercury vapour which adhered to the parts exposed to light, and finally the plate was 'fixed' in hyposulphite solution. Exposure times of several minutes were needed, even for a bright image.

The new invention was almost immediately applied to astronomy. Because of the abundance of light from the sun, it was the first astronomical object to be successfully photographed. This was done, by H. Fizeau and L. Foucault, using the Daguerreotype process, in 1845, with a resolution that is respectable even by modern standards. Sunspots were recorded very clearly, with umbrae and penumbrae, while the visual impression that the sun was less bright towards the limb (limb 'darkening') was confirmed.

The slow Daguerreotype process was replaced in time by the wet and dry 'collodion' processes, in which plates were prepared by an involved method which left them covered in a layer of collodion, i.e. nitrocellulose impregnated with silver halide salts. With the wet collodion process, the plates had to be prepared immediately before exposure, and used while still wet; dry collodion plates were at first much less sensitive, but still suitable for solar photography because of the large amount of light available. A number of successful plates were obtained, starting in 1854, some showing details of sunspots and the 'granulation', giving a mottled appearance to the sun's surface away from spots.

Carrington and others had been maintaining a visual record of sunspots at around this time, but it occurred to some that such a record could now be made photographically. In 1857, Warren de la Rue designed a telescope to do this for his private observatory near London. This *photoheliograph* (Fig. 1.12) consisted of a stopped objective lens and eyepiece to produce a solar image 10 cm in diameter on a photographic plate. At the request of the Royal Society, it was transferred to The King's Observatory, Kew (Fig. 1.13), where a regular programme of daily photography of the sun began. The short exposures needed were achieved with a focal plane shutter with a narrow slit. De la Rue supervised this programme until 1872, during which time several thousand solar photographs showing sunspots and faculae had been taken. At this point, the programme was taken over by the Royal Observatory, Greenwich. De la Rue's original photoheliograph continued in use till 1874 when it was replaced by a Dallmeyer instrument.

The Greenwich programme continued for over a century, with improvements and additions to the instrumentation, and a move of the whole Observatory to Herstmonceux in Sussex in 1949. For each day the sun was photographed, the total area of sunspots was estimated. This was done by comparing the sunspots on the photograph with a reticule of fine squares, counting the number of squares each spot covered. Corrections were applied for foreshortening, i.e. the apparent decrease in a spot's area near the solar limb. Areas were expressed (as they are still) in millionths of a solar hemisphere: one millionth equals about 3 million square kilometres. Photographs from other observatories (Mauritius, Melbourne in Australia, Harvard in the United States and Dehra Dun in India) were used to fill in gaps caused by cloudy weather at Greenwich. The time record of spot areas (Fig. 1.6) follows the Wolf relative sunspot number very closely, and so shows the eleven-year cycle.

Other observatories had begun their own photographic programmes, for example, that at Meudon Observatory, near Paris, founded in 1876 by Janssen. His impressive photographs (Fig. 1.14) succeeded in recording fine details of spots and individual photospheric 'granules' with angular sizes as small as 1 arc second (corresponding to 725 km on the sun).

George Ellery Hale's experiments in solar photography ushered in a new era in solar physics. In 1894, he had designed a solar telescope at Yerkes Observatory, near Chicago, which was fixed in a horizontal position, having sunlight fed into it by a rotating mirror system to compensate for the sun's movement across the sky. A later version of this telescope, the Snow telescope, was built in 1903. One of two concave mirrors, each with very large focal lengths, were used to produce images of the sun. The 'seeing' conditions, i.e. degree of atmospheric turbulence affecting image quality, were

Fig. 1.12 Photoheliograph built by Warren de la Rue for Kew Observatory, later used for sunspot photography at the Royal Observatory, Greenwich. (Reproduced by permission of the Trustees of the Science Museum, London)

Fig. 1.13 Engraving of Kew Observatory (The King's Observatory), originally built for King George III, and used as a solar observatory and laboratory for geomagnetism and meteorology by the British Association in the latter part of the nineteenth century.

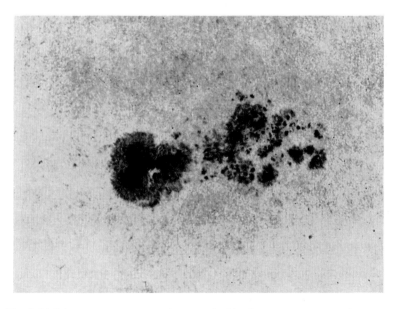

Fig. 1.14 A large sunspot group photographed by Janssen at Meudon Observatory in 1885.

judged to be inferior to a site in southern California, Mount Wilson, which Hale had been investigating and found to be near-ideal. The Snow telescope was accordingly transported there. Further experiments indicated that a telescope with a vertical arrangement instead of a horizontal avoided impairment of image quality because of heated air currents in the telescope. More particularly Hale found that raising the telescope as high as possible above the ground gave considerable improvements, since much of the external atmospheric turbulence arose within a few metres of the heated ground. The result was the 60-foot (18-metre) tower telescope of 1907, followed by a 150-foot (45-metre) tower telescope in 1912. These and the Snow telescope are shown in Fig. 1.15. The 60-foot telescope has a 30-cm-diameter lens with 18-m focal length, and has been used almost daily for taking white-light photographs of the sun. The many thousands of these images have recently been analysed in an extensive programme.

Much effort has gone into solving the problem of poor seeing, e.g. by seeking the best possible sites for observatories. Seeing 'monitors', in which different parts of a solar image are examined by photoelectric cells, were developed in the 1950s, and allowed photography to be carried out at the most favourable moments. More drastic solutions have involved balloons, both manned and unmanned. The most successful of the latter was the 1959 Project Stratoscope, directed by Martin Schwarzschild at Princeton University. This was a 30-cm Newtonian reflector mounted on a balloon which flew at a height of up to 24 000 m. The best of the images obtained showed details

Fig. 1.15 The horizontal Snow telescope and the two tower telescopes used by G. E. Hale at Mount Wilson, California. (Courtesy The Observatories of the Carnegie Institute of Washington)

of sunspots and granules only a fraction of an arc second across, and are still among the best ever produced.

1.7 Spectroheliography and solar magnetic fields

In 1869, Janssen suggested the principle of an instrument to observe the sun, not in 'white' light, but in isolated regions of the spectrum such as the strong hydrogen Hα line or calcium H and K Fraunhofer lines. The instrument would consist of a telescope forming a focused image of the sun on a slit. Light from this slit would pass through a prism and the dispersed light would be intercepted by a second slit placed at the position of an absorption line such as Hα. The primary slit, isolating a narrow region across the solar image, would then be moved across the whole sun and the second slit moved to maintain its location on the Fraunhofer line, and an image of the sun in the light of the Fraunhofer line would result. (For further details, see §10.1.) The *spectroheliograph*, the photographic version of this instrument, was eventually built by Hale in 1892. A slightly different instrument was devised by H. A. Deslandres at Meudon, called by him a *spectro-enrégistreur des vitesses*. The latter instrument maintained the primary slit in the one location across the sun while the observer searched for bright spectral features near particular spectral lines, displaced by the Doppler effect caused by the rapid motion of the emitting material on the sun.

Hale's instrument had the advantage of seeing the whole sun at any particular wavelength, and with it he was able to record not only the chromosphere and prominences off the solar limb (as had first been done spectroscopically by Janssen and Lockyer) but features on the solar disc. The first photographs were in the light of the calcium K line, located in the blue part of the spectrum; those in the Hα line, in the red, were not possible until the development of red-sensitive photographic plates a few years later. A mottled structure was visible, with bright patches which Hale called flocculi similar to the faculae visible in white light. He also recorded long, dark filaments on the disc, afterwards identified with prominences.

With the two tower telescopes at Mount Wilson, Hale made several important discoveries including one on the magnetic nature of sunspots. It had earlier been discovered, by P. Zeeman, that certain emission lines in laboratory spectra were split into two components when the emitting gas was in the presence of a strong magnetic field directed along the line of sight, and that these components had polarization properties (these are described in 3.6). C. A. Young had noticed in solar spectra that some Fraunhofer lines were doubled when a spectrograph slit admitted light from a section of the sun that included a sunspot. Hale investigated this further in 1908 with the 60-foot telescope and a large spectroheliograph, and found with a polarization analyser that the components were polarized, confirming the presence of large magnetic fields. Field strengths of a few tenths of a tesla were measured. He also noticed that the individual spots of pairs or groups – the westernmost ('leading') and easternmost ('following') – had opposite magnetic polarity to each other, i.e. spot pairs or groups had a 'bipolar' character. Further, the disposition of polarities was found to be opposite for northern and southern hemispheres. These observations were continued until 1913, the year of sunspot minimum, when spots of the new cycle started to appear at high latitudes. Hale discovered that the polarity sense of these new spot pairs was the reverse of the old-cycle spots in each hemisphere, so once again spot groups in either hemisphere had opposite polarities. A magnetic reversal in fact occurs when the new-cycle spots appear each sunspot minimum. Magnetically, then, the sunspot cycle has a period of 22 years rather than 11 years, i.e. after a space of 22 years, pairs or groups of spots repeat their magnetic polarities.

Hale found sunspots to be cooler (about 4200 K for a large spot) than the surrounding photosphere (6000 K), a fact already deduced from the identification in sunspot spectra of broad absorption features called bands, noted by Lockyer and Young, with molecules such as titanium oxide. The unspotted photosphere is too hot for such molecules to form, but they are able to exist at the cooler temperatures in spots.

At about this time, in 1908, John Evershed at the Kodaìkanal Observatory in India observed distortions in the shapes of Fraunhofer lines in spot spectra. They were due to Doppler shifts, which implied the existence of a flow of gas outwards from the spot centre, with speeds up to 2 km/s. It was later discovered from the calcium H and K lines (by C. St John at Mount Wilson) that the flow was *inwards* at higher altitudes, where the calcium lines are formed, so that there seemed to be a 'down-draught' at the centre of a spot.

The visual version of the spectroheliograph – the *spectrohelioscope* – was also the product of Hale's inventive mind, though it was not until the 1920s that it became a reality. The principle is identical to the spectroheliograph, but the primary and secondary slits are moved rapidly backwards and forwards so that the eye, unaware of this motion, sees an image of the sun in the light of the wavelength selected by the secondary slit. Hale very soon discovered that certain brilliant eruptions, already recorded with the spectroheliograph several years before in the light of the calcium K line, showed minute-by-minute movements and intensity changes that could actually be watched with the eye.

The story of these eruptions starts many years before Hale's observations. On 1 September 1859, Carrington and R. Hodgson observing with conventional telescopes (and therefore in 'white light') had witnessed the sudden appearance of an intensely bright patch of light within a large sunspot group. Within a matter of five minutes, the patch developed then faded from sight.

An intimation of the importance of the phenomenon was the fact that there were simultaneous changes in the earth's magnetic field and large displays of aurorae soon after (§§1.3, 1.9). It was the first recorded observation of a solar *flare*: flares are only rarely visible in white light, but are much more common when seen in the light of Hα and the calcium H and K lines. The first such observations were made in the 1870s by C. A. Young, who noticed that occasionally the central parts of these lines were bright instead of dark, forming a 'line reversal'. With a sunspot group focused on his spectroscope slit, and by opening the slit very wide, Young found that for one of these line reversals the bright core of the Hα line originated from bright filamentary structures near the sunspot group.

Hale's spectrohelioscope observations had likewise been made in the neighbourhood of a large sunspot group which was seen in January 1926. Several brilliant eruptions (then known as *éruptions chromosphériques* – the name flare is due to H. W. Newton) occurred when the spots were near the centre of the sun's disc, and were accompanied soon after by terrestrial magnetic field changes, as in the 1859 flare. Within a few years of these discoveries, spectrohelioscopes were operating at a number of observatories, including Arcetri (near Florence) and Greenwich, so that an almost continuous watch on flares and other chromospheric phenomena could be maintained.

1.8 The sun and relativity

One of the most far-reaching developments in physics since the time of Newton was the announcement by Albert Einstein of his theory of relativity in the early years of this century. The theory of general relativity, propounded in 1916, corrected Newton's law of gravitation, but the small differences between the old and new theories were extremely difficult to measure for relatively weak gravitational fields like those of the earth or even the sun.

An early test proposed for the theory of general relativity involved a very slight 'bending' of rays of starlight passing near the sun, as if light 'particles' had mass and were deflected by the sun's gravitational field. Even in Newton's theory there is a tiny deflection, 0.9 arc seconds, but it is precisely double – 1.8 arc seconds – for Einstein's theory. Such shifts could only be seen during the total eclipses when the sun (and therefore moon) happens to lie in a rich star field. The sun does actually pass through a star cluster, the Hyades, at the end of May each year, and by a fortunate coincidence a total eclipse at that time of year occurred in 1919, shortly after Einstein's theory had been published. The famous astronomer and popularizer of relativity, Sir Arthur Eddington, travelled to Principe, off the coast of West Africa, to observe the deflections of the Hyades stars, visible at eclipse totality. His measurements and those of observers despatched to Brazil on the same occasion fairly definitely confirmed the Einstein deflection. This was for many years unclear, but has been recently established by a re-examination of the original plates, one from the Brazil expedition giving 1.90 (standard error 0.11) arc seconds for the deflection. Subsequent eclipses up to recent times give similar results.

There was also an attempt to apply solar spectroscopy as a test of Einstein's theory. In this case, the theory predicts that a spectral line emitted by an atom near the sun's surface will have its wavelength increased very slightly. The so-called *Einstein shift* – only 0.0014 nm at the wavelength of the Hα line (656.3 nm) – was carefully searched for, in particular by C. St John at the Mount Wilson Observatory. (Allowance has to be made for the Doppler shift due to solar rotation.) The results were (and to this day remain) disappointingly equivocal; an increase of wavelength was found, but was most likely attributable to motions in the sun's lower atmosphere.

For relativistic effects on planetary orbits, see §7.1.

1.9   Solar–terrestrial connections

Throughout the nineteenth century, there was a growing realization that sunspot activity, the occurrence of aurorae visible in polar regions of the earth, and disturbances in the earth's magnetic field were all connected. However, these three phenomena were at first studied independently, and the relationships between them only found little by little. Having dealt with sunspot and auroral observations (§§1.2, 1.3) we now review early studies of the earth's magnetic field.

The use of compasses for navigation has a long history. In 1600, William Gilbert, physician to Queen Elizabeth I, discovered that it was because the earth is itself magnetized that compass needles point towards a particular direction: *Magnus magnes ipse est globus terrestris* (the earth's globe itself is a great magnet), as he put it in his account *De Magnete*. Now a compass needle is simply a small magnet free to rotate in a horizontal plane, and it points along the direction of local magnetic field lines, by convention towards the magnetic pole in the northern hemisphere. But also by convention, a magnetic field line's direction is defined to start from the north pole of a magnet and end at its south pole. Strictly, then, the magnetic pole in the northern hemisphere is a magnetic *south* pole (sometimes called a 'north-seeking' pole), and vice versa. The magnetic poles are some distance away from the geographic poles, and are not at opposite ends of the earth: the north-seeking pole is at present located at 77°N, 102°W (in the Queen Elizabeth Islands, north of Canada), the south-seeking at 65°S, 139°E (just off the Antarctic coast, south of Australia), while the line joining the magnetic poles is off-centre by 1200 km. The field strength at either pole is about 60 μT (i.e. 6×10^{-5} T). It was recognized very early on that the compass needle does not in general point at true geographic north. Thus, a compass needle at London pointed 11 degrees east of true north in 1580, but by 1660 pointed almost exactly at true north; by 1800 it pointed 24 degrees west. At present, the movement is in the reverse direction, being $4\frac{1}{2}$ degrees west in 1990. Voyages were taken to measure the angle of *declination*, or the angle between compass and true north; Halley had himself undertaken such a voyage in 1698–1700. As far back as the fifteenth century, R. Norman had noted that a compass needle left free to rotate in a vertical plane is inclined to the horizontal. The angle at which it is inclined (the angle of *dip*) is, like the angle of declination, subject to long-term, or secular, changes. Again, for London in 1990, the angle of dip is around 66 degrees and decreasing slowly.

In 1722, the English clock-maker George Graham discovered that there were short-lived variations in the magnetic field. Some of these were regular, occurring during the course of a day, but there were also irregular disturbances in which the angle of declination could vary by up to half a degree in only a few minutes. Celsius found in 1740 that these variations occurred simultaneously in two widely different locations, so the earth's magnetic field seemed to be disturbed in a general rather than a local way. While the cause of these disturbances was being debated, the German physicist K. F. Gauss had found that nearly all the field must have its origin in the earth's interior rather than its atmosphere. Present theories in fact hold that the field is generated by convection cells in the earth's liquid core constituting a dynamo. Gauss is remembered for his field strength measurements which he placed on a more satisfactory basis than before at a newly established observatory at Göttingen in 1834. Delicate magnetometers were set up to measure the three so-called elements of the earth's field, viz. the angle of declination and the horizontal and vertical components of the field strength (from which can be derived the angle of dip). Readings were made by eye, but a later (1847) invention by Charles Brooke made use of photographic recording of these elements for use at the geomagnetic observatory at Greenwich, just set up by the then Astronomer Royal, Sir George Airy.

While these discoveries of the geomagnetic field were being made, studies of aurorae continued, with increased understanding of their geographic distribution. There was an obvious increase of aurorae towards the poles (for instance, in the British Isles, an average of only six aurorae per year are visible in the south, but 60 per year in the far north). But it was found (by Muncke in 1833) that there was a zone of maximum auroral frequency near the poles beyond which there was a decrease. This zone (now called the *auroral oval*) was found, by the American Elias Loomis in 1860, to be roughly circular and 4000 km across, as was also illustrated by the map of Hermann Fritz (1881) which had contours of constant auroral frequency (or *isochasms* from Aristotle's reference to aurorae as 'chasms').

The connection between aurorae and the earth's magnetic field had been surmised by Halley when discussing the 1716 auroral display. It was indicated by observations of geomagnetic disturbances during aurorae by Celsius and Hiórter in Sweden. Another Swede, Wilcke, in 1770, found that rays in auroral displays lay along magnetic field lines. But the evidence now provided by Fritz's map of isochasms was overwhelming, as it clearly showed the auroral zone to be centred, not on the earth's geographic poles, but the magnetic poles. It was found that the zone had a similar shape to lines joining places with equal angles of magnetic dip.

The question of whether the sun might be the controlling influence on both the occurrence of aurorae and that of geomagnetic disturbances was considered, particularly after the discovery of the eleven-year sunspot cycle. Loomis and Fritz found aurorae to be correlated with sunspot number, and Edward Sabine discovered an eleven-year periodicity in magnetic measurements in Canada. The white-light solar flare seen by Carrington and Hodgson in 1859 (§1.4) was apparently related to geomagnetic disturbances and an auroral display: the magnetic records at Kew Observatory showed a disturbance that was simultaneous with the flare, while some 17 hours later there was a very large magnetic 'storm' and an aurora seen down to tropical latitudes. Despite this, Carrington was reluctant to associate these facts, though others were less cautious.

How the sun, 149 600 000 km away, could interact with the earth to cause magnetic and auroral effects was a formidable question. By the 1880s it was suggested that the sun emitted particles, which in 1896 the Norwegian K. Birkeland identified as beams of cathode rays (or electrons); his laboratory experiments indicated that a magnetic field causes such beams to be focused at a magnetic pole. He and his collaborator C. Störmer accounted for the geomagnetic disturbances and the geographical distribution of aurorae on the assumption that beams of electrons impinge upon the earth which was taken to be a uniformly magnetized sphere.

In 1904 E. W. Maunder distinguished between large, sunspot-associated geomagnetic disturbances and smaller ones without association. The large disturbances generally occurred when a sunspot group was just west of the sun's central meridian, while the smaller disturbances recurred at intervals of 27 days, equal to the sun's synodic rotation period. He concluded that there were indeed streams of particles emanating from the sun, which, for the large disturbances, were in the form of a short-lived blast travelling outwards from the sun in a curved path because of the sun's rotation. The smaller disturbances were puzzling up to quite recent times. J. Bartels called the supposed solar origin of the particle streams M-regions (M standing for 'magnetic effective'), but there was no particular solar feature that they could be associated with – indeed, the most well-marked disturbances occurred at solar minimum, when the sun was nearly spotless. Although the M-regions could not be identified, it was accepted that the streams emerged like water from a rotating garden sprinkler. Figure 1.16 shows the geometry imagined by Bartels giving rise to the recurring disturbance: a stream from an M-region on the sun passes by the earth each synodic rotation.

Investigating the connection between the sun and the earth's magnetic field was systematized early this century by devising a scale that indicated the degree of magnetic disturbance for each day. Several magnetic observatories around the world contributed, and a daily average, the character or C index, was compiled. The C index was used to illustrate the 27-day period of the small magnetic disturbances, which sometimes persisted for several solar rotation periods. This is shown by 'Bartels diagrams' in which the C indices are symbolically represented for each day in a synodic period, and with successive synodic periods arranged in a vertical sequence with time progressing downwards (Fig. 1.17). The patterns of symbols indicate the developing recurrent tendency of magnetic disturbances as the solar activity cycle approached a minimum.

The study of auroral and geomagnetic phenomena required simultaneous observations at various locations, and as early as 1882–83 an international collaboration was set up called the International Polar Year, to make observations at selected days of the month. A second such collaboration was organized in 1932, by which time radio observations had gained much importance. The International Geophysical Year – actually an 18-month period starting on 1 July 1957 – had wider intentions. It was due to coincide with a maximum in the sunspot cycle, which turned out to be the largest recorded up to that time, and among its aims was the study of the connection between solar flares, geomagnetic disturbances, radio fade-outs and particle emission from the sun. Apart from the much better understanding of the earth's environment and its connection with the sun, the period was made

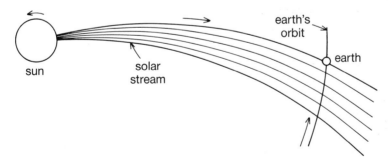

Fig. 1.16 Bartels's explanation for the recurrent magnetic storms near the time of solar minimum: a stream of particles emanates from the sun in a curved trajectory, encountering the earth every 27 days (the solar synodic period). (Adapted from Chapman and Bartels (1962), by permission of Oxford University Press)

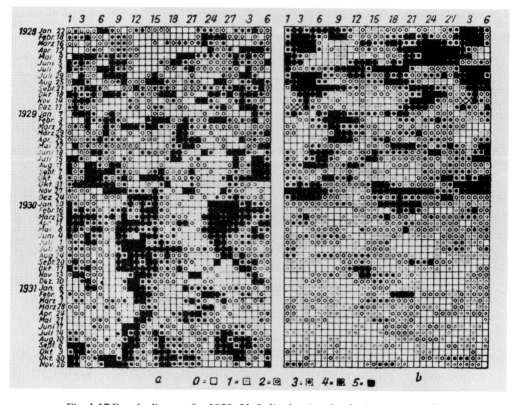

Fig. 1.17 Bartels diagram for 1928–31 (left), showing developing patterns of recurrent small magnetic disturbances as solar minimum approached (year of minimum was 1933). The right diagram shows the daily number of sunspots. (From Chapman and Bartels (1962), by permission of Oxford University Press)

memorable by the launch of *Sputnik I*, the first artificial satellite, by the Soviet Union in October 1957.

1.10 Measurements of solar energy

Although much of the sun's energy is radiated in the 'visible' range (about 400 to 700 nm), there are contributions from the infrared (wavelengths longer than 700 nm), and ultraviolet (shorter than 400 nm) spectral regions. The name 'solar constant' has been given to the amount of solar energy of all wavelengths received at the earth outside its absorbing atmosphere per unit area and per unit time, with the earth at its mean distance (exactly 1 astronomical unit) from the sun. As it is known to vary slightly, the name is not very apt, and we shall refer to it instead as the solar *irradiance*. Its SI unit is watts per square metre (W/m^2).

Attempts were made at measuring the solar irradiance as early as 1837. Some of the first experiments were based on the rise in temperature of some device on exposing it to direct sunlight. The amount of energy incident from the sun is then equal to the heat capacity of the device, measurable from other experiments, and the measured temperature rise. The experiment by the Frenchman C. S. M. Pouillet consisted of a blackened copper vessel containing water and with a thermometer inserted. Measurements were made of the temperature in the shade and then of its rate of increase on exposing the vessel to direct sunlight. The measured energy flow divided by the copper vessel's cross-sectional area gave the solar irradiance. Not all the sun's energy incident on the apparatus went into heating the copper vessel and its contents – some was reflected (despite the fact that the device was blackened), and some energy was absorbed by the earth's atmosphere. Pouillet's corrections for these led to the surprisingly precise value of $1260\,W/m^2$ for the solar irradiance; it compares with the modern value from spacecraft instruments of $1368\,W/m^2$. Pouillet's apparatus, called a pyrheliometer, was improved upon in later experiments, in particular by Secchi in the 1860s.

Further developments were made with electrical devices. Ångström in 1896 invented a compensating pyrheliometer, consisting of two metal plates, one exposed to sunlight and the other unexposed but heated to the same temperature by an adjustable electric current passed through it. The solar energy incident on the first plate could then be found from the fact that it was equal to the electric power dissipated in the second, which could be measured. In 1880, S. P. Langley developed an electrical instrument called a bolometer consisting of two thin, blackened platinum strips to form two arms of a Wheatstone bridge apparatus for measuring electrical resistances. One of the strips was exposed to sunlight, the other not. A change in the resistance of the exposed one produced a measurable change in the current passing through the bridge, which could then be related to the amount of incident solar radiation.

Substantial corrections had to be made to solar irradiance measurements to account for the absorption of solar energy by the earth's atmosphere. Langley had tried to minimize the correction by transporting his bolometer to the top of Mount Whitney (altitude $4400\,m$) in California. Mountain-top sites were chosen for an extensive solar irradiance measurement programme, begun in 1902, run by the Smithsonian Institution of Washington. Under the direction of Langley and later C. G. Abbot, the programme continued until 1960. The solar irradiance was routinely measured using a device consisting of blackened silver discs (Fig. 1.18). The rate of temperature rise on exposing a disc to sunlight was measured to give a relative measure of the irradiance. The silver-disc device had to be absolutely calibrated, and this was done using a separate instrument, the water flow pyrheliometer, built by Abbot. In this device, a blackened cone is exposed to the sun, absorbing practically all the

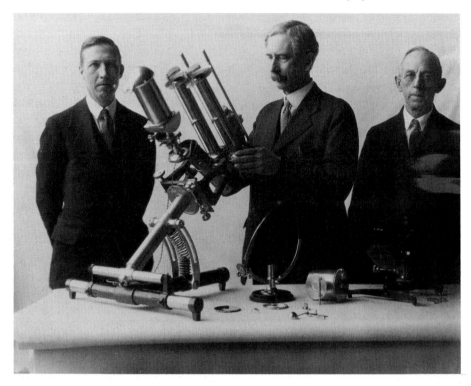

Fig. 1.18 L. B. Aldrich, C. G. Abbot and A. Cramer demonstrating instruments for measuring solar energy in the programme run by the Smithsonian Institution of Washington. The left-most instrument is a pyranometer to determine sky brightness, and on the right is a pair of silver-disc pyrheliometers for measuring the solar irradiance. (Courtesy Smithsonian Archives)

incident sunlight. Water flows in one end of a surrounding envelope, past a first platinum strip, then round the cone which is heated by the incident sunlight, and finally past a second platinum strip. Measurements of the change in the electrical resistance of the platinum strips and of the amount of water flow allowed the irradiance to be derived.

Recent analysis of the Smithsonian results indicates a mean value of the irradiance over the 1902–60 period of 1353 W/m², with variations of not more than 1%. Smaller variations may be attributable to imprecise corrections for atmospheric absorption: even at the mountain sites used in the Smithsonian programme, the earth's atmosphere absorbs at least a quarter of the incident energy.

The solar irradiance can be used to find how much radiation the sun emits at its surface (this calculation is done in §3.4). There were several attempts in the mid-1800s to derive how hot the sun's surface must be to emit the calculated amount of radiation, but the relation between temperature and radiant energy was very poorly understood then. A law proposed by Sir Isaac Newton led to calculated temperatures of several million degrees Kelvin, while an experimental law derived by P. L. Dulong and A. T. Petit gave only 2000 K! The discrepancy was eventually resolved by the 1879 discovery (due to J. Stefan and L. Boltzmann) that, for a perfectly absorbing body, the amount of radiant energy is proportional to the fourth power of absolute temperature. By the 1890s, using the Stefan–Boltzmann law, the solar surface temperature was derived to be about 6600 K, about the same as the modern value.

1.11 The nature of the sun

It was not until the mid-nineteenth century that solar scientists could be said to be even vaguely aware of the true physical nature of the sun, or of its surface features and chromosphere and corona. Sir William Herschel's conception of an inhabitable sun eventually fell into disrepute, Janssen, for instance, holding it to be 'ridiculous' by 1879. There was by then a general acceptance that, far from there being a dark, solid interior, the sun was entirely gaseous. A fundamental question now being asked was how did the sun shine, that is, what was the source of the sun's energy?

By the mid-nineteenth century, the earth was known to be vastly older than the six thousand years which Archbishop Ussher had estimated it to be from Old Testament accounts; the sun and rest of the solar system were presumably of a similar age. This meant that the sun had been radiating energy, possibly at more or less its present rate, for a very considerable time, so that the question of the energy source was a challenging one indeed. J. R. Mayer in 1848 proposed that a continual rain of meteorites might be this source, but the theory was easily refuted by the fact that the sizeable increase in solar mass predicted was not observed. A much more plausible idea was that of Hermann von Helmholtz in 1854 in which the gain of gravitational energy by a gradual contraction of the sun was balanced by the loss of radiant energy. The contraction needed was calculated to be less than what was measurable, so that there was at least no contradiction with observations. The calculated solar age by this process was around 20 million years, which at the time was considered accommodatingly large. However, by the end of the century, biological and geological evidence suggested terrestrial, and therefore solar, ages to be much larger still.

Theoretical studies, or 'models', of the internal structure of stars including the sun gave important understanding. Several investigators (Lord Kelvin, J. Homer Lane, R. Emden and others) had constructed models based on the known laws relating to perfect gases that allowed pressures and temperatures to be derived at various points in a star's interior. In these models, heat energy is transferred from the hottest region at the star's centre to its surface by convection. This was developed by Sir Arthur Eddington who assumed energy was transferred by radiation. Eddington calculated that the central region of the sun, given the known solar mass and radius, was at a temperature of about $10\,000\,000\,\mathrm{K}$ ($10 \times 10^6\,\mathrm{K}$) and a pressure of 10^{11} standard atmospheres (10^{16} Pa).

By the early 1930s, there was speculation (by R. d'E. Atkinson) that the sun's energy source was nuclear (or 'subatomic') in origin; the release of energy could result from either heavy-element atoms being broken apart or light-element atoms being fused together. Enough nuclear physics was understood by 1939 for Hans Bethe and Carl von Weizsäcker to propose that the basic process was in fact the fusion of four hydrogen atoms to form a helium nucleus; the calculated central temperature of the sun allowed such a reaction to proceed. The details of the nuclear reactions are discussed in Chapter 2.

Sunspots were no longer regarded as openings in a luminous layer of the solar atmosphere through which a solid interior could be viewed. However, exactly what they were remained a mystery. H. Faye imagined them to be places where hot gases from deep down rose to the surface, cooling as they did so and dissipating the material making up the photosphere. Spots were dark because the hot gases beneath the photosphere were supposed to be too hot to emit radiation. Warren de la Rue and his colleagues at Kew proposed in 1865 that on the contrary spots were where descending currents of cool gases occurred. Secchi and Lockyer thought that spots were formed by material falling back to the sun, having been ejected by surrounding bright faculae or high-latitude prominences. Meteorite impacts were suggested by Sir John Herschel, the eleven-year cycle supposedly arising from a swarm of meteorites in a highly eccentric orbit that intersected the solar surface. Hale in 1908 combined his discovery of the magnetic nature of spots with the observation of a vortex-like appearance round spots sometimes seen in Hα light in a theory in which the magnetic fields were generated by charged particles drawn into the vortex.

Two sunspot theories, both rather involved, held sway for several years until present ideas became acceptable (the latter being described in §6.5). In 1926, V. Bjerknes developed a vortex theory in which a pair or group of spots were produced at the points where a largely submerged vortex ring, circling the sun parallel to the equator, rose to pierce the photosphere. Without showing how, Bjerknes supposed that a magnetic field was generated 'hydrodynamically', i.e. by the motion within the vortex, the field direction depending on the direction of the vortex whirl. This gave rise to the opposite polarities of the leading and following spots of pairs or groups. Two vortex rings, north and south of the equator and having opposite vortical motions, explained the presence of spots in either hemisphere with opposite polarities, and a circulation of the vortex rings from pole to equator led to the eleven-year cycle. In the theory of H. Alfvén and C. Walén, a general magnetic field at the sun's centre was postulated, from which 'magnetohydrodynamic' waves were generated by a disturbance that recurred at eleven-year intervals. The waves were turned into pairs of small vortex rings on their way out, one propagating northwards, the other southwards, with opposite directions of motion. After several years, the vortex rings would reach the photosphere, and give rise to bipolar sunspot pairs or groups, with the opposite magnetic polarities as observed.

The nature of the solar chromosphere and corona was poorly understood until comparatively recent times. It was known that both were at higher temperatures than the photosphere. In the case of the chromosphere, this was deduced from the presence of the helium yellow line in the chromospheric but not photospheric spectrum, a higher temperature being needed to excite the line. The green and red coronal lines implied extremely high temperatures there – of the order of $1\,000\,000\,\mathrm{K}$. A. Unsöld and others considered the chromosphere to be made up of small prominences, and the higher chromospheric temperature was then explained in a theory of Alfvén by which prominences were the result of electrical discharges. Although the corona was very hot, it was also known to be extremely tenuous as comets occasionally passed through it without appreciable effects. As a result, a comparatively modest amount of energy could be supplied to it to explain the high temperature as loss of energy by radiation in such a rarefied atmosphere would be slight. A possible energy source considered was motions of the chromosphere or prominences.

1.12 Radio astronomy and the space age

The sun is a powerful emitter of radio waves, but its detection at these wavelengths was not made until several years after the 1931 discovery of radio emission from the Milky Way (i.e. the plane of our galaxy) by Karl Jansky. Searches for solar radio emission had in fact been made by Jansky and Grote Reber at metre wavelengths, but were unsuccessful owing to the fact that they were made at sunspot minimum, radio emission being very dependent on solar activity. The detection at metre wavelengths was in the event made accidentally, most probably by amateur observers in 1936, then by British Army radar operators in February 1942, the interference it was causing being mistaken for jamming of their equipment by the Germans. Analysing the evidence, James Hey concluded that the emission was in fact due to solar radio emission. During late February 1942, a large sunspot group was crossing the sun and there had been large flares associated with it.

Further investigations by Hey, Edward Appleton, Martin Ryle and others were made when large sunspot groups appeared in 1946 and 1947, using radio observations and Greenwich sunspot data. Radio emission was found to be associated with sunspots and to be highly directed, i.e. at a maximum when a sunspot group was near the centre of the sun's disc. Bursts of solar radio emission were also observed, and were found to be coincident with flares; this burst emission was not directed, i.e. bursts occurred regardless of the flares' locations on the sun's disc. The group at the Cavendish Laboratory, Cambridge, found from polarization evidence that the sunspot emission was due to moving electrons in a strong magnetic field.

In addition to the radio emission linked with solar activity, a steady component was recognized. This was discovered by G. C. Southworth in 1942 at centimetre wavelengths. The intensity of such emission could be related to the temperature of the emitting source, but it was found to be inconsistent with the 6000-degree temperature of the solar photosphere. Combining centimetric and metric observations of this steady component indicated that the source temperature increased with increasing wavelength from 10 000 K to about 1 000 000 K. In 1946, this was explained, by D. F. Martyn and V. L. Ginzburg, by the fact that the shorter-wavelength radiation arises from the chromosphere, where the temperature is about 10 000 K, and the longer from the corona, at about 1 000 000 K.

These very high temperatures mean that the chromosphere and corona radiate strongly in the ultraviolet part of the spectrum (wavelengths less than about 300 nm), and the coronal temperatures are so high that X-rays (wavelengths less than about 10 nm) are also emitted. This radiation is completely absorbed by the earth's atmosphere and so instruments designed to observe it must be flown on rockets or earth-orbiting satellites to altitudes where the absorption is reduced. Thus, progress in this area had to await the arrival of the space age. The rocket technology developed by the Germans during the Second World War was soon adapted for scientific purposes, and rapid advances in our knowledge of the sun at short wavelengths occurred in the decade following the war. Solar X-ray emission was observed by T. R. Burnight in a 1949 flight of a rocket carrying a camera with beryllium and aluminium filters. The photographic plate was found to be blackened on recovery, and since only X-rays were able to penetrate the filters used, X-ray emission from the sun had been positively identified.

The sun's ultraviolet spectrum is quite different from the Fraunhofer spectrum of the visible region. The continuum intensity decreases with decreasing wavelength beyond the violet, and, at wavelengths less than 208.5 nm, bright emission lines start to appear, rather like those in flash spectra during total solar eclipses. The solar continuum eventually fades to invisibility at a wavelength of around 140 nm. There is a very bright emission line at 121.5 nm – the Lyman-α line – due to the excitation of hydrogen atoms in the chromosphere. Rocket-borne instruments flown from 1946 onwards, developed by R. Tousey, W. A. Rense and others, were the first to detect ultraviolet wavelengths. Figure 1.19 shows the first solar ultraviolet spectrum obtained from an instrument built by Tousey's group at the US Naval Research Laboratory, flown on a captured German V2 rocket. Spectrographs on British and American rockets then began to reveal the presence of numerous other emission lines, some due to atoms stripped of several of the electrons that make up their normal retinue, indicating the high temperatures of the corona and even higher temperatures – up to about $4 \times 10^6 \, \text{K}$ – that occur in coronal active regions above sunspot groups in the photosphere.

In the 1960s, telescopes were built that could produce ultraviolet and X-ray images of the sun. Since glass absorbs in the ultraviolet, reflecting rather than refracting telescopes were developed and, for wavelengths shorter than about 100 nm, the optics had to be designed such that rays of light were incident at glancing angles off reflecting surfaces. Although the stabilization and accurate pointing of a rocket carrying such experiments had reached a high degree of sophistication, the brevity of a rocket flight (typically about five minutes long)

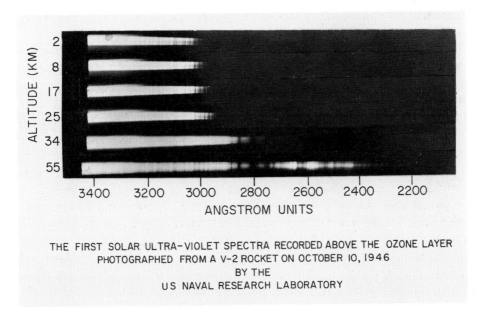

THE FIRST SOLAR ULTRA-VIOLET SPECTRA RECORDED ABOVE THE OZONE LAYER
PHOTOGRAPHED FROM A V-2 ROCKET ON OCTOBER 10, 1946
BY THE
US NAVAL RESEARCH LABORATORY

Fig. 1.19 The first photograph of the solar ultraviolet spectrum. It was obtained with an instrument built by the US Naval Research Laboratory on a V2 rocket launched in October 1946, and shows how, as the rocket altitude increases, the spectrum extends to progressively shorter wavelengths because of the decreasing atmospheric absorption. The broad dark lines at wavelength 280 nm (2800 ångström units, where 1 Å=0.1 nm) are due to the magnesium h and k lines. (See §4.4.) (Courtesy R. Tousey, US Naval Research Laboratory, Washington DC)

gave a considerable advantage to earth-orbiting satellites. The US series of *Orbiting Solar Observatories* (*OSOs*), consisting of a rotating wheel structure and a section permanently pointed at the sun, carried several such imaging instruments. The results showed that coronal active regions had much more intense ultraviolet and soft X-ray emission than 'quiet' regions because of their enhanced densities and temperatures.

Results from these instruments showed the great importance of solar flares at short wavelengths. At the time of a large flare, ultraviolet radiation but more especially X-rays showed large increases. Using spectrometers (i.e. spectrographs with electronic detectors of the radiation instead of photographic film) in which certain crystal material is used to disperse the radiation, US and Soviet instruments detected emission lines in the X-ray region due to atoms like iron with as few as two of the electrons left, the remaining electrons stripped off by the extremely high temperatures – up to 20×10^6 K – occurring in a flare. Gamma rays, at the extreme short-wavelength end of the electromagnetic spectrum, were detected for the first time by an instrument on *OSO-7* built by E. Chupp and collaborators during strong flares in August 1972, showing the presence of extremely energetic particles giving rise to nuclear reactions.

This completes our historical survey of solar observation. The rapid pace of discovery over the past 30 or 40 years has resulted in a much altered perception of the sun's nature, and the following chapters outline what our present understanding is and the problems that remain.

Chapter 2

The solar interior

In this and the following four chapters, we shall be studying in turn the chief parts making up the sun: the interior, surface layers, atmosphere and its extension the solar wind. We shall see what has been learned from observations and how these have been assembled to build our present understanding of the sun and its structure. First, then, we shall investigate what is known about what lies below the photosphere – the solar interior.

2.1 Structure of the interior and its energy source

The solar interior cannot be directly observed, as no radiation emerges from it – almost all comes from the photosphere. Nowadays, we have two important sources of observational evidence about the nature of the interior – neutrinos and solar oscillations – and we shall explore them later in this chapter. Even before these were available, however, it was possible to put together a picture, or 'standard model', of the sun's interior based on both measured quantities and laws of physics. The standard model is a mathematical description of, for example, the way pressure and temperature vary with the distance from the centre of the sun to its surface. The standard model has been refined with the new observational data, though there are still important unresolved issues, which again are dealt with later in this chapter.

The main observational facts the standard model uses are the measured solar mass (2×10^{30} kg), radius (696 000 km), energy output or *luminosity* (3.9×10^{26} W) and age (4.6×10^9 years). The first three quantities are fairly precisely determined in ways that are given in subsequent chapters; the age, the least well-known quantity, is deduced from datings of the oldest meteorites that have been recovered on the earth, and the assumption that the sun is of roughly the same age. There are many other observed facts that will affect the details of the model (such as the solar rotation rate) though not the substance.

The physical principles that the standard model also relies on can be briefly summarized as follows. Firstly, use is made of the fact that the sun is a sphere and in a steady state, i.e. neither appreciably expanding nor contracting. Secondly, it is assumed that energy is generated at the hottest region, the core, by nuclear reactions (or nuclear 'burning'), each involving the fusion of four protons to make a helium nucleus. Thirdly, it is assumed that the sun started its life by contraction out of a cloud of gas made up of mostly hydrogen and helium, the initial (or 'primordial') composition of the sun becoming altered in time by the steady increase of helium through the nuclear burning in the sun's core. Fourthly, the energy generated at the core is transferred by radiation except where convection is a more efficient process. We will shortly deal with these assumptions in some detail.

The standard model is actually a succession of models. The first model in the sequence is constructed by taking a mass of the primordial gas to simulate the sun at its birth as a star. The hydrogen at its core is allowed to burn, i.e. turn into helium, providing the energy the fledgling sun radiated and its heat that, in the form of gas pressure, supported the sun against gravitational collapse. More and more hydrogen turns into helium at the core, and successive models take into account the gradual depletion of hydrogen and formation of helium. After a time equal to the estimated age of the sun, the model should approximate the present characteristics of the sun, in particular it should have the radius and luminosity that are now observed. If this is not so, the sequence of models can be recalculated with adjustments made to two of the input quantities that are poorly known: these are the chemical composition of the primordial gas (in particular how much helium it had) and the nature of convective energy transfer in the outer layers of the solar interior. As has been indicated already, there is now observational evidence about the interior, and so the standard model can be tested.

We now describe the four main assumptions, given above, that the standard model makes. Firstly, the fact that the sun is very nearly a steady-state sphere of gas must mean that at every depth inside the sun the weight overlying material is exactly balanced by the pressure. Most of this pressure is due to the gas, though there is a small amount that must be taken into account due to radiation itself. The assumption is equivalent, to use a phrase in physics, to 'hydrostatic equilibrium'. Gas pressure itself is given by the 'perfect gas' laws, and is proportional to temperature multiplied by density.

The second assumption is that the energy generation at the core is by the fusion of four protons to form a helium nucleus. The dominant process consists of a chain of reactions called the proton–proton (pp) chain. Now, at low enough temperatures, hydrogen consists of a positively charged proton forming its nucleus with an orbiting, negatively charged electron; the charges exactly cancel to give an electrostatically neutral atom. Other atoms also have a retinue of orbiting electrons that balances the positive nuclear charge. At the centre of the sun, where the temperature is many millions of degrees Kelvin, countless violent collisions tear apart all hydrogen atoms as well as the much smaller number of other atoms to form separate protons, electrons and bare atomic nuclei, which all move with great speeds. These particles are available for nuclear reactions.

The first reaction in the pp chain involves the fusion of two protons to form a deuteron, which consists of a proton and neutron (a zero-charged particle with about the same mass as the proton). The two colliding protons must approach each other to within a tiny distance (10^{-15} m) to overcome the strong electrostatic force that tends to repel like charges, and simultaneously one of the protons must decay to a neutron and positron (positively charged electron). This highly improbable circumstance leads to an extremely low reaction rate – only one reaction per particle in about 14 thousand million years! However, there is a vast supply of protons available, so that in fact many such reactions occur. The second step in the pp chain is the fusion of a deuteron with another proton to form a nucleus of an isotope of helium, consisting of two protons and one neutron. The final step is the fusion of two such helium nuclei to form a nucleus of helium of the more common sort (also called an α-particle), having two protons and two neutrons. Other particles with very small or zero mass are also produced.

The reactions can be illustrated by equations in which the particles undergoing the collision are represented on the left-hand side and the products of the collision on the right. We use chemical symbols to denote the nuclei of the elements involved – H for hydrogen, He for helium, with D for the deuteron. Also, we indicate by superscripted numbers the mass of each of the nuclei in units of proton (or neutron) mass. Thus, ^{1}H denotes a proton, ^{3}He a helium nucleus with one neutron, ^{4}He one with two neutrons, and so on. The reactions, in the order just described, are:

$$^1\text{H} + {}^1\text{H} \rightarrow {}^2\text{D} + e^+ + \nu$$

$$^2\text{D} + {}^1\text{H} \rightarrow {}^3\text{He} + \gamma$$

$$^3\text{He} + {}^3\text{He} \rightarrow {}^4\text{He} + {}^1\text{H} + {}^1\text{H}$$

The symbol e^+ indicates a positron, ν a neutrino and γ a gamma ray. Note that, for each occurrence of the third reaction, the first two must have occurred twice. Taking this into account, we can add the numbers of particles on the left-hand and right-hand sides of the equations, cancelling out those occurring on both sides, to give the net result:

$$4\,{}^1\text{H} \rightarrow {}^4\text{He} + 2e^+ + 2\nu + 2\gamma$$

i.e. four protons fuse to make a helium nucleus.

The important point is that every completion of the pp chain results in the release of energy. This is as a consequence of the equivalence of mass (*m*) and energy (*E*), introduced by Einstein, given by $E=mc^2$, where *c* is the velocity of light in a vacuum (3.0×10^8 m/s). The masses of the particles can be conveniently expressed in atomic units, where the mass of the oxygen atom is 16. On this scale, the mass of a hydrogen atom is 1.008 13, so four such atoms have a mass of 4.032 52. However, the mass of a helium atom is only 4.003 86, leaving a difference of 0.028 66 units. Now, one atomic mass unit in SI units is 1.66×10^{-27} kg, so that this mass difference is 4.8×10^{-29} kg. Multiplying by the velocity of light squared (c^2) gives the energy liberated as a result of the mass difference, 4.3×10^{-12} J. Actually, a small amount of this energy (about 0.1×10^{-12} J) is removed by the neutrinos of the first reaction, leaving 4.2×10^{-12} J. It is this amount that is available for providing the sun's radiant energy, that is, for making it shine.

The third assumption of the standard model is that the sun contracted out of a cloud of gas, largely of hydrogen with some helium, and that its chemical composition changed as nuclear reactions proceeded. The sun's age (4.6×10^9 years) is much less than the estimated age of the universe, some 16×10^9 years. This is the time that has elapsed since the 'Big Bang' explosion, in which all the matter in the universe started a steady expansion which continues to this time as is evident from distant, 'red-shifted' galaxies. In the first few minutes of this expansion, the largely hydrogen gas was so hot that nuclear reactions occurred, turning some of the hydrogen into helium. At a much later stage, when galaxies and stars had been formed, more helium and heavier elements were synthesized at the centres of stars. Some of those stars then ended their lives explosively and in doing so returned their 'processed' gas to space. This is the gas out of which a late-generation star like the sun contracted. It was still largely composed of hydrogen, but helium made up about a tenth of it (by numbers of atoms), with very much smaller fractions of heavier elements. As already indicated, a major uncertainty in the standard model is exactly how much helium there was in the material out of which the sun was created.

The fourth assumption concerns the transfer of energy throughout the sun. Three main ways can be considered: conduction, convection and radiation. Conduction is thought to have very little importance in the interior. In the inner part of the sun, radiation is the dominant process. By it, energy diffuses outwards from the core by photons, the massless 'particles' of light. They move out only very gradually as they are continually scattered when encountering free electrons, protons and bare atomic nuclei. Every now and again, a proton or atomic nucleus will acquire one or more orbiting electrons, but near the sun's centre, it is so hot that these electrons are quickly stripped off again. Further out, the temperature has fallen to the point where appreciable numbers of atoms, particularly of heavier elements, have orbiting electrons. This simple fact has a profound consequence for the structure of the interior: because of the existence of atoms with some electrons, the outgoing radiation gets absorbed much more readily – the gas is suddenly more 'opaque'. The amount of radiative energy outflow is reduced, leading to the gas becoming unstable and to the formation of convection currents.

We can construct a thought experiment to illustrate convection. Suppose there is a small bubble of gas at a level inside the sun and we remove it upwards to a cooler level, allowing it to expand quickly enough to prevent any exchange of heat with its surroundings. On releasing the bubble, does the gas (*a*) sink back to its original position (i.e. is denser than its surroundings), or (*b*) continue to rise (i.e. is less dense than its surroundings)? In case (*a*), radiation will still be the dominant transfer process; but at the level where case (*b*) starts to occur, convection becomes more important. In the sun, this level is where the heavy-element nuclei start to acquire electrons, i.e. where the gas is suddenly more opaque, and is about two-thirds of the way out from the centre of the sun. There is still no robust theory of convective energy transfer, but instead it is often assumed that convection proceeds by cells ascending through a certain range (called the mixing length) before merging with their surroundings; one of the uncertainties in the standard solar model, mentioned earlier, is what size the mixing length is.

These points illustrate the assumptions made in calculating the standard solar model. Figure 2.1, from a recently calculated model, shows how temperature and density vary with distance from the centre of the sun. At the centre, the temperature is calculated to be 15.6×10^6 K and the density $148\,000$ kg per cubic metre. The third panel from the top in the figure shows the radial distribution of energy production (energy per unit volume) by nuclear reactions (almost entirely the pp chain). It shows how most of the energy is produced in a comparatively small region near the sun's centre. Also shown are the principal zones of energy transfer – the radiative zone (inner two-thirds) and the convective zone (outer one-third). Convection sets in where the temperature has fallen to about $1\,000\,000$ K. It is possible that giant elongated convective cells, shaped like bananas bunched round the rotation axis, continually transfer material from the bottom of the convective zone to near the surface. (These convective motions should not be confused with those of a much smaller scale nearer the photosphere, the 'supergranules', or the still smaller 'granules', discussed in §3.1.)

Some further points can be made about the convective zones. The occurrence of convective motions does not in any way upset the hydrostatic equilibrium which is assumed in the standard model; the motions in question are too slow (less than 0.1 km/s) for this to happen. However, the giant cells do effectively mix the gas in transferring it from one level to another, so that the chemical composition of the convection zone should be quite uniform. Nevertheless, the products of the nuclear reactions at the core, especially the

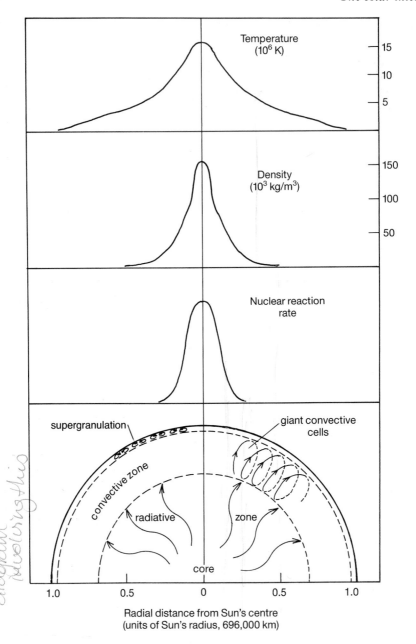

Fig. 2.1 Variations of temperature, density and nuclear reaction rate (energy per
unit volume) with radial distance, with sketch of the principal zones in the
solar interior. (Based on standard solar model of J. N. Bahcall (1989))

helium, remain where they were formed as they are out of reach of the convection cells. There is thus a steady build-up of helium and depletion of hydrogen at the core (the standard model indicating that the proportions of hydrogen and helium are 34% and 64% by numbers of atoms respectively).

Finally, we may ask how much longer the sun's hydrogen fuel will last. We can answer these questions by taking the observed energy output, or luminosity: this is 3.85×10^{26} W. As shown earlier, every pp chain of nuclear reactions liberates 4.2×10^{-12} J. Dividing this number into the luminosity gives the number of reactions occurring at the core per second: it is 9.2×10^{37}. Each reaction results in a mass reduction of 4.8×10^{-29} kg, as calculated earlier, so that the sun's mass is being consumed at the rate of 4.4×10^9 kg per second, or 1.4×10^{17} kg per year. This sounds like an alarmingly large amount! In fact, it is only 7×10^{-14} of the sun's present mass, or not much more than a millionth of the moon's mass. Even over the 4.6×10^9 years that the sun has been shining it would still only amount to 0.03% of its mass. Model calculations suggest that after another 5×10^9 years the sun will have depleted the reserves of hydrogen at its core, and when this occurs it will evolve into a 'red giant' (see §8.3).

2.2   The neutrino problem

Our present understanding is that the solar energy output is derived from nuclear reactions, almost entirely the pp reactions, at the solar core. Every second, some 1.8×10^{38} neutrinos are produced (two for each of the pp reactions), which are emitted in all directions. At the distance of the earth, some 6×10^{14} of them per second cross each square metre. Thus, a human body measuring 2 m high and $\frac{1}{3}$ m wide is crossed by 4×10^{14} neutrinos every second! There is, however, not the slightest interaction with any one of them. But if we could intercept a few neutrinos with some apparatus, we would have an excellent check on the conditions deep in the solar interior without actually being able to see it by conventional means.

The nuclear reactions of the pp chain are the most important for producing the sun's energy, though they are not the only ones occurring, nor are they the only ones that result in neutrinos. In a side-chain to the main pp reactions, for instance, one ^{3}He nucleus and one ^{4}He nucleus (α-particle) combine to form a beryllium nucleus, ^{7}Be, which may capture a proton to form a boron nucleus, ^{8}B. This unstable nucleus decays to form an ^{8}Be nucleus, a positron and a neutrino. This neutrino may have an energy of up to 2.2×10^{-12} J, much higher than the pp chain neutrinos whose energy does not exceed 7×10^{-14} J. However, the ^{8}B neutrinos are much rarer, only about two being produced for every 10 000 completions of the pp reactions. Nevertheless, detection of these neutrinos is of great interest as their numbers very sensitively depend on the sun's central temperature, far more so than the neutrinos of the main pp chain. Also, their greater energies make them much easier to detect.

Though neutrino detection is generally very difficult, it has been found that neutrinos with energies of more than 1.3×10^{-13} J have a slight but possibly measurable interaction with a common isotope of chlorine, in the ^{37}Cl form (consisting of 17 protons and 20 neutrons). This would make it suitable for detecting the ^8B neutrinos but not those of the main pp chain as their energies are less than 1.3×10^{-13} J. The interaction of a neutrino having sufficient energy and a ^{37}Cl nucleus results in an argon (^{37}Ar) nucleus and an electron. The ^{37}Ar form of argon is unstable, decaying back to a ^{37}Cl nucleus and a neutrino.

A vast experiment constructed 1.5 km underground in the Homestake gold mine in South Dakota has been operating for some years now, detecting ^8B neutrinos from the sun using these principles. It was set up by Raymond Davis and colleagues at the Brookhaven National Laboratory (Fig. 2.2). The detector consists of 400 cubic metres of perchloroethylene, a dry-cleaning solvent, held in a cylindrical vessel. It is located deep underground to minimize the effect of cosmic-ray particles which continually bombard the earth. Precautions are taken to reduce the effect of radio-activity in the surrounding rock, which is significant as the solar neutrino 'signal' is so small. The dry-cleaning solvent is left exposed to solar neutrinos for a period of about three months, during which time only a few of the immense number (about 10^{30}) of ^{37}Cl atoms are expected to interact with solar neutrinos. The ^{37}Ar atoms so formed are removed from the apparatus by purging with helium gas, and passed into a detector that counts the number of decays back to ^{37}Cl.

The number of reactions with solar neutrinos in such an apparatus can be calculated using the standard solar model. John Bahcall at the Institute for Advanced Study, Princeton, and colleagues have been calculating the expected reaction rate using increasingly sophisticated models. A unit defining the number of neutrino detections has been introduced (by Bahcall, partly in jest) known as the SNU (pronounced 'snoo'), where 1 SNU equals 10^{-36} interactions per second per atom. A first calculation, in the 1960s, gave the expected neutrino detection rate as 30 SNUs, but this has been much reduced, with estimates over the past few years ranging from 6 to 13 SNUs. A recent (1988) estimate is 7.9 SNU, which is considered unlikely to be wrong by more than 33%. The chief uncertainties in these calculations include those in the standard solar models as well as nuclear rates: it is judged that, if the measured rate is outside the error range, 'someone has made a mistake'.

Great concern has been caused as the neutrino interaction rates measured at the Homestake experiment have averaged only 2.2 SNUs, with a deviation of 0.3 SNU, for a large number (almost 80) of 100-day runs between 1970 and 1988. The experimental number of neutrino interactions is, then, only

Fig. 2.2 The ^{37}Cl neutrino detector in the Homestake Mine in South Dakota, 1500 m below ground level. Raymond Davis, the principal scientist of the experiment, is on the catwalk above the tank which contains 400 cubic metres of perchloroethylene. (Courtesy Brookhaven National Laboratory)

about a third of the calculated number, and the discrepancy is well outside both the uncertainty of the calculated result and the experimental deviations. The solar neutrino 'problem' is the still unexplained difference between observed and expected neutrino interactions. There is no reason to doubt the validity of the Homestake results, but it is desirable to have confirmation from a completely different apparatus. This has recently been provided by the Japanese Kamiokande II experiment. Unlike the Homestake experiment which counts neutrinos from all directions (it is assumed that they are entirely solar in origin), Kamiokande II is a scattering experiment that actually gives the direction of neutrino arrivals, so is able to sort out those neutrinos in the direction of the sun from any uniform background. It was, incidentally, successful in detecting neutrinos from the 1987 supernova in the Large Magellanic Cloud, a nearby galaxy.

The neutrino problem is 'unexplained' only in the sense that there is no consensus as to the cause; astronomers and physicists have come up with plenty of possible reasons for the discrepancy, some more or less improbable. Those connected with the sun, i.e. that involve possible inaccuracies in the standard solar model, are generally attempts at cooling the interior so that the production of neutrinos is reduced to the observed level without affecting the amount of supplied energy; the energy escaping as radiation from the sun's photosphere is always taken to be equal to that generated in the solar interior. One suggestion that has been made involves the existence of a type of elementary particle to explain a large amount of hidden mass that astronomers have found evidence for; this is the weakly interacting massive particle (WIMP). It has been suggested that large numbers of WIMPs were produced in the early universe, and that the sun captured some of them. If so, they could carry some of the energy from the hot core to a region a little further out, so that the number of neutrinos is reduced to the observed amount without reducing the total energy output. There is, unfortunately, no direct evidence of the existence of WIMPs, so the whole idea is conjectural, and there is some recent indication that the standard model without WIMPs is in fact correct; we shall deal with this in 2.3. Another suggestion put forward is that the sun has a rapidly rotating core with strong magnetic fields, which would also reduce the neutrino flux without reducing the energy output, but there are now observations indicating that this is not the case. A third suggestion involves the mixing of material in the sun's core, which would temporarily alter the neutrino rate. As mentioned in §2.1, mixing is not thought to occur near the sun's core as energy is transferred by radiative rather than convective means there, but mixing 'episodes' may take place. This is possible without there being an effect on the observed solar luminosity as radiation takes several million years from its generation at the core to its emergence at the photosphere. However, exactly why periodic mixing occurs is still to be explained.

Other solutions to the problem are related to particle physics rather than the sun. Very little is known about the neutrino, even whether it has mass; it appears to be like the photon, i.e. massless and travelling at the speed of light, though it is quite possible that it does have a very small, as yet unmeasured, mass. One explanation for the solar neutrino problem put forward is that the neutrino, once created at the sun's core, decays into another particle that cannot be detected by present experiments on its way to the earth. But if the neutrino travels at the speed of light, it takes not more than about eight minutes to reach the earth, and so the decay must occur in this time. This idea has now ceased to be acceptable since neutrinos were detected from the 1987 supernova in the Large Magellanic Cloud, and they must have been travelling for thousands of years. However, it is possible that neutrinos 'oscillate' from one form to another. The neutrinos of the pp reactions are known as electron neutrinos (their presence was originally invoked in explaining the decay of neutrons to protons and electrons). There are also neutrino types associated with two other particles – the muon and the tauon – and these would not be detected by neutrino experiments. If the neutrino does have a small mass, the electron neutrino could be transformed to one of the other two types between the sun and earth, and if there is perfect mixing, the fact that the observed neutrino captures are about a third of those expected is then explained. It has been recently suggested that this mixing would be assisted when neutrinos pass through the material making up the sun, the neutrino 'picking up' some mass on its way; this has become known as the MSW effect (after the initials of the authors who first proposed it).

An intriguing possibility is that the neutrino capture rate as measured with the Homestake detector varies in anti-correlation with the sunspot number – fewest events are detected at the time of most numerous spots. This has been concluded by some from results since 1979 (Fig. 2.3). Some have dismissed this as a statistical fluke, though there may be a reason for the neutrino modulation. If the neutrino has both a mass and a magnetic field associated with a spin, it is possible that the neutrino may perform a spin 'flip' in the presence of the sun's magnetic fields which are enhanced at times of high solar activity. The spin-flipped neutrino will not be detected by the Homestake and other experiments, so a reduction in the number of captures is expected. We must wait and see whether the anti-correlation indicated by Fig. 2.3 continues to hold up.

The great interest in the solar neutrino problem has led to several proposals for experiments that are either larger than those currently operating or of a different type. Large detectors using gallium are being built, one in Italy, the other in the CIS; they will detect the neutrinos of the main pp reactions, which are less energetic than the ^{8}B neutrinos. Another detector that will do this using indium is being developed in Europe. A major new detector using 'heavy water' (D_2O instead of H_2O), a collaboration of Canadian, US and British groups, is under construction in Ontario. This will observe neutrinos both by capture and by neutrino-scattering using a variety of detectors. The next decade or so may see much improved observational evidence about neutrinos, though it is less certain that we shall have a convincing explanation of the solar neutrino problem in this time.

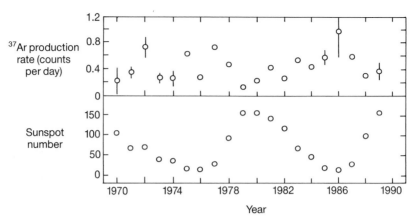

Fig. 2.3 Yearly averages of the neutrino capture rate (as measured by the number of ^{37}Ar atoms produced) at the Homestake Mine experiment and of the sunspot number, showing possible anti-correlation. (Adapted from Bieber *et al.* (1990))

2.3   Solar oscillations

Apart from solar neutrinos, we have further observational evidence about the sun's interior that may be used as a test of the standard solar model. This is in the form of oscillations at the sun's photosphere that are the surface manifestation of subtle seismic activity occurring throughout the solar interior. And just as terrestrial seismology provides information about the structure of the earth, so the new subject of *helioseismology* gives solar astronomers a means of studying the solar interior. However, the earth's structure – a solid crust and mantle with a core that is liquid in its outer part and possibly solid at its centre – is quite different from the sun which is entirely gaseous. This means certain wave motions – specifically shear waves – that propagate in the earth do not do so in the sun. The wave motions that are observed are of the nature of sound (or pressure) waves, though it is possible that those due to gravity are present.

Solar oscillations were first detected in 1960 by R. B. Leighton, R. W. Noyes and G. W. Simon at the Mount Wilson Observatory. They measured small displacements in the wavelengths of absorption lines in the photospheric spectrum to derive velocities by the Doppler effect. This was done at several points over the solar disc. What was found was a pattern of temporary oscillatory motions at various locations; the period of the waves was always about five minutes, and the maximum velocities seen (velocity 'amplitudes') were about 0.5 km/s towards or away from the observer (i.e. lines shifted to the violet or red end of the spectrum respectively). The pattern, having appeared at some point, would persist for about half an hour (or about six cycles of the wave motion), then die away, but a similar pattern would then be in progress elsewhere. Figure 2.4 shows similar observations of such motions taken over a narrow strip of the sun's disc 60 000 km long in a period of about 100 minutes.

Although not understood at the time, it was later suggested, then confirmed, that these surface motions were due to sound waves excited and trapped in the solar interior. To see this, we will first examine the propagation of sound waves.

Sound waves, having been excited by some disturbance, propagate through a gas as a progression of parcels of compressed and rarefied gas. The pressure of the gas acts as a restoring force which tends to bring back the gas to its undisturbed state. Just as light waves undergo reflection at a surface, sound

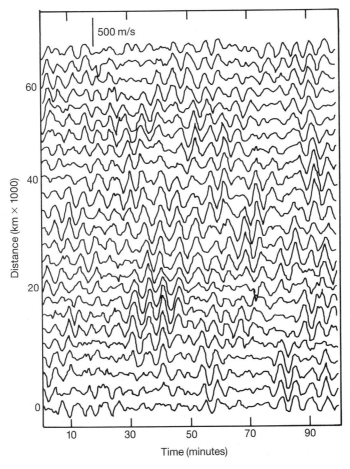

Fig. 2.4 Five-minute oscillations revealed by plots of velocity at various locations in a 60 000-km-long strip of the sun's surface over a period of 100 minutes. The scale of the velocity amplitudes is indicated at the top of the figure. (From Howard, Tanenbaum and Wilcox (1968))

waves too may be reflected from a boundary (Fig. 2.5a). An acoustic or 'resonance' cavity may be formed when a sound wave travels repeatedly (is 'trapped') between two solid boundaries (Fig. 2.5b). The result is a wave motion that progressively builds up and fades away as a whole, with 'nodal' points where the displacement and velocity are always zero. This motion is called a standing wave or *mode*. The condition for a mode is that its wavelength must be twice the distance between the boundaries (the fundamental mode), or equal to submultiples $(1, \frac{2}{3}, \frac{1}{2}, \frac{3}{4}$, etc.) of the fundamental wavelength (known as overtones or harmonics).

Acoustic cavities exist in the solar interior, even though there are no solid boundaries. A sound wave that propagates into the solar interior from the surface at an angle (Fig. 2.6a) will travel in an arc. The reason is that the speed of sound is greater at greater depths in the sun where it is hotter (it is proportional to the square root of the absolute temperature). Thus, the wave front (plane perpendicular to the direction of the wave) has its deepest part travelling at a greater speed than the shallowest part; the wave front is therefore turned (or 'refracted') until the wave is again directed towards the sun's surface. The solar analogue of the acoustic cavity has the sun's surface and bottom of each of these arcs as boundaries. The solar analogues of wave modes are those waves that are trapped between these boundaries; they are called p modes, 'p' standing for pressure which is the restoring force. A mode traces out a sequence of arcs below the sun's surface, as Fig. 2.6a indicates, producing oscillations at the surface that can be observed by their Doppler

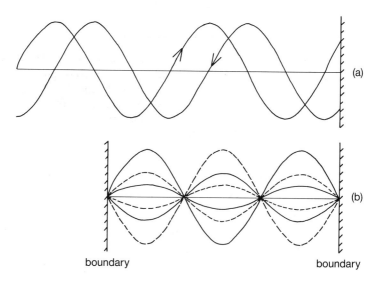

Fig. 2.5 Wave motion to illustrate formation of solar oscillations: (a) wave being reflected from a single boundary; (b) standing wave formed by wave reflected at two boundaries located at wave nodes.

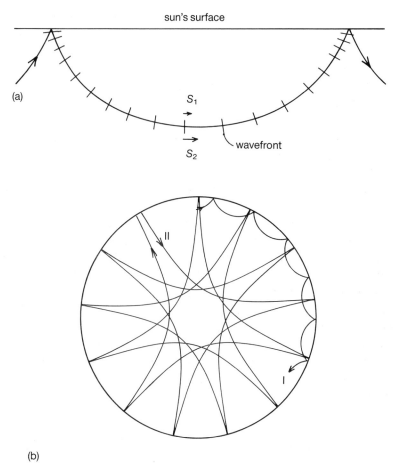

Fig. 2.6 (a) Wave being reflected at two points in the solar surface. The short lines along the wave's motion represent wave fronts, which on propagating downwards are turned round by the fact that the speed of sound is larger at larger depths, e.g. is larger for point S_2 than S_1. (b) Wave modes penetrating to two depths in the sun, both large fractions of the sun's radius, so forming global oscillations.

shifts. A horizontal wavelength (value of wavelength measured parallel to the sun's surface) and a period can be defined for each mode, both of which could in principle be determined from the observed oscillations. For each mode, there is a whole number of wavelengths (called the mode order) along the path of propagation between the surface and the bottom of the path and back to the surface again. For a particular horizontal wavelength, there is a fundamental mode, i.e. mode of order one, having one wavelength between successive reflections at the surface, and overtones which are shorter-period modes, penetrating deeper the greater the order.

These points can be illustrated by a power spectrum in two dimensions that is a graph of observed wave period (or its inverse, frequency) against horizontal wavelength. The 'degree' l is sometimes used instead of horizontal wavelength, and is equal to the solar circumference divided by the horizontal wavelength. Such a graph can be constructed from measurements of Doppler velocities as a function of position on the sun and time. Figure 2.7 shows a

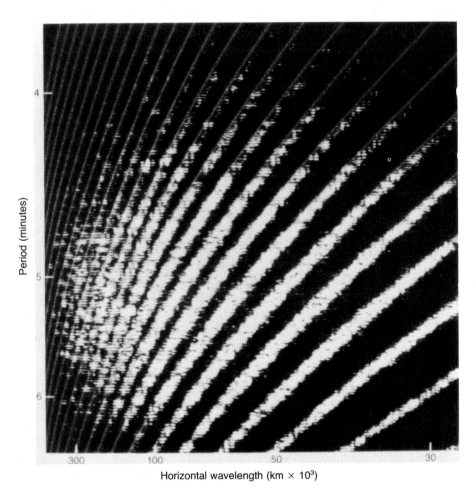

Fig. 2.7 Power spectra of observed solar oscillations, shown by the bright curves in this plot of wave period against horizontal wavelength. Only those waves with certain combinations of period and horizontal wavelength are trapped inside the sun, as indicated by the thin white lines, based on calculations with a standard solar model having a helium abundance of 25% by mass. The agreement of the observed spectra with the theory curves shows that the sun's interior does behave as an acoustic cavity and that the solar model is basically correct. (From Leibacher *et al.* (1985), adapted from Duvall and Harvey (1983); courtesy National Solar Observatory, Tucson, Arizona)

graph that was produced in this way for a period of a few days in 1981. Several million modes are present on the sun at any time, but the strongest oscillations can be picked out as a pattern of bright ridges, each of which belongs to a particular mode order (orders greater than two are shown in the figure). While the velocity amplitudes of an individual mode are tiny (less than about 20 cm/s) and well below the observational limit, the observed oscillations have amplitudes of up to 0.5 km/s because of a random superposition of the many modes present. The strongest oscillations have periods averaging about five minutes, i.e. a frequency of 3.3 millihertz (mHz), and have a very wide range of mode degree (from very low values to about a thousand). The corresponding horizontal wavelengths are from 4000 km to hundreds of thousands of kilometres.

Modes having intermediate values of degree – from, say, 50 to 150 – penetrate only the upper tenth of the sun's radius. But when the mode degree is very low, the horizontal wavelength and penetration depth are large fractions of the solar circumference (equal to 4 370 000 km), and the solar curvature must be taken into account. The paths of propagation in the solar interior of two such modes are illustrated in Fig. 2.6b. Now, a mode with degree less than about 100 travels around the sun in a comparatively short time (a few hours), but such modes are estimated to persist for times much longer than this, so for them there is a further restriction: only those modes that 'constructively interfere' with themselves around the solar surface exist, i.e. a whole number of horizontal wavelengths for such modes must fit into a solar circumference. For these 'global' oscillation modes, the nodes, i.e. where the observed surface velocity is always zero, are in the form of circles round the entire sun. The number of these nodal circles corresponds to the degree l of the mode. Figure 2.8 shows a computer-generated solar image for a single trapped acoustic wave.

Extreme cases of very-low-degree modes are those with l equal to 0, for which the whole sun pulsates in and out in phase, or l equal to 1, in which one hemisphere expands in phase while the other contracts in phase. The lower the degree, the greater is the fraction of the entire sun that moves in phase. Observing the sun's integrated light, i.e. without forming an image of the sun, is therefore a way of separating these modes from those of higher degree. A spectrum of the sun's integrated light is formed and the Doppler displacements of certain absorption lines are measured over a period of time. As individual oscillation modes have small velocity amplitudes, very careful measurements must be made. The technique used by researchers at the University of Birmingham, who are pioneers in such observations, involves the precise comparison of the wavelength of a solar absorption line due to potassium with its laboratory wavelength. Observations are continued over the whole time the sun is observable at mountain-top observatories in Hawaii and Tenerife and a station in Australia. Combining the sets of observations can

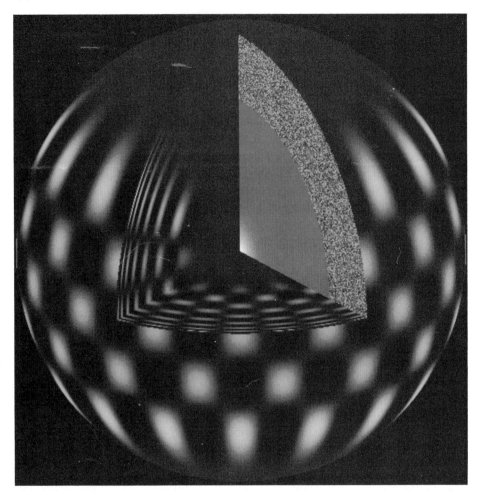

Fig. 2.8 A computer-generated image of the sun showing zones of gas that are
approaching (dark spots) and receding (light spots) from the observer when
a three-dimensional sound wave resonates inside the sun. An observer
would be able to detect this by the oscillations produced by the wave at the
solar surface. The cut-away shows the wave motion propagating in the
solar interior, and the extent of the convection (outer one-third) zone.
(Courtesy J. W. Leibacher)

give continuous coverage of the sun for days at a time. Researchers at Nice
University have a similar technique using one of the sodium D lines. They
have sited their equipment in the Antarctic from where, during the southern
summer, observations can continue for extended periods of time.

These investigations have shown the existence of several oscillation
periods, all in the region of five minutes. They occur as sharp peaks when the
data are plotted as a power spectrum, a one-dimensional version of Fig. 2.7.
The University of Birmingham measurements for an extended period in 1981

are plotted in this way in Fig. 2.9. The peaks occur as closely spaced pairs which, by comparison with calculations, are identified as having degree *l* equal to 0 and 2 or 1 and 3. The frequency spacing of the 0 to 2 peaks is only 9.5 µHz, and of the 1 to 3 peaks 15 µHz. These extremely small frequency differences need very long stretches of data to resolve, hence the endeavour to get the longest possible coverage from several sites.

Standard solar models, mentioned in §§2.1 and 2.2, may be used to calculate the characteristics of solar p mode oscillations which can then be compared with observations. This is because standard models give the temperature structure of the sun and therefore the speed of sound waves traversing the interior; from this information, frequencies (or periods) and horizontal wavelengths (or degrees) of modes can be predicted. This was done for Fig. 2.7, for example, where the thin white lines represent calculations from a standard model. They closely match the observed oscillations that show up as the bright ridges, but there are differences which are more than the estimated errors of the observations, suggesting the need for refinements. Another test of the standard-model calculations is the splitting of the low-degree peaks in the integrated-sunlight data (Fig. 2.9). This is of interest as these low-degree modes give information about the solar core to which they penetrate. As mentioned, this splitting is 9.5 µHz for the peaks with *l* equal to 0 and 2. Until recently, standard-model calculations gave this splitting as 10.6 µHz, which is considered significantly larger as the observations are so precise. These observations also seemed to point to needed refinements in the standard model.

As already indicated, the standard model can be adjusted by changing two of the input quantities, the primordial helium abundance and the nature of convective energy transport. It was found, for example, that on increasing the helium abundance, the calculated splitting of the *l*=0 and 2 peaks can be made to agree with the observed amount. There was concern that doing this worsened the neutrino problem since more helium would enhance the cal-

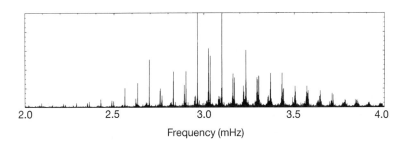

Fig. 2.9 Power spectrum of global oscillations obtained from velocity data using integrated (unimaged) sunlight over three months in 1981 from University of Birmingham instruments at its stations in Tenerife and Hawaii. (Courtesy G. R. Isaak)

culated neutrino emission, already much more than that observed. Several
suggestions or contrivances were proposed as ways of avoiding this dilemma
(see 2.2). However, recent standard models seem to remove the need for this
as agreement between observed and calculated splittings now appears to have
been achieved. There is, apparently, no longer any need for WIMPs or
mixing episodes, and the standard model seems to be a satisfactory description of solar oscillations if not the observed neutrino emission.

Attempts have been made to measure the rotation of the sun's interior
using solar oscillation data. It is most likely that oscillations of intermediate
degree especially will show this since a rotating interior will slightly alter their
frequencies. The results so far indicate that the rotation rate decreases as
depth increases, though this may not apply to the core itself, and that there is
a latitude dependence. This is surprising since it had been assumed that the
sun originally rotated much faster than at present, and that the interior might
still be rotating at the original rate.

Other oscillation modes are expected to propagate in the solar interior in
which buoyancy rather than pressure is the restoring force. These are the g
modes, 'g' standing for gravity. They can propagate in a medium where there
are no convection currents, i.e. in the sun's radiative zone. Since this is deep
down, it might be thought that g modes would not have observable effects at
the surface, but in fact their amplitudes are such that this may be so. They are
believed to have much longer periods than the p modes, of the order of an
hour. Great interest has centred on a possible oscillation with period equal to
160 minutes, noted for some years. However, the fact that 160 minutes is
exactly a ninth of a day has given rise to a suspicion that some terrestrial effect
is responsible.

The burgeoning subject of helioseismology has led to a considerable number of plans for future instrumentation to make further observations. Net-
works of observatories round the world will enable long stretches of data to be
acquired so that searches for long-period oscillations can be made and power
spectra refined. The University of Birmingham's stations in Tenerife, Hawaii
and Australia have already been mentioned. The Nice group is leading a
project called the International Research on the Interior of the Sun (IRIS),
which will be a network of instruments observing the spectrum of integrated
sunlight, like those used in the Antarctic. The network, already partly set up,
will operate in Chile, California, the CIS, Morocco, Tenerife, Australia,
China and Nice itself. An instrument for observing oscillations over an entire
image of the sun has been developed for the Global Oscillation Network
Group (GONG), funded by the US National Science Foundation. It will
gather oscillation data from 62 500 'pixels' or small areas covering the solar
disc using precise measurements of Doppler shifts in a solar absorption line.
Some six instruments are planned in a world-wide network. Spacecraft
instruments have also made helioseismology studies by measuring the time

variations of total solar 'irradiance', or amount of energy received (see §9.1). They include the ACRIM instrument on the NASA *Solar Maximum Mission* (*SMM*) which operated between 1980 and 1989, and an instrument on the Soviet Mars probe *Phobos*. The low-degree p modes seen in Doppler measurements of integrated sunlight were observed by both these instruments. A helioseismometer based on the same principles as the integrated-sunlight instrument of the Nice group will be part of the package of instruments on board the joint US and European *SOHO* (*Solar Heliospheric Observatory*) spacecraft, to be launched in the mid-1990s. It will be placed in an orbit about an equilibrium point between the earth and sun, and will be able to view the sun uninterruptedly for indefinite periods of time. It will thus be able to search for long-period g modes. An immensely improved amount of high-precision data can therefore be expected in the next few years.

2.4   The origin of the sun's magnetic field

Magnetic fields play a dominant part in giving rise to solar activity. Thus, sunspots are locations of strong magnetic field, and sunspot pairs or groups are basically bipolar, with the polarities showing reversals every 22 years, or two sunspot number cycles (§1.7). There are also numerous other features in the photosphere, chromosphere and corona dealt with in Chapters 3–6 which are also associated with fields.

Observations of these magnetically associated features indicate that they are locations where the field is raised to the sun's surface from the interior. It is possible that when the sun was first formed by contraction of a gas cloud, a 'primordial' magnetic field was dragged in with the gas so that the interior still contains this field. The idea that the surface field arises from this primordial field, most of which must lie deeply buried in the sun's radiative zone, has been vigorously defended over the years by, most notably, J. H. Piddington. But it is widely accepted that there is a mechanism for the continual regeneration of magnetic fields from motions of the hot gas making up the convection zone. This is known as the *dynamo* mechanism.

By the action of a dynamo, the kinetic energy of an electrically conducting body is converted into magnetic energy. Familiar examples in everyday life are the dynamos or alternators attached to motor cars. The basic principle can be briefly stated as follows. When a loop of wire is allowed to rotate in the field of a permanent horseshoe magnet, a current is induced. If instead of a single loop an entire coil is used, the induced current is large enough to be useful, and may be led away (via slip-rings and carbon brushes) to supply lights and other electrical devices. Alternatively, the permanent magnet may be replaced by an electromagnet, i.e. a piece of soft iron magnetized by a coil of wire surrounding it (the 'field' winding) carrying a current. Some of the current induced in the moving coil may be fed back to the field winding, increasing the current flowing in it and so increasing the field of the electromagnet. The energy of the rotating coil's motion, according to the dynamo principle, is converted to the energy of the magnetic field produced in the electromagnet.

Although the solar interior does not have solid wires or solid magnets to constitute a dynamo, it is composed of a highly electrically conducting gas, made up of freely moving charged particles (electrons, protons and ions). Such a gas is called a *plasma*. A consequence of the high electrical conductivity is that any magnetic field is 'frozen' into the plasma, a concept introduced by Hannes Alfvén, the Swedish Nobel laureate who is one of the leaders of the subject. Thus, when the plasma moves, e.g. by convection, the field is constrained to move with it. It has been found that, starting with a small 'seed' magnetic field, certain motions of the plasma can generate much larger fields, i.e. the solar interior could act as a dynamo. More particularly, these motions could give rise to fields that periodically oscillate, so possibly explaining the solar activity eleven-year (or magnetic 22-year) cycle.

Several theories of how the solar dynamo mechanism operates have been proposed. Generally, the essence of the problem is, starting from a magnetic field geometry in which lines of force are parallel to lines of longitude (a *poloidal* field), how to generate a field with lines orientated parallel to lines of latitude (a *toroidal* field), as sunspot fields appear to be; then to have a feedback process that regenerates the poloidal field from the toroidal. This would account for the oscillatory nature of the sunspot cycle. Another feature that needs to be explained is the 'butterfly diagram' of sunspots, i.e. the migration of spots towards lower latitudes during the course of a cycle, also known as Spörer's law (§1.2).

An early, semi-observational model by H. W. Babcock in 1961, since modified by R. B. Leighton and others, accounts for the main features of the solar 22-year magnetic cycle by frozen magnetic fields and non-uniform (or differential) solar rotation – the rotation period is about 25 days near the equator but about 27 days for high-latitude spots (§1.2; see also §3.2). Babcock's picture was of an initially poloidal field, with the field lines just beneath the photosphere (see Fig. 2.10). By differential rotation, these field lines, being attached to the solar interior gas, are drawn out in longitude. In a time of about three years, they are wrapped round the sun about five turns, forming a spiral pattern. It is from these bands of field lines that the sunspots are pictured as emerging. They are considered to be due to previously submerged bundles of field lines which, by turbulent motions, become twisted into rope forms and rise to the surface by magnetic buoyancy: this is more fully described in §6.5. This emergence of field produces sunspot pairs or groups or simple bipolar magnetic regions without spots. The parts of the stretched-out field lines with maximum strength move progressively nearer the equator, so explaining Spörer's law of spot migration. The regeneration of the poloidal

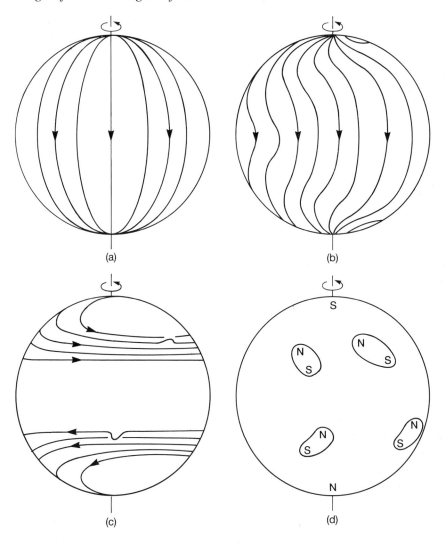

Fig. 2.10 Simplified version of Babcock's scheme for generation of sunspot
magnetic fields. In (a), the sun's magnetic field, present just below the
photosphere, is entirely poloidal. In (b) and (c) this field is deformed into
a partly toroidal field by the sun's differential rotation. At the time of
sunspot maximum (c), active regions are formed as the field breaks
through the photosphere to form bipolar groups with opposite polarity in
N and S hemispheres, as is observed. In (d), in the waning stage of the
sunspot cycle, the 'following' polarity areas of active regions move
polewards, neutralizing an already existing polar field (not indicated in a–
c) which was previously of opposite polarity (polarities are indicated by 'N'
and 'S' symbols). Thus, the poloidal field of (a) is regenerated, but with
opposite polarity, so explaining the 22-year magnetic cycle of the sun.

field from the toroidal is explained in the model by the observational fact that spot groups are generally tilted with respect to lines of latitude, with leading parts nearer the equator than following parts. The leading parts of groups in the northern and southern hemisphere, having opposite polarities, neutralize each other's magnetic field as they approach the equator. Meanwhile, the following parts migrate towards the poles where their fields begin to neutralize the weak polar field, eventually replacing this field. The last stage in the sequence is a magnetic configuration in which the field is completely poloidal but with polarity opposite to the initial poloidal field. From this, a toroidal field can be regenerated by differential rotation, and so the sequence can begin all over again, but with the spots and other bipolar regions having reversed polarity.

An objection that has been raised to the Babcock and similar models concerns the location of the field lines in the solar interior. Babcock and others took them to be just below the photosphere, i.e. in the upper part of the convection zone. But the buoyancy of the magnetic-field rope structures, tending to bring them to the surface to cause spots and other features in the model, is now known to occur rather too readily: the problem is to find a way of stopping all the field from being brought up to the photosphere. The field lines in later models are originally sited in the lower part of the convection zone or even just below it, in a region where downward-moving convection cells 'overshoot' slightly into the radiative zone; they are brought up with rising cells of gas.

Other models have been proposed, but most use Babcock's idea of differential rotation to generate a toroidal field from a poloidal. A number of models, starting with that of E. N. Parker in 1955, have considered the motion of convection cells to be the mechanism for generating the poloidal field back from the toroidal field. According to this, the bands of toroidal field are disturbed by the rising and sinking convection cells; thus, a cell of rising gas (see Fig. 2.11) takes the field lines upwards while cyclonic (or 'Coriolis') forces produce a twist. These cyclonic motions are the same as those producing cyclonic rotation of winds round a region of low pressure in the earth's atmosphere; in the earth's northern hemisphere the pattern is an inward spiral with an anti-clockwise sense, while in the southern hemisphere, it is an inward spiral of clockwise sense. Thus, looking down on the sun at the rising cell, the field lines as they rise are twisted in an anti-clockwise sense in the northern hemisphere, clockwise in the southern. This twisting of field lines tends to be counteracted by the action of sinking cells which produce opposite twists. However, there is a net twist due to the expansion of rising cells but contraction of sinking cells. The result is the generation of small sections of poloidal field from the toroidal field bands. These small sections are then supposed to be linked together to form a larger-scale poloidal field. This is stretched by differential rotation, tending to reinforce the original field at the

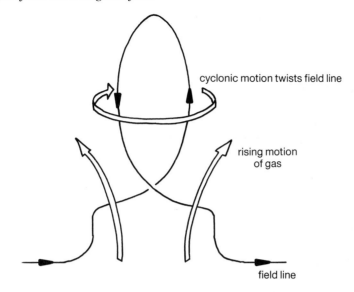

cyclonic motion twists field line

rising motion of gas

field line

Fig. 2.11 A magnetic field line is carried upwards by a rising cell of gas, then turned by cyclonic motion. The clockwise sense shown is appropriate for the southern hemisphere of the sun.

equatorward edge of each of the toroidal bands but cancelling the field at their poleward edges. A migration of the toroidal field bands results, so explaining the decreasing latitude of spots during the eleven-year cycle. Meanwhile, a new band of toroidal field of opposite polarity is created at high latitude to continue the process.

A major and still unanswered objection levelled against this mechanism is that as a consequence of the rising and sinking cells the rotation rate should increase with increasing depth. In fact, helioseismology results (§2.3) indicate the reverse, and these are supported by some theoretical considerations.

A second objection has been raised by T. G. Cowling, not only against Parker's mechanism but against all dynamo models. This concerns the way in which small sections of poloidal field are linked to form a larger-scale field. The mechanism involved is called turbulent diffusion. In fact, this results in the mixing of elements of magnetic field, not joining them up, which can only be done through the diffusion of field (or equivalently the reconnection of field lines). This is dealt with in §6.5 in discussing solar flare theories, but meanwhile it can be briefly said that, in the highly conducting gas of the solar interior, reconnection proceeds extremely slowly, perhaps too slowly to explain the evolution of the sun's eleven-year cycle.

These problems are still being tackled, but it has led some to reject the essential idea of the dynamo mechanism, the regeneration of fields from motions of gas. Piddington has championed the position that the sun's captured primordial field is the origin of that seen at the surface. In his picture, a poloidal field in the radiative zone produces a toroidal field with one orientation for 11 years, the reverse for the succeeding 11 years; this field at some stage in its development becomes so strong that it bursts through the convection zone to form the surface field. The theory has received much criticism (e.g. what is the origin of the toroidal field oscillation?). On the other hand, while the objections to the dynamo mechanism lack convincing answers, it cannot be said that the sun's magnetic field is completely understood.

Chapter 3

The solar photosphere

In our imaginary journey outwards from the solar interior, we have arrived at the solar photosphere. Up till now, we have referred to the photosphere – loosely – as the sun's surface, though it is in fact a very thin layer of the sun's atmosphere. It is from here that the bulk of the energy, mostly in the form of visible and infrared light, is radiated: hence the name photosphere or 'light sphere'. In this chapter, we will describe the fine features of the photosphere, the radiation emitted from it and what information has been gained from analysis of its spectrum.

3.1 The solar granulation and other fine structure

Even a modest-sized telescope reveals that a fine granulation is visible all over the photosphere, especially near the centre of the disc. It has been variously called 'mottling' and 'rice grain patterns'. The individual features making up the pattern are generally known as granules. Telescopic observations are hampered by the blurring due to turbulence in the earth's atmosphere or in the telescope itself, but at the occasional moments of good seeing the granulation appears with startling clarity. To record features during such intervals, bursts of photographs may be taken and those of best quality selected later. Some excellent photographs have been obtained this way at the Pic du Midi Observatory (Fig. 3.1). A recent development involves the use of a video camera and the automatic selection of the recorded images with a seeing monitor. High-resolution images have also been acquired with instruments above much of the earth's atmosphere: these include the Project Stratoscope telescope, mentioned in §1.6, and the Solar Optical Universal Polarimeter (SOUP) on *Spacelab 2*, launched on the NASA Space Shuttle in 1985. The smallest features recorded measure 0.25 arc seconds (175 km on the sun).

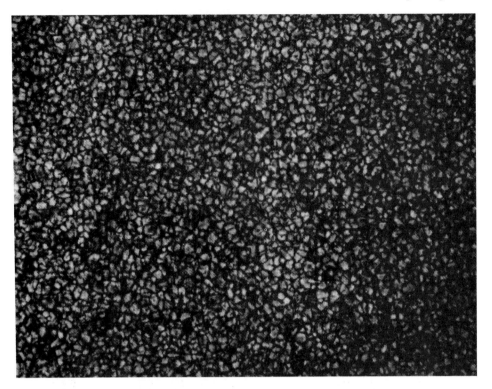

Fig. 3.1 Granulation observed at the high-altitude Pic du Midi Observatory.
(Courtesy R. Müller, Observatoire de Pic du Midi)

A 'typical' granule, if such can be defined, measures about 1.5 arc seconds
(1100 km) across, though it is not clear whether there is a definite size 'scale'
for granules since they seem to be steadily more numerous the smaller they
are. The larger granules are bright, polygonal areas separated by darker
channels, called intergranular lanes; a typical distance between two granules
is about 1.9 arc seconds (1400 km). The smaller ones appear less regularly
shaped. Their appearance is altered near sunspots, becoming lengthened
when in contact with the penumbral boundaries of spots, and occur (as
'umbral dots') within spot umbrae (§6.1). It has been claimed that granules
are on average smaller at sunspot maximum than at minimum. The brightest
part of a granule is generally about 30% brighter than the intergranular lanes;
this translates, by the Stefan–Boltzmann law, to a temperature difference of
400 K.

Granule lifetimes (including their development and decay stages) average
about 18 minutes, with the largest granules lasting the longest. A granule is
born out of the fragments of a previous granule that has broken up, or else
from the merging of neighbouring granules. They usually expand, then end
their brief existence by fragmenting, merging with other granules or merely
fading. For some bright granules, often with a dark centre, the fragmenting

may be more dramatic, the granule being described as 'exploding'. Movies from the SOUP instrument suggest these are quite common events. A horizontal flow pattern seems to be associated with an exploding granule, covering about 5000 km and lasting several tens of minutes, and this has been called 'mesogranulation'.

Granules are associated with upward motions and intergranular lanes with downward motions, with a difference of velocities of perhaps 2 km/s. This is revealed by spectrograms in which solar features are imaged along the length of the entrance slit. Figure 3.2 is an example of such a 'wiggly line' spectrogram; the granules, which show as bright streaks across the spectrum, are

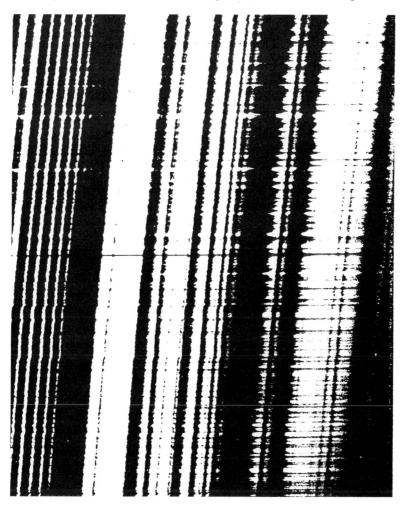

Fig. 3.2 'Wiggly-line' spectrogram. Granules appear as bright lines across the spectrum, and the displacements of the Fraunhofer lines show that they are rising cells of gas. Other 'wiggles' in the lines are due to the five-minute oscillations and the supergranulation flow. (Courtesy National Solar Observatory/Sacramento Peak)

associated with approach velocities (Fraunhofer lines Doppler-shifted to the violet).

There is another aspect of solar granules shown by observations made with the telescope at the Swedish Solar Observatory, using electronic imaging under excellent seeing for a period of more than an hour. During this time granules were seen to spiral and disappear into a vortex, some 5000 km in diameter. The granules' shapes became altered along the direction of the spiralling motion. As only a small area of the sun was examined, there could be many thousands of such vortices, and their presence may be important in a wider context (§5.7).

It seems likely that granules are rising convection cells of hotter gas and intergranular lanes descending currents (or 'down-draughts') of cooler gas. There are strong horizontal flows from the centres of granules towards the intergranular lanes. It has been pointed out that, if the depth of a granule is about the same as its size (say, 1000 km), granules arise at about the level below which hydrogen begins to be ionized. This may be important as a driving mechanism for granules since energy is released when hydrogen atoms recombine. But recent computer simulations suggest that granules are surface phenomena, with the hot ascending currents moving through a height range of only 100 km before releasing their energy as radiation. A small part of this radiated energy is absorbed by the upper photosphere, which is as a result hotter than it would otherwise be. The competitive processes of convective cooling and radiative heating lead, in this model, to large temperature fluctuations over small distances which in turn give rise to the observed granules.

A much larger-scale but more subtle convection occurs on the sun, known as *supergranulation*. It was first recognized (by A. B. Hart in the 1950s) from the Doppler shifts of Fraunhofer lines which indicated horizontal flows occurring over tens of thousands of kilometres. The images produced by Leighton, Noyes and Simon (who had also discovered the five-minute oscillations: §2.3) showed that there were cell structures, about 30 000 km across and lasting a day or so, revealed by an outward, almost horizontal flow of material from the centre of a cell to its sides, with velocities of 0.4 km/s. There is a weak upward flow at the centre of each cell, and a downward flow of about 0.1 km/s at its edges. They show up in cancelled 'Dopplergrams' (images that show areas of approaching or receding velocities as light or dark – see §10.1) most clearly when near the solar limb, as there the larger horizontal flows are most nearly in the line of sight (Fig. 3.3).

The improved resolution of solar photographs in recent years has resulted in the identification of a very fine bright structure in spectroheliograms taken in the light of weak Fraunhofer lines or in the wings of the $H\alpha$ line, both of which are formed in the photosphere. This consists of tiny bright points ('filigree') strung along the dark lanes between granules, with frequently a

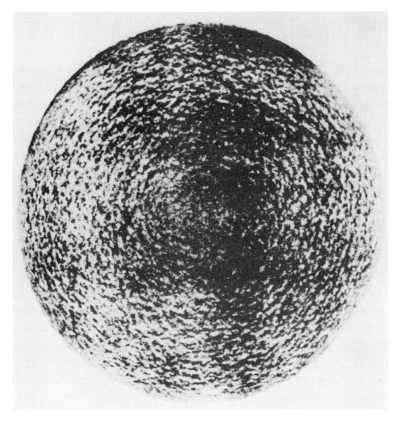

Fig. 3.3 Slight, near-horizontal motions at the boundaries of supergranules are revealed by this 'photographically cancelled' image, produced by the superposition of spectroheliograms in the red and violet wings of the Hα line. Areas approaching the observer appear light, those receding dark. (Courtesy R. B. Leighton, Mount Wilson Observatory)

clustering to form linear structures ('crinkles'). The smallest elements are perhaps 150 km in size (0.2 arc seconds), lasting for about 20 minutes. They are a few hundred degrees Kelvin hotter than the surrounding photosphere, and are associated with high magnetic field (§3.8). A 'photospheric network' composed of filigree and crinkles (called collectively 'network bright points') is recognizable, coinciding with the boundaries of the supergranule cells.

Connected with the network bright points are *faculae*. The most conspicuous are seen in the neighbourhood of sunspot groups – they are discussed in §6.1 – though others occur at very high latitudes and are therefore known as 'polar faculae'. Both spot-associated and polar faculae are associated with high magnetic fields and both vary in number over the course of the solar cycle; however, spot-associated faculae are most numerous at sunspot maximum whereas polar faculae are most numerous on the rising phase of the sunspot cycle.

3.2   The solar differential rotation and large-scale flows

As indicated in §§1.2 and 2.4, observations of sunspots show that the solar photosphere rotates *differentially*: spots near the equator rotate in periods that are about two days less than those at highest latitudes, about 40° north or south. However, there is not only a latitude dependence as certain sunspot characteristics affect rotation rates. Very long-lived, regular-shaped spots, for example, rotate relatively slowly, but rapidly developing spots rotate faster. Certain 'proper motions' of spots occur, such as the separation of the components of a spot pair shortly after their birth (§6.1), and these affect the observed rotation rates.

The rotation period can in principle be found spectroscopically, by observing the Doppler shifts of Fraunhofer lines for different points along the sun's limb – points along the west limb, for example, are receding from the observer because of solar rotation, and so give rise to a red shift of a spectral line. The largest velocity, at the equator, is about 2 km/s, giving a shift of only 0.003 nm at a wavelength of 500 nm. Finding the differential rotation rates from such Doppler measurements was until fairly recently not nearly as accurate as from sunspots. But instrumental developments have resulted in considerable refinement. The present technique uses a 'Doppler compensator', a very sensitive spectrograph in which the Doppler shift is precisely measured by electronically comparing the light intensity at positions either side of the unshifted position of a Fraunhofer line. One of the problems encountered is that there are several small-scale motions in the photosphere which also give rise to Doppler shifts, e.g. supergranulation, but these can be approximately allowed for by averaging over long periods.

Table 3.1 summarizes current information about the sun's differential rotation. It gives synodic and sidereal rotation periods (derived from formulae expressing the rotation rate per day as a function of latitude) from both the Doppler compensator method and sunspot motions. For periods derived from spot motions, values are given for all spots and (in parentheses) those for large, regular spots with a slower rotation. The periods for spots were taken from an analysis of the Greenwich photoheliograph results over the period 1874 to 1976.

The faster equatorial rotation has been ascribed to motions on a large scale from the solar poles to the equator, a so-called 'meridional flow'. Very small velocities are thought to be involved, perhaps only a few metres per second. At present, there is no observational confirmation, from either sunspot motions or from Doppler measurements, of the type of flow that could maintain differential rotation. There is, however, a motion of large-scale magnetic field patterns from sunspot latitudes towards the poles (§3.8).

An interesting discovery by R. Howard and B.J. LaBonte in 1980, in a synoptic programme of Doppler measurements at the Mount Wilson

Table 3.1. *Solar rotation periods for different latitudes*

Latitude	Rotation period (days) from Doppler method		Rotation period (days) from sunspots	
	Sidereal	Synodic	Sidereal	Synodic
0°	25.6	27.6	24.7 (25.1)	26.5 (26.9)
10°	25.7	27.7	24.9 (25.2)	26.7 (27.1)
20°	26.0	28.0	25.3 (25.6)	27.2 (27.6)
30°	26.6	28.7	26.0 (26.4)	28.0 (28.4)
40°	27.7	30.0	26.9 (27.3)	29.0 (29.5)
50°	29.3	31.9		
60°	31.4	34.4		
70°	33.6	37.1		
80°	33.5	39.3		

Sources: Data from Snodgrass (1984) and Balthazar *et al.* (1986).

Observatory, is that of 'torsional oscillations'. These consist of zones of alternating fast and slow rotation which move from the solar poles to the equator over a 22-year-long period. The exact nature of the zones is still being debated (the difference in rotation speeds between the fast and slow zones is only 3 m/s). However, there is clear evidence of an association with sunspot activity; such activity appears to occur at the poleward edges of the fast zones, at least when the fast zones arrive at latitudes of about 30° north or south. The fast zones appear to start at polar latitudes several years before the spots of each new cycle appear, when the spots of the previous cycle are at maximum development.

It is possible that surface velocity patterns exist corresponding to the giant 'banana-shaped' convection cells (§2.1). Such patterns would show as longitude-dependent flows. The measurements of such flows are much more difficult to make than those of differential rotation, and so far searches for them have been without definite success.

3.3 The solar diameter

The diameter of the sun, i.e. of the photosphere, is found from the angular diameter as seen from the earth and the distance to the sun. If the angular diameter, θ, is expressed in radians, the linear diameter, in km, is $r\theta$ km, where r is the earth–sun distance. More usually, the angular diameter is expressed in arc minutes or arc seconds. The angular diameter at mean earth–sun distance (one astronomical unit) is almost exactly 32 arc minutes, or just over half a degree.

Measurements of the angular diameter have been made for over 300 years, and analyses of the data over this long time-span have led to claims that the size varies, either steadily or depending on solar activity. Several methods have been used, including direct measurements with telescopes, timings of total eclipses and timings of occasions when Mercury transits the sun.

One of the oldest series of measurements is that due to Jean Picard at the Paris Observatory. He measured the vertical diameter using a telescope of long focal length fitted with a micrometer and the horizontal diameter by timing the transit of the leading and following edges of the sun as it moved across the field of view of a meridian telescope by the earth's rotation. The timing method was continued until 1719 by Picard's successors. The data at face value would seem to indicate a larger sun for the period of the Maunder minimum (1645–1715) but the reality of this is debated as instrumental errors, though commendably small for the time, may account for the supposed change.

There are more recent series of measurements of the vertical diameter made at various observatories using meridian circle telescopes. Some cover long periods – e.g. Paris (1837–1906), the Cape Observatory, South Africa (1834–87), and most notably those made between 1836 and 1953 at the Royal Greenwich Observatory, where horizontal diameters were also measured. Analysing the Greenwich data, J. A. Eddy and A. A. Boornazian in 1979 made the claim that the sun was contracting at the rate of 0.1%, or one arc second in the angular diameter, per century – a contraction considered to be very large indeed. However, the reality of the alleged variations may be questioned as the measurements show occasional discontinuities associated with changes of telescope, observing method or even the person making the measurement. These doubts have led some to disregard these direct measurements in favour of other methods.

The timing of total eclipses of the sun gives a very accurate measure of the solar angular diameter. The instants when totality starts or ends can be precisely timed as the presence of even the smallest 'Baily's bead' (§1.5) renders the eclipse non-total. Edmund Halley's observations of the 1715 eclipse (§1.5) gave the first reliable estimate of the angular diameter – 32′ 00″ arc, or 1920 arc seconds – by this method, which is very close to the present value. There are a few other accurate timings of eclipse totality durations made since then, all of which give the sun's angular diameter to within two arc seconds of 1920 arc seconds. Figure 3.4 shows these measurements (open circles with error bars) plotted against time since 1715.

The method of timing transits of Mercury is in principle very similar. When Mercury transits the sun, it is visible as a tiny black disc, some 7 arc seconds in diameter. Transits of Mercury are not particularly frequent – about 14 in a century – but they have been well observed and measurements of their durations (timed from the internal contacts of the limbs of the sun and Mercury) give angular diameters comparable in accuracy to those obtained

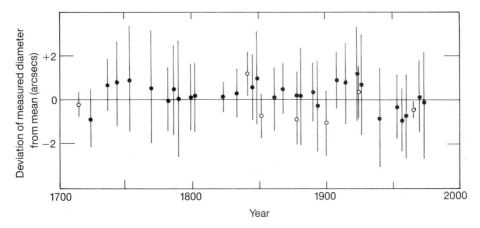

Fig. 3.4 Measurements of the sun's angular diameter since 1715, shown as a plot of deviations from the mean value of 31′59.26″ arc. Measurements made from timings of total solar eclipses are indicated by open circles, those from timings of Mercury transits by filled circles. Error bars represent standard deviations of measurements made during each event. (Courtesy J. H. Parkinson *et al.*; © 1980 Macmillan Journals Ltd)

from eclipse timings. These are added to the graph in Fig. 3.4 for transits of Mercury seen between 1723 and 1973.

Figure 3.4 shows that there is little evidence of a changing solar diameter over the past 250 years or more. The study by J. H. Parkinson and colleagues, on which the figure is based, failed to find any detectable change of diameter over the 250 years of measurements. The difference between this result and that of Eddy and Boornazian is explained by the inaccuracies in the direct diameter measurements used in the latter analysis. The mean value of the angular diameter from the Parkinson *et al.* analysis is 1919.2 arc seconds, giving a linear radius of 695 970 km.

For a possible difference between equatorial and polar diameters and its implication for general relativity, see §7.1.

3.4 Radiation from the photosphere: the temperature minimum

Radiation from the photosphere spreads over a wide range of wavelengths, from the ultraviolet (down to about 140 nm), through the visible wavelength range (400–700 nm) to the infrared (wavelengths of a few micrometres). To show its distribution with wavelength, we use a quantity called *spectral flux*, equal to the solar energy per unit time in a unit wavelength interval received at the distance of the earth but outside the earth's atmosphere. Figure 3.5 shows

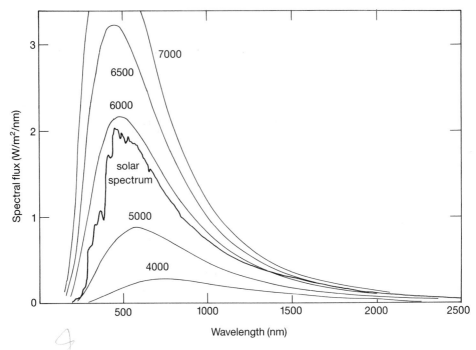

Fig. 3.5 Distribution of photospheric radiation with wavelength (the spectral flux
from the whole sun received outside the earth's atmosphere by an area of 1
square metre in a wavelength range of 1 nm). It is compared with that of
black bodies at different absolute temperatures (indicated in K). (From
Labs and Neckel (1968))

the spectral flux distribution over the range from the ultraviolet (wavelength
of about 200 nm) to the 'near' infrared (wavelength up to 2500 nm); the SI
unit of spectral flux used is watts per square metre per nanometre
($W/m^2/nm$). The solar curve, which is compared with some other curves to
be explained shortly, is based on measurements from the earth (corrected for
extinction by the earth's atmosphere) and from rocket-borne instruments.
The measured spectral fluxes were made over 10-nm wavelength intervals,
and so the fine details of all but the strongest Fraunhofer lines (which would
show up as dips in the curve) do not appear. The maximum spectral flux
occurs at about 460 nm, i.e. in the blue part of the visible spectrum, but as can
be seen this is far from well defined. Although the peak is in the visible range,
the total amount of visible-wavelength radiation is rather less than half the
total, as can be roughly judged by the area below the curve in the wavelength
range 400–700 nm. Slightly more than half the radiation lies in the infrared
region, while less than a tenth lies in the ultraviolet.

We may compare the solar spectrum with an ideal case met with in thermal
physics. Kirchhoff's law (§1.4) states that objects which are good absorbers of

radiation are, for the same temperature conditions, good radiators. If, at a particular temperature, an object is what is called a *black body*, i.e. one that perfectly absorbs all radiation incident on it, it radiates the maximum amount for that temperature. A surface coated with lamp-black sometimes serves as the physicist's approximation to a black body, or better a cavity with sides coated with absorbing material and closed except for a small aperture through which its radiation may be observed. The amount and distribution of radiation depend on the absolute temperature of the black body, and can be exactly calculated. The smooth curves in Fig. 3.5 are distributions of spectral flux for black bodies at particular absolute temperatures.

Two notable properties of black-body radiation curves can be illustrated from Fig. 3.5. Firstly, the amount of radiation over all wavelengths (i.e. the area under each curve) rapidly increases with absolute temperature: by the Stefan–Boltzmann law (§1.10), the amount is proportional to the fourth power of temperature. Secondly, the maximum radiation for a curve occurs at a progressively shorter wavelength for increasing temperature. The exact relation is that the wavelength of maximum spectral flux is inversely proportional to absolute temperature, this being known as Wien's law.

There are marked differences between any one of the black-body curves and the solar curve, though there is a crude similarity of the solar curve with that of a black body at about 6000 K. This is very roughly the temperature of the photosphere. Another estimate of the sun's temperature can be made by summing the solar spectral flux over all wavelengths and seeing at what temperature a black body must be to emit the same amount. This estimate of temperature is called the *effective* temperature, and is widely used by astronomers to characterize stars as well as the sun.

The sun's effective temperature can be calculated from the amount of solar radiation received at the earth. The area under the solar curve of Fig. 3.5 gives the total amount of solar energy received per unit time and per unit area at the distance of the earth. This quantity is the solar irradiance (§1.10). Modern spacecraft measurements (§9.1) give the solar irradiance as 1.368 kilowatts per square metre. Since the number of square metres on a sphere with radius r, where r is the sun–earth distance in metres, is $4\pi r^2$, the number of kilowatts the entire sphere receives is $1.368 \times 4\pi r^2$. With $r = 149\,600\,000$ km, this works out to be 3.85×10^{23} kilowatts, which must equal the power the sun radiates, for there is negligible absorption of radiation between us and the sun. This radiated power is known as the solar *luminosity*. Thus, 3.85×10^{23} kilowatts are radiated by the entire solar surface which has an area $4\pi R_\odot^2$, where $R_\odot$, the solar radius, equals 696 000 km. The power per square metre of solar surface can thus be calculated, and is found to be $63\,200$ kW/m^2. Now the radiant energy per unit area of a black body, by the Stefan–Boltzmann law, is σT^4 watts per square metre, where σ is the Stefan–Boltzmann constant, equal to 5.67×10^{-8} SI units, and T is the absolute temperature (in K). Equating σT^4 to $63\,200\,kW/m^2$ gives T, the value of the sun's effective temperature, equal to $5778\,K$.

Effective temperature is a convenient means of comparing the sun and other stars: thus, hot white stars like Sirius or Rigel have an effective temperature of about $10\,000\,K$, while for a 'red giant' star like Aldebaran it is around $3000\,K$. It also gives a very rough idea of how hot the photosphere is, since the radiation involved in the calculation we have just done nearly all comes from there.

A remarkable property of the photosphere is how opaque it is. This is especially surprising in view of the fact that even at the deepest photospheric levels the gas is very tenuous: its density is only one five-hundredth of that of the earth's atmosphere at its base, and its pressure one-tenth. A small part (15%) of this 'opacity' is accounted for by the absorption at discrete wavelengths by atoms of various elements, revealed by the absorption lines in the solar spectrum. There is also continuous absorption of radiation over the entire visible range of the sun's spectrum. The atomic process giving rise to this was identified by R. Wildt in 1939. In the photosphere and upper levels of the solar interior there is an abundant supply of electrons, stripped from the atoms of various elements. A sodium atom, for instance, in its neutral state, has a single loosely bound electron which is easily removed by a collision with another particle. On the other hand, a hydrogen atom is better able to retain its single electron, since a relatively large amount of energy is required to remove it. Occasionally, one of the supply of free electrons liberated, e.g. from sodium, combines with a neutral hydrogen atom to form a *negative* hydrogen ion, i.e. a proton with two orbiting electrons, having a net negative charge of one unit. Although this process per atom is infrequent, there are plenty of negative hydrogen ions present at any time because of the overwhelming abundance of hydrogen in the sun. Once such an ion is formed, one of its electrons can be easily removed again by a photon having an energy that can be any value within a very wide range: in the process, the photon is absorbed, and its energy imparted to the once-again liberated electron. Thus, negative hydrogen ions are excellent absorbers of continuous radiation in the photosphere, from the ultraviolet into the infrared, and they account for most of the photosphere's large opacity.

Related to the photospheric opacity is the small height extent of the photosphere, i.e. the solar limb appears extremely sharp. Part of the reason for this is the rapid decrease of gas pressure with increasing height. It can be assumed, as is done for the solar interior, that the photosphere is in hydrostatic equilibrium (§2.1), i.e. gas pressure balances the weight of overlying material. The consequence is that gas pressure falls 'exponentially', i.e. is proportional to $e^{-h/H}$, where h is the height and H is a constant known as the pressure *scale height*, and is a measure of how fast the pressure fall-off is. It is equal to the height range over which the gas pressure falls to $1/e$ (or 37%) of its original value. In kilometres, the scale height is $0.03 \times T$, where T is the absolute temperature. At the base of the photosphere, the temperature is 6400 K, so the scale height works out to be a little less than 200 km. This is only 0.03% of the solar radius. Seen from the earth, such a scale height would subtend less than 0.3 of an arc second. The height extent of the photosphere is more exactly determined by the density of negative hydrogen ions, which determine the opacity, and this density falls off even more sharply. The reason for this is the rapid decrease in the numbers of free electrons which are required in the formation of the ion from hydrogen.

We now discuss how radiation reaching an observer on the earth emerges from the photosphere. Radiation reaching this observer from some particular point on the sun starts off at some deep part of the photosphere and, on the way out, its intensity is reduced as it undergoes some absorption through the partly opaque photosphere. On the other hand, as all photospheric layers are radiating, emission processes occurring along the line of sight will add to the intensity of radiation. The absorption and emission processes result from transitions between energy states of individual atoms making up the photosphere. In addition to these processes, radiation may suffer scattering off these atoms, so for a particular line of sight radiation may be added to or subtracted from the original beam of radiation by these scattering processes. We now define some important quantities widely used in this context. Firstly, the *absorption coefficient* of the gas is defined to be the fraction of radiation subtracted from a beam of radiation by absorption or scattering per unit length (say, 1 m) of the beam. Correspondingly, the *emission coefficient* is the amount of radiation, to include scattering processes, emitted by the gas per unit volume of the gas (e.g. 1 cubic metre). Lastly, the emission coefficient divided by the absorption coefficient gives what is called the *source function*. All three of these quantities vary with wavelength, and indeed may be completely different in different parts of the spectrum, e.g. from the visible to ultraviolet regions.

A quantity often used by astronomers is *optical depth*. The optical depth of a slab made up of partly opaque material is the physical depth of the slab (e.g. in metres) multiplied by the absorption coefficient. The slab's optical depth is defined to be one if a beam of light passing through the slab has its intensity reduced to $1/e$ (37%) of its incident intensity. More generally, the absorption coefficient varies along a line of sight, so the optical depth is the sum of all the contributions made by small length elements, each having its particular absorption coefficient. Thus, looking into the solar atmosphere from the earth, the optical depth along the line of sight increases steadily until we can see no further.

To a good approximation, the solar radiation that we see emerges from a layer in the atmosphere where the line-of-sight optical depth is two-thirds. The height in the solar atmosphere to which this level corresponds depends on what part of the solar disc is viewed (see Fig. 3.6). An observer looking at disc centre (point C in the figure) sees radiation that is emitted at the very base of the photosphere, but moving away from the disc centre, the level where the radiation emerges corresponds to an increasing height in the solar atmosphere because of the increasing slantness of the line of sight. Eventually, at the limb, radiation emerges from a level that is several hundred kilometres above the photosphere's base (point L).

Over the height range of the photosphere, the temperature decreases from about 6400 K at the base to 4400 K at the top. Beyond this level, which we will define to be the base of the solar chromosphere, temperature increases again, so that there is a *temperature minimum* region. In visible light, the point L in Fig. 3.6 (i.e. from where radiation emerges at the limb) is at a level which is just beneath the temperature minimum, so we see a less hot part of the atmosphere than at sun centre (point C). Assuming the radiation to have approximately the wavelength dependence of equivalent black bodies, and recalling both the Stefan–Boltzmann and Wien laws, we see that the radiation from the limb must be both less intense and somewhat redder. Thus, there is a decrease of solar intensity towards the limb, or a *limb darkening*, very noticeable in whole-sun photographs. Measurements of limb darkening pro-

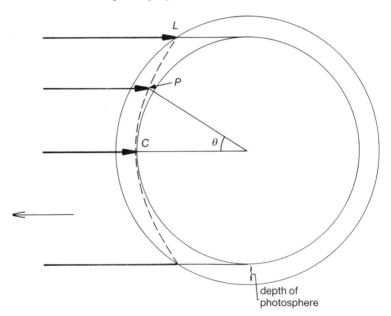

Fig. 3.6 When the sun is viewed in white light, radiation essentially emerges from a point P in the photosphere whose height depends on its angular distance θ from the sun's centre; at disc centre, radiation emerges from point C, at the photosphere's base, and at the limb, it emerges from point L, several hundred kilometres above the photosphere's base, near the temperature minimum.

vide information about the temperature variations over the extent of the photosphere, and are used as input for models of the solar atmosphere. We shall deal with these in §4.4.

In an ideal case, in which there is a complete interaction between the atoms of a gas and the radiation they produce, the source function is the same as the black-body radiation function which can be exactly found from the gas temperature. The real situation for the solar photosphere is not quite this since radiation does not completely interact with the gas but rather some escapes to space. However, an approximation to the ideal case exists, known as local thermodynamic equilibrium (LTE). Although there is a variation of temperature within the photosphere, the LTE approximation amounts to the source function being the same as the black-body radiation curve at the *local* value of temperature. An observer measuring the spectral shape of the sun's radiation – averaged over the solar disc – would obtain precisely this black-body radiation if for all wavelengths it emerged at the same level in the sun's atmosphere. This is the equivalent to the absorption coefficient being constant with wavelength, the so-called 'grey atmosphere' approximation. The solar radiation curve departs from that of a black body, however, and the reason is the fact that the absorption coefficient varies with wavelength. The general wavelength dependence of the absorption coefficient is, as indicated earlier, largely determined by negative hydrogen ions. Over the visible range, the absorption coefficient steadily rises with wavelength, reaches a maximum at a wavelength of about 800 nm, then decreases again, at least up to about 1600 nm. Now, for wavelengths of around 800 nm, where the absorption coefficient is relatively large, the radiation emerges from a higher level in the photosphere where it is cooler, and therefore the intensity of this radiation is smaller than the grey-atmosphere case. Conversely, where the absorption coefficient is relatively small, i.e. at wavelengths appreciably shorter or longer than 800 nm, radiation emerges from a lower level where it is hotter, and so is more intense than the grey-atmosphere case. This can be partly seen from Fig. 3.5, where the solar spectral flux crosses over black-body curves for progressively higher temperatures above wavelengths of about 1000 nm. It is not obvious at all for very small wavelengths. This is because the negative hydrogen ion does not entirely determine the absorption coefficient; hydrogen in its usual atomic form (one proton, one electron) now plays a part, but more particularly the numerous Fraunhofer lines are important. We now turn to the question of their formation.

3.5   The Fraunhofer lines

Before discussing the details of the solar spectrum and the origin of the Fraunhofer lines, we will outline the way quantum mechanics explains how the patterns of bright, or emission, lines are produced by atoms of an incandescent gas. By far the simplest atom is hydrogen, with its one proton and orbiting electron, and the simplest spectrum is the one hydrogen atoms produce. In Bohr's quantum-mechanical picture (§1.4), the electron in the hydrogen atom moves round the proton in specific orbits that are circular or elliptical. The smallest allowable orbit is circular with a radius of 0.053 nm; this orbit is labelled $n=1$, the number n being called the principal quantum number. Succeeding orbits, with $n=2, 3, 4$, etc., have radii that are 4, 9, 16, etc., times as large, i.e. they increase as the square of n. In its lowest-energy or 'ground' state, a hydrogen atom has its electron in the $n=1$ orbit. The atom is 'excited', i.e. raised to a higher-energy state, when the electron jumps from a close-in orbit to one further out; this excitation may take place either because of a collision with a particle such as a free electron or absorption of a photon. The atom may be 'de-excited' by the electron jumping from an orbit with higher n value to one with a smaller n with the emission of a photon.

Figure 3.7 illustrates the first few orbits of the electron. Photons having particular energies are emitted when the electron jumps from $n=2, 3, 4$, etc., orbits to the lowest, $n=1$ orbit; and photons with different (much smaller) energies are emitted when the electron jumps from $n=3, 4, 5$, etc., orbits to the $n=2$ orbit, and so on. The result is that a gas containing a large number of hydrogen atoms at a sufficiently high temperature (to ensure frequent particle collisions) will emit photons of particular energies, i.e. radiation at discrete wavelengths corresponding to the various electron jumps or 'transitions'. The

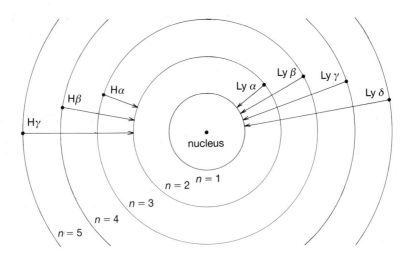

Fig. 3.7 Simplified model for electron orbits round a hydrogen nucleus (proton), based on the Bohr model for the hydrogen atom. Orbits are not to scale: $n=1$ has radius of 0.053 nm; $n=2$ orbit four times this radius; $n=3$ orbit nine times, etc. Spectral lines in the Lyman series and Balmer series are emitted by electrons making the transitions indicated.

energy released when an electron makes a transition from an $n=3$ to $n=2$ orbit corresponds to the wavelength of the Hα line, 656.3 nm, and an entire series of lines – the *Balmer* series – is formed by similar transitions from higher-n orbits to the $n=2$ orbit, members of which include Hβ at 486.1 nm ($n=4$ to $n=2$), Hγ at 434.0 nm ($n=5$ to $n=2$), and so on. All these lines are in the visible range of wavelengths. Equally, de-excitation can occur to the ground state, or $n=1$ orbit, forming a series of lines, called the Lyman series, in the ultraviolet; this series includes the Lyman-α line at 121.5 nm ($n=2$ to $n=1$), Lyman-β at 102.5 nm ($n=3$ to $n=1$), Lyman-γ at 97.2 nm ($n=4$ to $n=1$), and so on.

The simple spectrum of hydrogen is readily explained by the principles of the quantum theory. Helium, in a neutral state, has two orbiting electrons, and can be excited when one of the two electrons is taken to a higher orbit. The spectrum as a result is rather more complicated than hydrogen, though there are patterns of lines not unlike those of the line series in the hydrogen spectrum. Singly ionized helium (we will denote it by He^+, the superscripted '+' indicating the charge of the ion) is akin to hydrogen, as it has only one orbiting electron, so the spectrum has striking resemblances. The differing nuclear charge – two instead of one – results in line wavelengths that are almost exactly one-quarter of those of hydrogen: thus, Hα in ionized helium is 164.0 nm compared with 656.3 nm. Fully ionized hydrogen (i.e. with its one electron removed) and helium (with both its electrons removed) do not produce line spectra at all.

It takes a certain number of electrons to 'fill' an orbit, and the electrons of inner, filled orbits do not generally make transitions. Sodium is a case of some interest. It has 11 electrons, but 10 are in filled orbits with $n=1$ and 2. The remaining electron is in an outer ($n=3$) elliptical orbit (labelled $3s$) which may be excited to a circular $n=3$ orbit (labelled $3p$). Now, an electron behaves as if it were spinning, and associated with this spin is a tiny amount of energy. Thus, when a sodium atom is de-excited, i.e. the outer electron makes the transition from a $3p$ to $3s$ orbit, it can either retain its spin direction or reverse it. The photons produced for the two cases have energies that are slightly different because of the electron spin energy, so that a gas of sodium atoms emits two lines, very close in wavelength. These are the famous D lines, at 589.0 and 589.6 nm, forming a doublet in the yellow part of the spectrum. Another well-known doublet, the H (396.8 nm) and K (393.4 nm) lines in the violet, is formed by transitions in ionized calcium (Ca^+). In this case, the transitions involve a single outer electron making a transition between $4p$ and $4s$ orbits.

Heavy atoms with very many orbiting electrons, then, have spectra that are not as complex as might be imagined, since the electrons in already filled orbits do not participate in transitions. However, some elements (notably iron) do have spectra that consist of hundreds of lines over a wide wavelength range, making the identification of lines in the spectrum of the sun, where some of these elements are abundant, an involved undertaking.

This explains the formation of emission-line spectra characteristic of incandescent gases in the laboratory and certain astronomical objects like gaseous nebulae and the sun's chromosphere and corona. But the visible spectrum of the solar photosphere consists of a continuous background with numerous *dark* lines – the Fraunhofer spectrum. Figure 3.8 shows the visible spectrum, from red to violet wavelengths. These are at the wavelengths of emission lines produced by incandescent gases, so there are Fraunhofer-line equivalents of the sodium D lines, ionized calcium H and K lines, and hydrogen Balmer lines, for instance. The Fraunhofer spectrum is in fact like the one produced by an incandescent solid source shining through a container of cold gas, as in the experiment that Kirchhoff had performed (§1.4). The hot source gives a continuous spectrum (or 'continuum') and the gas absorbs at the same wavelengths as the emission lines it produces when hot. The absorption coefficient must therefore sharply increase at the wavelengths of these lines, but be very small elsewhere.

In applying this to the solar spectrum, we follow the same reasoning as was used in §3.4 to explain limb darkening and the departures of the solar spectrum from that of a black body. We use the fact that most Fraunhofer lines are formed in the photosphere, and that the photospheric temperature decreases with height. At the wavelength of a Fraunhofer line, the absorption coefficient is high, so the radiation comes from the top of the solar photospheric layer where it is cooler. Conversely, in a line-free part of the solar spectrum, the absorption coefficient is low so that radiation comes from near the base of the photosphere where it is hotter. As a result, the spectral flux at the position of the line is lower than the neighbouring continuum, i.e. the line appears relatively dark. The Fraunhofer lines, then, are dark as a consequence of the decreasing temperature from the base to the top of the photosphere. Notice that this is not quite the same as the older notion of a specific cool 'reversing' layer above the hotter photosphere (§1.5).

Some years ago, the atomic processes that resulted in the formation of the solar Fraunhofer lines were considered to be of two main types. In the first, line photons are repeatedly absorbed and re-emitted in different directions by atoms as if the photons were being scattered by the atom. The scattering is 'coherent' if the scattered photons have identical energies, and 'incoherent' if, as is more generally the case, there are slight changes in their energies owing to the motions of the absorbing and re-emitting atoms. The scattered photons are eventually absorbed by, e.g., a negative hydrogen ion which is able to absorb photons with a wide range of energies. In the second process, a line photon can be removed by pure absorption, in which it is absorbed by an atom which is raised from a lower state to a higher one, but which emits a completely different photon on de-exciting to a lower state different from the first.

Fig. 3.8 Spectrum of the sun obtained with the University College London Echelle Spectrograph. Each strip (or 'order') covers roughly 5 nm of the spectrum, with wavelengths increasing from left to right in an individual strip, and from bottom (violet) to top (red). The total wavelength range is about 423 to 657 nm. The Hα line appears twice, in the second order from the top (far left) and third order (far right). The sodium D lines in the yellow also appear twice (12th order from top, to left, 13th order, far right). The Hβ line appears as a prominent line a little above and to the right of the centre of the image (33rd order). In the bottom two orders, two extremely broad dark features mark the ionized calcium H and K lines. (Courtesy F. Diego and D. Walker)

This picture of line transfer is not too inaccurate for many of the photospheric lines. All the weak and medium-strength Fraunhofer lines have their origin in the photosphere, as do the outer parts (or 'wings') of the very strong lines like Hα and the H and K lines. But this picture is inadequate in dealing with the formation of the centres or cores of strong lines which have their origin much higher in the atmosphere – in the chromosphere. We will discuss the formation of these lines in §4.2.

As already indicated, the solar photospheric spectrum contains lines of not only neutral atoms but also ions. Up till now, the notation for particular ions used has been the chemical symbol and superscripted pluses – e.g. Ca^+ for singly ionized calcium. We will also adopt a widely used spectroscopic notation of Roman numerals in which the *spectrum* due to neutral atoms of an element is designated by the element's chemical symbol followed by 'I', that of the singly ionized atoms by 'II', and so on; so the H and K lines due to the *ion* Ca^+ are part of the Ca II *spectrum*.

The wide variety of the strengths of the Fraunhofer lines is apparent from Fig. 3.8. The fact that the hydrogen Balmer lines are conspicuous is understandable in view of the overwhelming abundance of hydrogen; but helium, the second most abundant element, is almost absent in the visible spectrum. On the other hand, by far the strongest lines of all – the H and K lines – are due to calcium, which is about a million times less abundant than hydrogen and indeed less than many other elements. Although the abundance of an element plays a part in the relative strength of the Fraunhofer lines due to that element, a much larger part is played by *temperature*.

Temperature affects what is called the ionization state of the photosphere. From the base of the photosphere to its top, the temperature varies from 6400 K to 4400 K. At such temepratures, calcium atoms are almost all in the singly ionized state (Ca^+) rather than the neutral state since it is comparatively easy to remove one of the two outer electrons of the neutral atom. It needs much more energy to remove the single electron of hydrogen, however, so this element is nearly all in a neutral state, with (as mentioned before) a small proportion in the negative ion state (H^-). Even more energy is required to remove one of the two electrons of neutral helium, so helium is mostly in a neutral state in the photosphere.

Although for a particular atom or ion there are many energy states the electrons may take, nearly all atoms will be in the ground state. Thus, hydrogen atoms are normally in the configuration in which their electrons are in the $n=1$ orbit, and Ca^+ ions normally have their single outer electron in a $4s$ orbit. From the ground state, atoms or ions can most easily give rise to lines in the next highest energy states, e.g. the $n=2$ orbit for hydrogen or $4p$ orbit for Ca^+. In the case of pure absorption, described above, the atom or ion will absorb photons corresponding to a *resonance* line. Ca^+ ions can very easily do this, since relatively low-energy photons are needed for the $4s$ to $4p$ excitation, and there are many such photons in the photospheric radiation. But there are no photospheric photons energetic enough to excite the Lyman-α line, so it is not seen as a Fraunhofer line. However, a tiny proportion of neutral hydrogen atoms in the photosphere exist, not in their ground state, but with the electron in the $n=2$ state. It takes a relatively low-energy photon to excite such atoms further, to the $n=3$ state; in doing so the Hα absorption line is formed. Thus, a combination of excitation and ionization conditions, both determined by the photosphere's temperature, give rise to the strong Ca II H and K lines and rather less strong hydrogen Hα line, even though the element calcium is much less abundant than hydrogen.

In Table 3.2, the strongest Fraunhofer lines in the solar spectrum are listed, from near-ultraviolet to infrared wavelengths. The emitting atom or ion, and the line wavelengths and strengths, indicated by their equivalent widths (which will be defined in §3.6), are given, as are common designations for some of the lines, many based on Fraunhofer's original classification.

Table 3.2. *A selection of strong Fraunhofer lines*

Wavelength (nm)	Designation	Emitting atom or ion	Equivalent width (nm)
279.5	k	Mg^+	} 2.2
280.2	h	Mg^+	
285.2		Mg	1.0
373.5	M	Fe	0.31
393.4	K	Ca^+	1.9
396.8	H	Ca^+	1.4
410.2	h (Hδ)	H	0.34
422.6	g	Ca	0.15
434.0	G' (Hγ)	H	0.35
438.4	d	Fe	0.11
486.1	F (Hδ)	H	0.42
518.4	b_1	Mg	0.16
589.0	D_2	Na	0.08
589.6	D_1	Na	0.06
656.3	C (Hα)	H	0.41
849.8	⌈ Calcium	Ca^+	0.13
854.2	⌡ 'infrared	Ca^+	0.36
866.2	⌞ triplet'	Ca^+	0.27

Source: Data based on Allen (1973).

3.6   Broadening and splitting of spectral lines

It will be seen that some of the stronger lines in the solar spectrum of Fig. 3.8 have a definite width – they are not infinitely narrow in wavelength. This is particularly true of the H and K lines in the violet, which are about 2 nm wide. Figure 3.9 is a graph of spectral flux against wavelength in the neighbourhood of a hypothetical but typical line. The intensity falls from the continuum level to a minimum which is above the zero level, then rises again, forming a 'profile' that is generally symmetrical. As a measure of the strength of a Fraunhofer line, we use the *equivalent width*, i.e. the width (in nm) of a rectangular strip that removes the same amount of radiation from the continuum as the line itself. Equivalent width is useful in several different contexts, though it does not give information about the kind of profile a line has.

There are several possible line broadening mechanisms. The first is that due to the motion of emitting atoms. By the Doppler effect, the wavelength of light emitted by a source approaching an observer is decreased by an amount that depends on the speed of the source with respect to the observer; and conversely the wavelength is increased when the source recedes from the observer. The atoms making up the solar atmosphere are in a state of agitation depending on the temperature of the gas. For instance, in the lower photosphere, where the temperature is about 6000 K, hydrogen atoms move around with speeds averaging about 10 km/s. There is a distribution of speeds along a line of sight, with some atoms receding, others approaching, and some moving transversely, i.e. with zero line-of-sight velocity. The light such atoms emit will therefore be Doppler-shifted towards both the red and

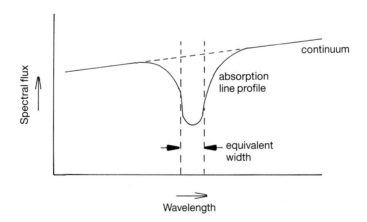

Fig. 3.9 A hypothetical section of the solar spectrum to illustrate the equivalent width of a Fraunhofer line. The equivalent width is the width of a rectangular section that removes the same amount of radiation from the continuum as the real line.

violet ends of the spectrum. The result is that any line will have its profile broadened. We will do a simple calculation to find out how much the broadening will be. Consider the Hα line, with a normal (or 'rest') wavelength of 656.3 nm, emitted by hydrogen atoms moving at speeds averaging 10 km/s. To an observer on earth, the Hα line is red-shifted or violet-shifted by a wavelength that equals, as a fraction of its rest wavelength, the speed of the atom expressed as a fraction of the speed of light (300 000 km/s). Thus, the wavelength shift amounts to 656.3 nm×10/300 000=0.02 nm. The motions of the atom are completely random, so there are as many atoms approaching at a particular speed as receding. Although the atoms move in all possible directions, only that component of their velocity along the line of sight actually gives rise to a Doppler shift. The net result is that the spectral line, instead of being sharp, is blurred or broadened in wavelength to a bell-shaped, or 'Gaussian', profile. Figure 3.10a shows such a profile for an emission line, but the foregoing applies equally to Fraunhofer lines. The line width for so-called *thermal Doppler* broadening would seem to be an imprecise quantity as the profile extends out indefinitely in wavelength from the centre, but a width can be defined, e.g. the full width of the emission-line profile at 1/e (37%) of the

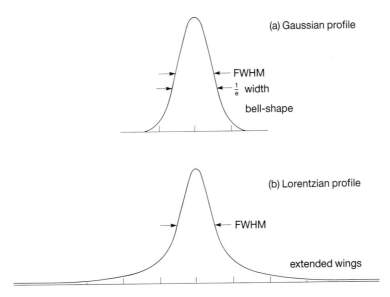

Fig. 3.10 Emission line profiles. (a) A line is broadened into a Gaussian (bell-shaped) profile when the emitting atoms have random motion giving rise to a range of Doppler shifts. (b) A line profile determined by natural broadening or by collisional broadening has a Lorentzian profile, with extended wings. The two profiles in this figure have identical widths at half their maximum intensities ('Full Width at Half Maximum' or FWHM). The width of the Gaussian profile can also be defined by the 1/e width.

maximum intensity; the shift of 0.02 nm, corresponding to 10 km/s just cal-culated, is in fact half of this width.

The second mechanism is connected with the amount of time an atom spends in its upper energy state. By a quantum-mechanical principle known as Heisenberg's uncertainty principle, this state is not infinitely sharp in energy but has a definite width, the amount of which depends on how quickly the atom leaves the state. Thus, an atom making a transition from this state to the lower state emits a photon with a small range of energies. The spectral line so formed is said to have *natural* broadening, or to be broadened by *radiation damping*. Figure 3.10b shows the profile, called a 'Lorentzian' pro-file, of an emission line with natural broadening. The line intensity only slowly falls away from line centre compared with the Doppler-broadened profile, forming extended (or 'damping') wings to the profile.

For certain lines, *collision* broadening is important. It is due to the effect that moving charged particles have on the light emitted by an atom. These particles do not collide with the emitting atom in a billiard-ball sense, but pass near enough that the atom comes under the influence of the tiny electric field which each moving charged particle carries. This electric field produces a momentary perturbation of the orbiting electrons of the atom as each charged particle dashes past. As these collisions are random, the perturbations are random and so any emission line is broadened. In the case of the Balmer lines, for instance, the broadening is greatest for lines emitted by transitions from high-n orbits to the $n=2$ orbit. Collision broadening is important for the wings of many spectral lines emitted in the solar atmosphere, and increases for increasing gas density, since then the chances of collisions between particles and emitting atoms become greater.

Magnetic fields also affect line profiles, a process which is important for the sun because of the presence of large fields near sunspots and other features. In this case, spectral lines are split into components rather than broadened, though the components may be so close together that their profiles already broadened by the Doppler or collision mechanisms overlap to form a profile that in effect has an extra broadening mechanism. The splitting of the line due to magnetic fields in this way is known as the *Zeeman effect*.

To illustrate the Zeeman effect, imagine an atom experiencing a strong external magnetic field. An orbiting electron of the atom, with its negative electric charge, gives rise to a tiny electric current, and this current has an associated magnetic field with direction perpendicular to the plane of the electron's orbit. It is the interaction of this field with the external field that produces a slight addition to or subtraction from the normal energy carried by photons that form the spectral line. The greater the external field strength, the larger is this added or subtracted energy. The result is that there are two components, called the σ components, of the line displaced to either side of the normal line wavelength, whose separation is proportional to the field strength. When viewed along the field lines ('longitudinal' field case), only the σ components are visible, but when viewed perpendicular to the field lines ('transverse' field case), the two σ components are visible together with another (the π component) at the undisplaced line position, giving rise to a triplet of lines. Figure 3.11 shows the two cases. The above refers not only to emission lines, which was how the Zeeman effect was originally found, but also to the solar Fraunhofer lines.

Both the σ and π components have a special property known as *polarization*. Now, light can be considered as a series of vibrations of wave-like or sinusoidal motion; the distance between successive crests is equal to what we have been calling the wavelength. According to Maxwell's theory of light, these vibrations are electromagnetic in character, with electric and magnetic fields varying sinusoidally in planes perpendicular to one another, and propagating in a vacuum at the speed of light. If these planes remain fixed in space, the light is said to be *plane-polarized*: the plane of polarization is identified with that of the electric field vibrations (Fig. 3.12a). Most light (e.g. direct sunlight) is *unpolarized*, and can be thought of as having planes of vibrations that change extremely rapidly – typically, in intervals of a hundred-millionth

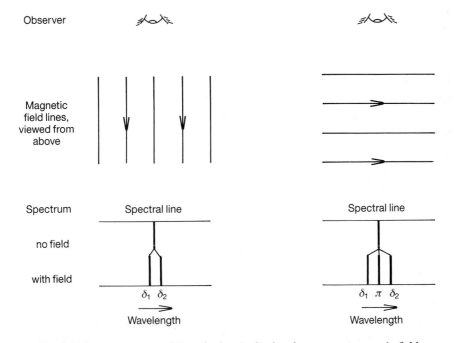

Fig. 3.11 Zeeman pattern of lines for longitudinal and transverse magnetic field. When the line of sight is parallel to the field lines (longitudinal case), a spectral line is split into two σ components; when the line of sight is perpendicular to the field lines (transverse case), it is split into three components (two displaced σ and one undisplaced π).

of a second – so that an observer viewing such light sees vibrations in all possible directions (Fig. 3.12b). Another special case of polarization is when two rays of plane-polarized light travel together with their planes perpendicular to one another and with their wave motions a quarter of a wavelength out of phase (Fig. 3.12c). The combination of these two wave motions is obtained by a vector addition of each of the two on a screen imagined to be travelling with and perpendicular to both rays. An observer viewing the combined beam sees the plane of vibration trace out a circle on the screen as it moves along; the light is said to be *circularly* polarized. If, to the observer, the plane of polarization rotates in a clockwise sense, the light is right-hand circularly polarized; if in a counter-clockwise sense, the light is left-hand circularly polarized.

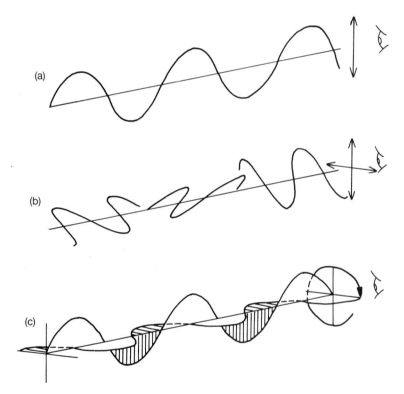

Fig. 3.12 Three special cases of light polarization. (a) Plane-polarized light arises
when the electric field vibrates in a single plane; such light is represented
by the double-headed arrow. (b) Unpolarized light occurs when segments
of light waves occur in the line of sight with randomly orientated planes
(each segment has many millions of cycles, not the few shown); such light
is represented by the crossed double-headed arrows. (c) Circularly
polarized light occurs when two plane-polarized rays are a quarter of a
wavelength out of phase, so that the vector sum of the wave motions
rotates as the waves propagate.

With the Zeeman effect in the longitudinal case, the two σ components are circularly polarized, with the plane of polarization rotating in opposite directions. In the transverse-field case, when all three components are visible, the two σ components are plane-polarized and so is the undisplaced π component, but with the plane of polarization perpendicular to that of the σ components. In general, one views the field in a direction that is neither along nor perpendicular to its direction. The resulting Zeeman pattern is still three components, with the undisplaced π component plane-polarized and the two σ components circularly polarized, one in a right-hand sense, the other left-hand.

These are the chief line broadening and splitting processes that operate in the solar atmosphere. In addition, a line profile is broadened simply by the finite size of the spectrograph slit and sometimes other 'instrumental' causes. Instrumental broadening can generally be taken account of if the characteristics of the spectrograph are known. The profile can then be analysed to give the relative contributions of Doppler and collision broadening for lines not sensitive to magnetic fields. For Zeeman-broadened or split lines, polarization measurements can give the magnetic field present. In §§3.7 and 3.8 we show how line profiles are analysed to give important physical information about the photosphere.

3.7 The photosphere's chemical composition

An important piece of information the Fraunhofer lines contain is the chemical composition of the photosphere. By this we mean the relative numbers of atoms due to the various elements present, or alternatively the element 'abundances'. The standard means of doing this is the *curve of growth*, originally due to M. Minnaert, which is based on the way Fraunhofer line strengths 'grow' as the number of absorbing atoms they are due to increases along a line of sight. The curve of growth is a plot (usually logarithmic) of line strength, defined by its equivalent width, against the number of absorbing atoms.

To see the form of the curve of growth, imagine first that there are only a few absorbing atoms of an element along a line of sight. A Fraunhofer line due to these atoms is relatively weak, and its equivalent width is mostly due to the central part of the line profile, which has a thermal Doppler form (see Fig. 3.13). The equivalent width grows linearly, i.e. in proportion to the number of absorbing atoms. As the number of absorbing atoms increases, the line fully absorbs all the continuum at its centre, i.e. it becomes 'saturated'. Adding more absorbing atoms to the line of sight at this stage does not increase the equivalent width appreciably, so the curve of growth flattens out considerably.

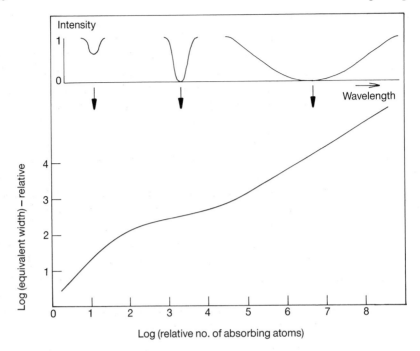

Fig. 3.13 Solar curve of growth, i.e. a graph of relative equivalent width against the
relative number of absorbing atoms, both plotted logarithmically. At the
top of the graph are representations of line profiles appropriate to the
various regions of the curve of growth – weak lines near the bottom left-
hand corner, intermediate-strength lines in the middle and strong in the
upper right-hand part.

Eventually, the line wings due to radiation damping or collision broadening
start to add to the equivalent width, so the curve of growth again steepens, but
is less steep than the first rise.

A theoretical curve of growth can be plotted on the assumption of some
model for the solar photosphere. It may then be compared with a curve of
growth that is experimentally obtained. The latter consists of points, each of
which corresponds to a line of some element. For each of these lines, an
equivalent width can be measured, so defining the y-values of these points.
Now, the x-values of each of the points are strictly equal to the number of
absorbing atoms multiplied by what is called the oscillator strength, a measure
of how probable is the upward (or absorption) jump that an electron must take
in the atom to form the line. An oscillator strength of one means the absorption transition is very probable, but an oscillator strength equal to a tiny
fraction of one means that the transition has very low probability. The number
of absorbing atoms, the unknown quantity we are seeking (since it gives the
abundance of the element), is the same for all lines chosen. Thus, the x-

values of the plotted points can be set to the oscillator strengths alone. These can be found experimentally or from atomic calculations.

The points constituting the experimental curve of growth can be made to lie on the theoretical curve of growth by adjusting the horizontal scale. This gives the number of absorbing atoms and so the element abundance. A set of element abundances can be found by forming experimental curves of growth using the Fraunhofer lines due to each element.

There are several complications in this method, one being that the temperature of the part of the photosphere where the lines are formed should be known in order to determine what fraction of the element is neutral or ionized. Lines due to, e.g., neutral iron give the number of neutral-iron atoms present, but to get the total number of iron atoms, the proportion in the ionized state has to be known. Also, the curve-of-growth method obviously does not work for elements that do not give rise to Fraunhofer lines: these include the 'noble' gases helium, neon and argon, all of which need much higher temperatures than those in the photosphere for significant excitation.

Oscillator strengths may, as indicated earlier, be determined by either laboratory measurements or calculations using quantum mechanics. Modern values are fairly accurate for most values unless extremely small. However, as recently as 20 years ago, laboratory measurements of the oscillator strengths of neutral iron were a factor of about ten in error, leading to an iron abundance that was a tenth of the true value. Improved techniques led to a revision to the present value.

Table 3.3 is a list of the most abundant elements in the sun. The abundance of an element is expressed as a fraction *by number* (i.e. not mass) of that of hydrogen. Photospheric values are given, derived by methods described here. So, too, are values appropriate to the solar corona, measured from emission lines: the method is described in §4.4.

The most striking feature of Table 3.3 is the overwhelming abundance of hydrogen, the lightest element, which makes up 91% of all atoms. Helium, the next lightest, cannot be directly determined for the photosphere, but it most likely accounts for about 9%. The sum total of all heavier elements accounts for only 0.1%. These abundances represent the primordial material out of which the sun was originally formed. (Although hydrogen is being converted to helium and some other elements by nuclear reactions at the sun's core, this has no effect on observed abundances as the core material is not mixed with the rest of the solar interior, and so the newly synthesized elements do not appear at the photosphere or corona.) This primordial material was processed by both the Big Bang 'fireball', in which some of the original hydrogen was converted to helium, and at the centres of first-generation stars which subsequently exploded or otherwise returned their contents to interstellar space.

Table 3.3 indicates that certain elements are more abundant in the photo-

Table 3.3. *The chemical composition of the sun*

Element	Symbol	Atomic number	Abundance relative to hydrogen Solar photospheric	Solar coronal
Hydrogen	H	1	1.000	1.000
Helium	He	2	(0.1)	(0.1)
Carbon	C	6	4.9×10^{-4}	2.4×10^{-4}
Nitrogen	N	7	9.8×10^{-5}	3.9×10^{-5}
Oxygen	O	8	8.1×10^{-4}	(2.5×10^{-4})
Fluorine	F	9	3.6×10^{-8}	–
Neon	Ne	10	(1.2×10^{-4})	(3.5×10^{-5})
Sodium	Na	11	2.1×10^{-6}	2.7×10^{-6}
Magnesium	Mg	12	3.8×10^{-5}	3.7×10^{-5}
Aluminium	Al	13	3.0×10^{-6}	2.7×10^{-6}
Silicon	Si	14	3.5×10^{-5}	3.9×10^{-5}
Phosphorus	P	15	2.8×10^{-7}	–
Sulphur	S	16	1.6×10^{-5}	(8.6×10^{-6})
Chlorine	Cl	17	(3.2×10^{-7})	(6.4×10^{-7})
Argon	Ar	18	(3.9×10^{-6})	2.1×10^{-6}
Potassium	K	19	1.3×10^{-7}	–
Calcium	Ca	20	2.3×10^{-6}	2.9×10^{-6}
Titanium	Ti	22	1.0×10^{-7}	–
Vanadium	V	23	1.0×10^{-8}	–
Chromium	Cr	24	4.7×10^{-7}	–
Manganese	Mn	25	2.8×10^{-7}	–
Iron	Fe	26	4.7×10^{-5}	3.9×10^{-5}
Cobalt	Co	27	8.3×10^{-8}	–
Nickel	Ni	28	1.8×10^{-6}	2.2×10^{-6}
Copper	Cu	29	1.6×10^{-8}	–
Zinc	Zn	30	4.0×10^{-8}	–

Notes:
Elements are included that have a photospheric abundance of at least 10^{-8} of that of hydrogen. In general, elements with smallest abundance are least well determined.

Individual elements:
He. No photospheric value available. Coronal abundance is uncertain estimate from analysis of prominence spectrum; approximately the same as a preliminary value from the *Spacelab 2* CHASE experiment.
O. Coronal abundance could be a factor three higher (C. Jordan, unpublished).
Ne. No photospheric value available; value quoted is for 'Local Galactic' in Meyer's work. Coronal abundance could be a factor three higher (C. Jordan, unpublished).
S. Coronal abundance could be a factor two higher (C. Jordan, unpublished).

sphere than in the corona, though some believe that the possible observational errors make these differences insignificant. The elements in question include carbon, nitrogen and oxygen. All these elements are relatively difficult to ionize, so exist in the photosphere in mostly neutral form. Elements like sodium and calcium are easier to ionize, and exist in the photosphere mostly in an ionized form. Those who maintain the photospheric and coronal abundance differences to be real argue that there may be a diffusion process tending to remove ions but not neutral atoms from the photosphere to the corona; thus calcium and sodium may be partly transferred from the photosphere to the corona by this means, but not carbon, nitrogen or oxygen, which are therefore more abundant in the photosphere.

3.8   The photospheric magnetic field

The photospheric magnetic field is measured by the Zeeman splitting of certain photospherically formed Fraunhofer lines. The largest field strengths – 0.4 T – occur in sunspots. Fields exist elsewhere, and indeed it is likely that the entire solar surface is pervaded by at least a very weak field. Weak fields do not give rise to a splitting of lines into separate Zeeman components – rather, a broadening is observed. The line-of-sight or longitudinal field can then be found using the polarization properties of the two blended σ components. The instrument that uses this principle is the Babcock magnetograph, introduced in 1953. The complete field – both its strength and direction – may be found from the longitudinal field component and the transverse field component, the latter being measured from the linear polarization of the central (π) component of the Zeeman-broadened line. The instrument that does this is the vector magnetograph. Further details about these instruments are given in §10.1.

Maps of the solar longitudinal fields – 'magnetograms' – are produced at a number of observatories. Daily full-disc maps from the Kitt Peak, Arizona, station of the National Solar Observatory appear in the US National Oceanic and Atmospheric Administration's (NOAA) bulletin *Solar-Geophysical Data*; Fig. 3.14 is an example. With the Kitt Peak magnetograph, field strengths as small as 0.05 mT in small area elements (about 5 arc seconds, or 3600 km

Notes to Table 3.3 – *contd.*
Cl. Photospheric abundance from HCl molecular lines in a sunspot spectrum. Coronal abundance from an X-ray flare spectrum (Phillips and Keenan, 1990).
Ar. No photospheric value available: value quoted is for 'Local Galactic' in Meyer's work.

Sources: Data mostly from Meyer (1985) and Grevesse (1984).

square) can be measured. In Fig. 3.14, white areas denote regions of north (also called 'positive') polarity, i.e. where field lines are directed towards the observer, black areas regions of south ('negative') polarity, i.e. field lines directed away from the observer, and grey areas those in which measured fields are equally distributed between the two polarities. Boundaries between regions of positive and negative polarity, where the longitudinal field is zero, are known as 'magnetic inversion' lines (sometimes, less accurately, neutral lines).

A sunspot group generally appears on such magnetograms as a bipolar magnetic area, with the leading spot having the largest field strength (about 0.4 T) of one polarity and the following spots slightly weaker fields of the

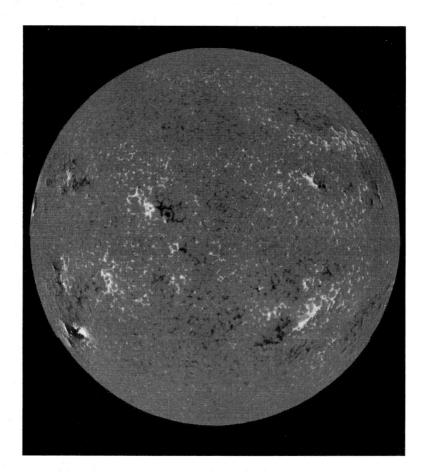

Fig. 3.14 Magnetogram of sun from Kitt Peak National Observatory for 18 July 1990. White areas denote areas of positive (or north) polarity, black areas negative (or south) polarity, and grey areas are those where opposite polarities are equally distributed. (Courtesy National Solar Observatory/Kitt Peak)

opposite polarity. Their dispositions obey Hale's polarity laws (1.7), i.e. the leading spot of a group in one hemisphere has the opposite polarity to the leading spot of a group in the other hemisphere, with the polarities reversed for successive sunspot cycles. For well-developed sunspot groups, there can be great magnetic complexity, and the bipolar structure may be hardly recognizable (see §6.1). As well as sunspot groups, there may also be other bipolar groups without spots, either the remnants of old spot groups or developing regions that will turn into spot groups. Collectively, all these bipolar regions (including the spot groups) are called 'active regions', and are confined to latitudes between about 35° north and south; we will deal with them more fully in §6.2.

In addition to the active regions, there are many very small bipolar magnetic areas without spots, and having a wider distribution in latitude than spot groups, though still having some concentration at lower latitudes. They even appear when the sunspot cycle is near minimum. They have lifetimes of less than a day, and are hence known as ephemeral regions. Like the active region bipolar groups, they obey Hale's polarity laws.

The small-scale magnetic field is also associated with the filigree (§3.1), which occurs where the field is particularly strong (about 0.1 T). There are also small clumps of field concentration (strength several mT) distributed round the boundaries of supergranules. They are coincident with structures observed in the chromosphere forming a 'chromospheric network' (§4.3). Very detailed studies with video-magnetographs by Sara Martin, Harold Zirin and colleagues at the Big Bear Observatory in California show that magnetic field exists even within the supergranules where small clumps of weak (1 mT) field, called 'intranetwork field', move radially out from near the supergranule centres towards the network field. This field may be either of the opposite polarity to the network field towards which it moves, in which case there is a 'cancellation' of the two fields, or of the same polarity, in which case there is a 'coalescence' of the two fields. In this way, the network field is continuously changing, either strengthening or diminishing with time.

The sun's large-scale magnetic field shows recognizable patterns and periodic behaviour. There is, for instance, a large-scale weak field at the solar polar regions. Hale claimed to have measured the strengths of such fields to be 5 mT, but this is now known to be incorrect, the field strengths being no more than 0.6 mT and probably much less. Nor is the field 'global', as Hale and earlier investigators had thought, but simply confined to the poles. This field reverses every 11 years, as do the sunspot fields, but the time of reversal is approximately at the peak of the sunspot cycle, i.e. when there are most spots. The field reversals in north and south hemispheres are not simultaneous – the reversal in one hemisphere may be up to a year earlier than in the other.

There is an intimate connection between the polar fields and the spot-associated field at lower latitudes when averaged over long periods. As sunspot groups start to decay, the areas of magnetic polarity associated with the following spots move to higher latitudes, often merging with each other to form 'unipolar' regions. Meanwhile, the field associated with the leading spots remains in equatorial latitudes. This fact was used by Babcock in his model for the solar dynamo (§2.4). At the rising stage of the sunspot cycle, the field at each of the poles has the opposite polarity to the unipolar areas approaching them, and it is then that the following-polarity areas have slowest drifts towards the poles, 5–10 m/s. After sunspot maximum, when the polar field has the same polarity as the unipolar areas, the flow of following-polarity areas is increased, with drifts up to 20 m/s. Possibly the motion of the unipolar areas is impeded in the rising stage of the sunspot cycle because their field can only merge with the polar field by a process of cancellation.

One's impression is that the sun is much more 'magnetic' when there are many spots than at solar minimum when there are few or none. The truth of this can be investigated by calculating the total magnetic 'flux' over the sun's surface, obtained by multiplying the area of small elements by the measured field strength in each area (regardless of polarity) and taking the sum over the solar surface. It is found that the total flux at sunspot minimum is between a third and a half – much more than would appear to be the case – of that at sunspot maximum. An additional impression that magnetic flux is concentrated in equatorial latitudes is borne out by measurements that indicate the average polar flux is only a sixth of the equatorial.

Chapter 4

The solar chromosphere

The chromosphere is a narrow part of the solar atmosphere above the photosphere not normally visible without special observing techniques or unless
there is a total solar eclipse. Its main characteristics are a rise of temperature
with height and complex structure which is constantly changing. In this
chapter, we will survey observations of the chromosphere that have been
made over a wide range of wavelengths and the deductions made from them
about the chromosphere's nature.

4.1   The chromosphere in profile

At the precise moment when totality begins or ends during a total eclipse, i.e.
when all the photospheric light is just blotted out, the chromosphere suddenly
appears as a red-coloured crescent or ring with irregular outer edge. More
closely examined, the edge is found to be made up of numerous fine jet-like
structures. The life-history and structure of these *spicules* (the name is due to
W. O. Roberts) can be examined outside the limited time of a total eclipse
using a filter with a narrow spectral band-pass, set at the wavelength of the
Hα line. Figure 4.1 is such an Hα 'filtergram' of a region of the limb. An
individual spicule is revealed to be a narrow column, a few hundred
kilometres in diameter, ascending almost radially into the corona with velocities of about 30 km/s, and attaining an altitude of about 9000 km. They last
approximately 15 minutes, ending their brief lives by fading from view rather
than descent; the ascent takes about 90 seconds. There may also be other
motions such as oscillations and a rotation about their long dimensions.
Spicules are very numerous indeed: seen on the solar limb, there is at least
one every few hundred kilometres, with a strong tendency to be clustered;
such clusters have been called 'porcupines'.

Below an altitude of about 1500 km, the chromosphere is more continuous in nature, though to call it a 'layer' of the solar atmosphere is not accurate. Above about 5000 km altitude, the chromospheric spicules appear to co-exist with the very hot gas – at temperatures of about 2 000 000 K – making up the corona. This is indicated by the presence of the coronal emission lines which occur down to 5000 km. Figure 4.2 shows a hypothetical section through the chromosphere to illustrate this: this figure will be discussed in more detail later.

The photospheric Fraunhofer spectrum is, at the moment when a total solar eclipse begins, suddenly replaced by an emission-line or *flash* spectrum. This spectrum undergoes rapid changes in the following few seconds as the moon covers up more and more of the chromosphere which emits the spectrum. Slitless spectrographs have commonly been used to form the spectrum, the chromosphere itself acting as a slit. The spectrum appears as a series of thin crescents, each one marking the wavelength of an emission line. The strongest emission lines are generally the familiar hydrogen Balmer lines and the ionized calcium (Ca II) H and K lines. Figure 4.3 is a flash spectrum during a total eclipse in 1970, covering the wavelength range from the $H\alpha$ to $H\beta$ lines. Very high members of the Balmer series have been recorded – lines corresponding to electron transitions between an $n=2$ orbit and one with n as high as 34. There are also many lines due to metals, such as magnesium, calcium, iron and nickel, as well as lines of these elements in a singly ionized form. They are relatively easy to excite, and so are emitted in the coolest and lowest part of the chromosphere.

The emission lines of the flash spectrum have their dark-line counterparts in the photospheric spectrum. There are, however, significant differences

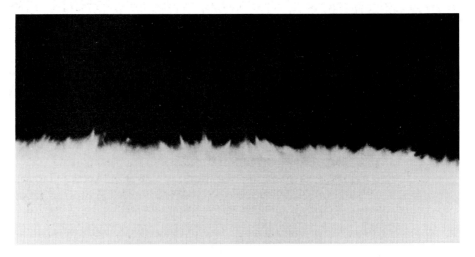

Fig. 4.1 Spicules on limb photographed in a band centred on the $H\alpha$ line (656.3 nm). (Courtesy National Solar Observatory/Sacramento Peak)

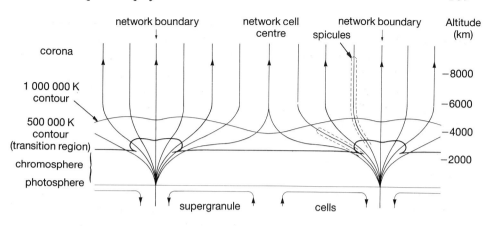

Fig. 4.2 Hypothetical section through chromosphere, showing the network structure and supergranules. Contours of temperature are shown for 500 000 K (taken to be the transition region) and for 1 000 000 K. The height scale on the right of the figure is only approximate. Spicules are represented by the dashed lines, the magnetic field lines by the arrowed lines: the field is swept horizontally towards the supergranule edges at low altitudes, but higher up the field lines spread out to form a canopy. (After Gabriel (1976))

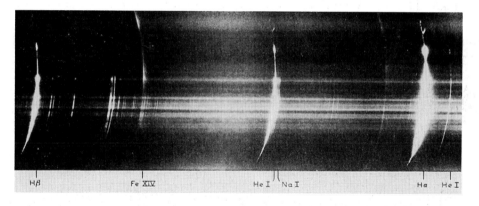

Fig. 4.3 Flash spectrum obtained shortly after the start of totality during the solar eclipse on 7 March 1970. Both chromospheric and coronal lines are visible: at the extreme left is the Hβ line (wavelength 468.1 nm), and at the extreme right the Hα line (656.3 nm). Between are the coronal Fe XIV green line (530.3 nm) and the chromospheric line of helium (587.6 nm) very near the sodium D lines (589.0 and 589.6 nm) in the yellow. The spectrum was taken with a 30-cm objective and a spectrograph with very wide slit, so that effectively the sun's limb acted as the slit whose crescent-shaped image is formed by each of the spectral lines. (Courtesy M. Kanno, Kwasan Observatory, Kyoto University, Japan and *Sky and Telescope*)

between the strengths of the chromospheric emission lines and those of the Fraunhofer lines. The flash-spectrum lines vary in intensity relative to each other in any case, depending on how much of the chromosphere is covered up by the moon. A most noticeable difference of the flash spectrum is the presence of helium lines, e.g. the yellow 587.6-nm line; these lines are absent in the Fraunhofer spectrum. The reason is that the hot gas of the chromosphere is able to excite helium atoms but the photospheric gas is too cool to do so. The excitation is by collisions with free electrons, and the gas temperature has to be about 20 000 K before there are enough electrons with the required energy. Such temperatures occur only in the upper chromosphere, so this is where the helium lines are emitted. Lines of singly ionized helium, He^+, also occur, and a still higher temperature is needed both to perform the ionization, again by electron collisions, and the excitation to the first excited levels of He^+.

The flash spectrum is made up of not only emission lines of varying strength but also a faint continuum stretching across the entire visible range. It is due to photospheric light scattered by free electrons in the chromosphere, a process called Thomson scattering. There is an additional continuum visible at wavelengths less than 364.6 nm, forming an 'edge' at this point, very near where the higher members of the hydrogen Balmer series of emission lines crowd together. This *Balmer* continuum is formed by the recombination of free electrons with protons to form neutral hydrogen atoms; the electrons are captured into the $n=2$ orbit, emitting photons as they do so. These photons, unlike those for line radiation, are not quantized, so their energy may take any value above an amount corresponding to the energy difference between an $n=2$ orbit and one with infinite n (i.e. when the electron just becomes free). This energy corresponds to a wavelength of 364.6 nm, so the continuum is emitted over wavelengths less than this amount.

4.2   The chromospheric Fraunhofer lines

Studying the chromosphere in profile during an eclipse or with $H\alpha$ filters gives only limited information – it does not, for example, allow comparison of features like spicules with the photospheric magnetic field as revealed in magnetograms. Chromospheric details on the disc can be observed, however, with spectroheliograms made in the light of the $H\alpha$ and Ca II H and K lines. The reason is that, whereas weak and medium-strength lines are photospheric in origin, these particularly strong Fraunhofer lines are formed over a much larger range of heights, including the chromosphere. In this section, we will outline the processes of the formation of these lines.

The older, 'classical' way of dealing with how Fraunhofer lines were formed relied on, among other things, the assumption of local thermodynamic equilibrium (LTE), by which the source function is equal to that describing black-body radiation (also known as the Planck radiation function) at the local value of temperature (§3.4). The intensity at the centre of a weak line is, by this treatment, approximately equal to the Planck function at an optical depth of two-thirds, and since this level is higher and cooler than that where the neighbouring continuum is formed, the line appears dark. With very strong lines, however, the assumption of LTE breaks down. The cores of these lines are formed above the height of the temperature minimum, where the local value of temperature, i.e. that describing the speeds of atoms, ions and electrons, may be quite different from that characterizing the radiation in which the atoms and ions in the chromosphere are immersed. This radiation is largely from the intensely bright photospheric layers, and has a characteristic temperature (the 'radiation' temperature) of about 6000 K. Take the core of the calcium K line, for example. It is formed at a 'height' (i.e. above the photosphere's base) of some 1900 km, where the temperature describing the motion of particles is a few hundred degrees higher than 6000 K. Now, both radiation and particle collisions play a part in how the line is formed, and so the different local and radiation temperatures must be taken into account. In other words, the source function can no longer be considered equal to the Planck function, but must be evaluated taking the individual atomic processes that are involved in forming the line.

The source function is defined by the emission coefficient (given by the atomic processes that create line photons) divided by the absorption coefficient (i.e. atomic processes that destroy line photons) (§3.4). Firstly, there are collisional processes that both create and destroy photons. For the $H\alpha$ line (formed by transitions between the $n=3$ and $n=2$ orbits of hydrogen atoms), line photons may be created by electrons colliding with hydrogen atoms resulting in the excitation of their electrons from $n=1$ orbits to $n=3$ orbits, followed by spontaneous transitions from $n=3$ to $n=2$ orbits. But $H\alpha$ line photons may equally be destroyed through the collison of electrons with hydrogen atoms already in the $n=3$ state, taking the atoms to a lower state without photon emission. Then there are 'photoelectric' processes in which photons of the photospheric radiation field likewise create $H\alpha$ photons or destroy them through the excitation or de-excitation of hydrogen atoms. Finally, there are scattering processes (already described in §3.5), though the effect of these is neither to add nor to take away line photons: they merely give rise to a diffusion of the line photons.

The chromospheric lines can be broadly classified accordingly as the collisional processes or photoelectric processes dominate. This depends largely on the particular atom or ion. It turns out that the $H\alpha$ line is photoelectrically controlled, while the Ca II H and K lines are collisionally controlled. Thus, the $H\alpha$ line intensity is governed more by the photospheric radiation than the local value of temperature, which determines the collisional processes. Going outwards from the photosphere into the chromosphere, there is an increasing 'leak' of photospheric radiation to space which results in the Hα source function decreasing outwards. Using the rough approximation that the line intensity is equal to the source function at two-thirds optical depth, we see that the line intensity steadily decreases towards its core which is formed at the highest level in the atmosphere, corresponding to the mid-chromosphere (about 1500 km altitude). Figure 4.4 shows the inner part of the profile of the Hα line.

On the other hand, the line source function for the H and K lines more nearly reflects the local value of temperature. As already indicated, the temperature rises with height from the temperature minimum out into the chromosphere. This results in an increase of the H and K line source function, at least over a certain height range. Eventually, at a height of about 1900 km, the source function again decreases owing to the effects of the scattering terms in the source function. The effect on the line profile of the rise of the source function at chromospheric heights is the presence of two

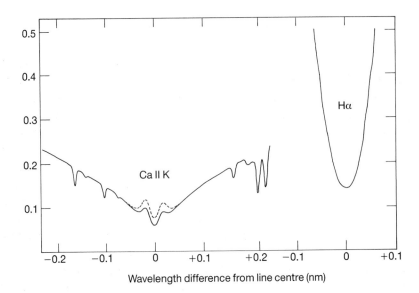

Wavelength difference from line centre (nm)

Fig. 4.4 Profiles of the chromospherically formed ionized calcium (Ca II) K line (left) and the hydrogen Hα line. The wavelength scale (in nm) is centred on the line centre in each case (393.4 and 656.3 nm respectively) and the spectral flux (vertical) scale is in fractions of the continuum value. For the Ca II K line profile, the solid line is that for integrated sunlight near sunspot minimum in 1977, the dashed line near sunspot maximum in 1979. The emission peaks in the K (and H) lines are the only evidence for a chromospheric rise in temperature in the visible spectrum. (Adapted from White (1964) and White and Livingston (1981))

bumps on the line profile near its core (see Fig. 4.4). For the K line, these emission peaks are called K2V (on the violet side of the line core) and K2R (red side). (The H2V and H2R peaks are similarly defined for the H line.) Their prominence is strongly related to solar activity (Fig. 4.4 shows both quiet-sun and active-sun profiles) since when the sun is active there is an increase in the collisional terms of the source function. The 'self-reversed' core, sometimes called K3, is formed highest in the chromosphere (altitude of about 1900 km, where the source function again decreases). The wings, called K1, are formed near the temperature minimum. They extend over such a wide wavelength region (392–395 nm) that many other, quite unrelated and much narrower, lines occur superimposed on the K line profile.

In summary, the Hα and calcium H and K lines and indeed some other strong Fraunhofer lines are partly formed in the chromosphere, so by viewing the sun at or near the cores of these lines we obtain an image of the chromosphere at a particular level. We now turn to what the appearance of the chromosphere is.

4.3   The chromosphere in spectroheliograms

Hα and calcium H and K line spectroheliograms of the sun show that the chromosphere is a highly non-uniform, structured region of the solar atmosphere. We can view different levels of the chromosphere simply by adjusting the spectroheliograph or 'tuneable' filter so that images at different parts of the line profile are produced. Thus, for the Hα line, moving the wavelength from the line core to the wings produces images that are successively of the mid-chromosphere (altitude 1500 km) down to the photosphere. Images at the wavelengths of the Ca II K2 emission peaks and the K3 core also show the mid-chromosphere, and the K1 wings the region of the temperature minimum. One complication is that the fine structures of the chromosphere may show rapid changes, and consequently their emission is Doppler-shifted towards the violet if viewed at the centre of the solar disc since then their velocities are towards the observer. A spectroheliograph or filter tuned to a particular part of the line profile may then not pick these features out because of their Doppler shifts.

Figures 4.5 and 4.6 are full-disc spectroheliograms made at the centres of the Hα and the Ca II K lines. They were taken on the same day as the magnetogram of Fig. 3.14. They illustrate vividly the difference from white-light images, i.e. of the photosphere. The most conspicuous features of the spectroheliograms are the bright patches called *plages* in the vicinity of sunspot (or active) regions, particularly in the K line image. Also noticeable are the long dark structures called *filaments*. Just visible in the Hα image of

Fig. 4.5 is a sunspot and near it are fine dark lines, known as *fibrils*. We shall discuss these more fully when describing the active sun, in Chapter 6.

The entire chromosphere – 'quiet sun' regions and active regions – is made up of bright clumps that form a pattern known as the *chromospheric network*. The quiet-sun network is very faint, but becomes enhanced near plages. An individual network cell measures about 30 000 km across and lasts a day or so. The bright structures are the so-called flocculi or 'coarse mottles' which show up more clearly in the Ca K image of Fig. 4.6. A coarse mottle in K line

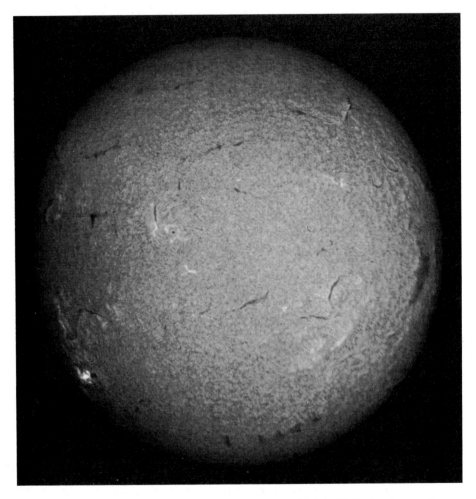

Fig. 4.5 Whole-disc Hα spectroheliogram taken on 18 July 1990, showing small active regions (bright areas). A sunspot is visible as a dark spot near disc centre, and there are several filaments, which are the long dark features. This spectroheliogram was taken on the same date as the magnetogram of Fig. 3.14, so comparison with the solar surface magnetic field can be made. (Courtesy National Solar Observatory/Sacramento Peak, Sunspot NM)

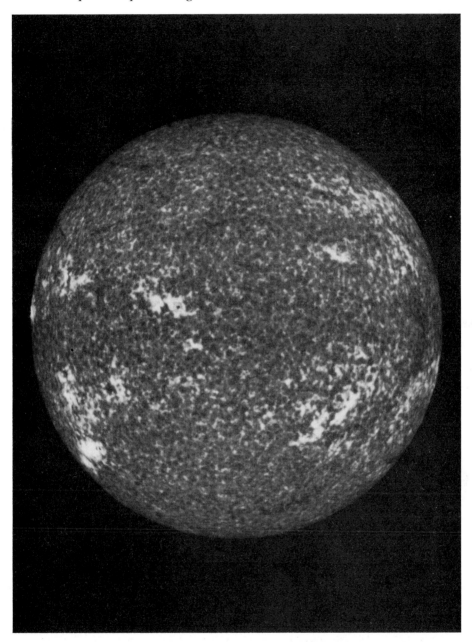

Fig. 4.6 Whole-disc spectroheliogram in the core of the Ca II K line taken on 18 July 1990 (the same date as in Figs. 4.5 and 3.14). The chromospheric network, marked by the coarse bright mottles, shows more clearly than in the Hα spectroheliogram, as do the active regions, but the filaments are barely visible. (Courtesy National Solar Observatory/Sacramento Peak, Sunspot NM)

images is composed of several bright 'fine mottles', about 7000 km long and 700 km in diameter, lasting about a day, and showing both upward and downward velocities. The fine mottles appear as both bright (if low-lying) and dark (if higher than about 3000 km) when observed at the centre of the Hα line, but are always dark in the Hα red wing. Most likely the latter are identifiable as spicules when viewed at the limb (§4.1). Figure 4.7, a photograph towards the limb taken in the red wing of Hα, shows how the fine dark mottles cluster together to form bushes, which are probably identifiable as the porcupine clusters of spicules on the limb. Positive identification is not possible as individual features on the disc fade long before solar rotation carries them to the limb.

Seen near the centre of the solar disc at the Hα line core, the fine dark

Fig. 4.7 Spectroheliogram taken in the red wing of the Hα line of a sunspot region and surrounding area near the solar limb. In the quiet areas, spicules show as fine dark mottles pointing towards the limb. In the active region, the small bright points are Ellerman bombs (§6.2). (Courtesy Big Bear Observatory)

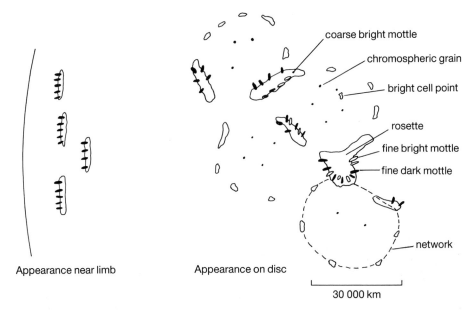

Appearance near limb   Appearance on disc

30 000 km

Fig. 4.8 Schematic diagram of quiet-sun chromospheric structures seen in Hα and Ca II H and K line images. The coarse bright mottles outline the chromospheric network, and fine dark and bright mottles emanate from the coarse mottles. At the intersection of two or three network cells, the fine mottles form a rosette pattern. Chromospheric grains are small dark features within network cells visible in the violet wing of the Hα line, while in Ca K bright cell points are small bright points that repeatedly brighten and fade. Near the limb the fine dark mottles seen in Hα point towards the limb.

mottles form radiating patterns called rosettes, with bright mottles at the centres of the rosettes and between the dark mottles. Inside network cells, visible in the violet wing of Hα, are small dark points about 1000 km in diameter known as 'chromospheric grains', which are associated with upward motions. At the wavelength of the calcium lines, there are small bright points that brighten and fade repeatedly in a few minutes, called 'bright cell points'. Their relation to the Hα grain is unclear. Figure 4.8 schematically shows these chromospheric network features as they appear at disc centre and near the limb.

The chromospheric features and photospheric magnetic field (see §3.8) are related on both small and large scales. The network structures in particular correspond to almost vertical fields with strengths of about 1–2 mT and much higher (several tens of milliteslas) at the intersections of several cells, where the Hα and K coarse mottles are particularly bright. The clumps of intranetwork field within network cells discussed in §3.8 may be associated with the bright cell points. As already indicated, this field appears at the centres of supergranules, and moves radially outwards to the supergranule edges where it either cancels or coalesces with the network field. The chromospheric network structures as a result are weakened or strengthened in intensity.

On a larger scale, strong fields present in and and around sunspots are related to the plage structures, while magnetic inversion lines separating large unipolar magnetic areas of opposite polarities are marked by the filaments.

4.4   The chromosphere in non-visible wavelengths: the transition region

Although much information about the chromosphere may be obtained from images made at the wavelengths of lines in the visible spectrum, there is no indication of the connection between the chromosphere and the overlying, much hotter corona. This connection can be studied by observing the sun in the ultraviolet part of the spectrum (so requiring space-borne instruments since such wavelengths are absorbed by the earth's atmosphere) and in short-wavelength radio waves.

Below a wavelength of about 400 nm, in the violet, the solar continuum decreases steadily in intensity. There are two strong Fraunhofer lines in the ultraviolet due to singly ionized magnesium, at wavelengths of 280.2 and 279.5 nm, called the Mg II h and k lines since the electron transitions involved are similar to those for the Ca II H and K lines. They, like the calcium lines, are collisionally controlled and have emission cores indicating the presence of the chromospheric rise of temperature; Figure 4.9 shows the core of the k line. Below about 140 nm, the continuum fades to invisibility, and at 121.5 nm is the intense Lyman-α line of hydrogen, formed by $n=2$ to $n=1$ transitions of hydrogen's single electron; Figure 4.9 shows the profile of this line. The line emission is formed purely by free-electron collisions which excite hydrogen atoms in their ground state to the $n=2$ state. Radiation has no contribution at all, since there are effectively no photons having the appropriate energy to excite to the $n=2$ state. It is calculated from its intensity that the region where Lyman-α is formed must be at a temperature of about 20 000 K.

The many other emission lines present are also excited by free electrons, though unlike Lyman-α many are optically thin, i.e. a line photon once emitted freely travels out from the sun without suffering any absorption or scattering. Many of the most intense lines are due to very highly ionized atoms. A doublet of lines at 155 nm, for instance, is due to three-times ionized carbon (C^{3+}), which can only be formed in a region where the temperature is 110 000 K. This is much higher than the chromosphere though still lower than the corona, at about 2 000 000 K. The lines are in fact emitted where the temperature rises extremely rapidly with height, a region of the atmosphere known as the *transition region*. Other prominent emission lines are due to even more highly ionized atoms and arise further up the steep

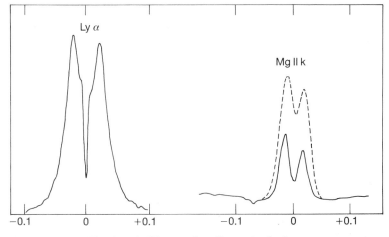

Fig. 4.9 Profiles of chromospherically formed ultraviolet lines: the hydrogen
Lyman-α line (left) and the ionized magnesium (Mg II) k line. The
wavelength scale (in nm) is centred on the line centre in each case
(121.5 nm and 280.2 nm). The profiles are dominated by emission peaks,
which are activity-dependent (the dashed line for the Mg II k line
indicating the profile when the sun is active). (After Lemaire *et al.* (1978)
and Lemaire and Skumanich (1973))

gradient of temperature, while still others are emitted at temperatures of
above 1 000 000 K and thus in the corona itself. Figure 4.10 is a section of the
quiet-sun ultraviolet spectrum from 45 to 140 nm, and shows lines due to
O VI (temperature 320 000 K), Ne VII (500 000 K) and Si XII (2 000 000 K).
Table 4.1 is a list of some of the more prominent ultraviolet lines with
approximate values for their temperatures of formation. The use of Roman
numerals to denote the spectra emitted by ions, introduced in §3.5, is continued and extended here (the 155-nm carbon lines, for instance, being part
of the 'C IV' spectrum).

Although the photospheric continuum ceases to exist at wavelengths
shorter than about 140 nm, there is another continuum – the *Lyman* continuum – which is formed by recombination of electrons with protons to form
neutral hydrogen. As the Balmer continuum is formed by recombination of
these electrons to the $n=2$ orbit (§4.1), so the Lyman continuum is formed by
recombination to the $n=1$ orbit. The edge is at 91.2 nm, corresponding to the
transition of an electron from a state in which it is just free to the bound $n=1$
state.

The ultraviolet lines of the chromosphere, corona and transition region tell
us a great deal about the structure of the solar atmosphere. Early spacecraft
observations did not resolve individual solar features but in time a clear
understanding of where the emission originated at particular temperatures

Table 4.1. *Some prominent ultraviolet emission lines*

Wavelength (nm)	Identification	Temperature of formation (K)
17.1	Fe IX	900 000
17.48	Fe X	1 100 000
18.04	Fe XI	1 300 000
19.5	Fe XII	1 500 000
21.9	Fe XIV	1 800 000
23.4	Fe XV	2 000 000
24.9	Ni XVII	2 500 000
25.1	Fe XVI	2 200 000
25.6	He II	50 000
26.3	Fe XVI	2 200 000
26.5	Fe XIV	1 800 000
28.4	Fe XV	2 000 000
30.4	He II	50 000
33.5	Fe XVI	2 200 000
36.8	Mg IX	1 000 000
46.5	Ne VII	500 000
55.4	O IV	160 000
62.5	Mg X	1 100 000
99.7	C III	70 000
103.2	O VI	320 000
117.6	C III	70 000
120.6	Si III	30 000
121.5	H I[a]	20 000
124.0	N V	200 000
133.5	C II	20 000
139.4	Si IV	60 000
140.3	Si IV	60 000
153.3	Si II	14 000
155.0	C IV	110 000
164.0	He II	50 000
165.7	C I	<10 000
167.0	C I	<10 000
180.8	Si II	14 000
181.7	Si II	14 000

Note:
[a] Lyman-α of neutral hydrogen.

began to emerge with the images obtained from instruments on spacecraft. Far-reaching advances were made when the NASA *Skylab* mission – an earth-orbiting manned laboratory – was flown in 1973–74, carrying ultraviolet and X-ray instruments on board. One of these – an imaging ultraviolet

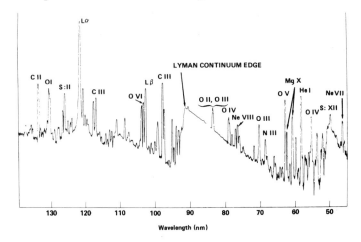

Fig. 4.10 A section of the solar extreme ultraviolet spectrum – from 45 to 140 nm – obtained by the Harvard College Observatory spectrometer on *OSO-6*. The ions responsible for some of the more intense lines are identified: they range in temperature from the chromospheric hydrogen Lyman lines to the coronal Mg X (nine-times ionized magnesium). The emission 'wedge' with peak flux at 91.2 nm is the Lyman continuum. (Courtesy R. Noyes, Harvard College Observatory and NASA)

instrument – showed that the upper chromosphere (as revealed by Lyman-α line emission) and the transition region (as revealed by, e.g., the O IV line at 55.4 nm) were confined to the network that is observed in Ca II K line spectroheliograms. Figure 4.11 illustrates this, showing six *Skylab* images of a quiet-sun region measuring 220 000 km square. The network is steadily less sharply defined as the temperature increases, and is blurred almost beyond distinction for the image in the Mg X line (at 62.5 nm), which is formed at 1 100 000 K and therefore in the corona. The network, then, is not traceable above the transition region. This can be understood with reference to Fig. 4.2 earlier; the chromospheric and transition-region line emission arises sufficiently deep down in the atmosphere that the magnetic field lines at the supergranule boundaries are still tightly bound together, but the Mg X line emission region, in the corona, is just above where the magnetic field spreads outwards to form a canopy and so its emission is not spatially confined at all.

The *Skylab* ultraviolet observations have been used to build models for the solar chromosphere, in which the temperature variation with height in the atmosphere is specified for particular solar features. In the models of J. E. Vernazza and colleagues, these features include the interior of a network cell, network elements away from plage areas (i.e. unenhanced) as well as bright network elements. With a trial temperature curve, the procedure is to find, using an elaborate non-LTE calculation, the emergent ultraviolet spectrum in the wavelength range 40–140 nm – the principal emission lines and also the

Lyman and other continua. The calculated spectrum can then be compared with that from the *Skylab* instrument and any differences reduced as much as possible by adjusting the temperature–height curve. Figure 4.12 is the calculated curve for the 'average quiet sun' by this method. It shows the temperature variation with height from the base of the photosphere to the start of the transition region, about 2300 km higher. The temperature minimum at 500 km height is evident, as is the rise above this level to the chromosphere and the extremely rapid rise at a height of 2000 km or so. The regions where various parts of the line profiles of the Hα, Ca K, Mg k and Lyman-α lines are formed are indicated on the figure.

Ultraviolet line intensities may be analysed to give element abundances in the chromosphere and transition region, and so allow comparison with those from the photosphere, obtained as we saw in §3.7 from curve-of-growth methods. Firstly, we know (from atomic calculations) the temperature at which a particular line is formed (actually, the line is formed over a small range: the temperatures in Table 4.1 are the mid-points of these ranges). Secondly, atomic calculations can give us the rate at which a line is excited by electron collisions for that temperature, so we can work out how much line emission emerges from a unit volume, say one cubic metre. If this is com-

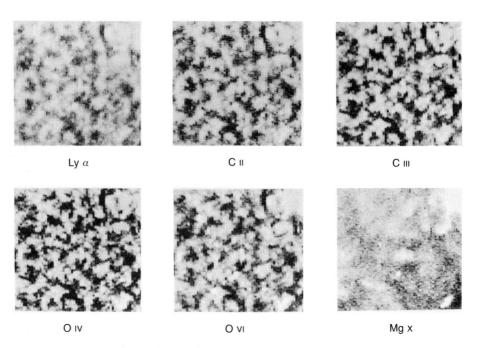

Fig. 4.11 Six *Skylab* images of a quiet-sun region taken in ultraviolet spectral lines in a temperature range from the chromospheric Lyman-α to the coronal Mg X. The chromospheric network becomes progressively more blurred. (Courtesy R. Noyes, Harvard College Observatory and NASA)

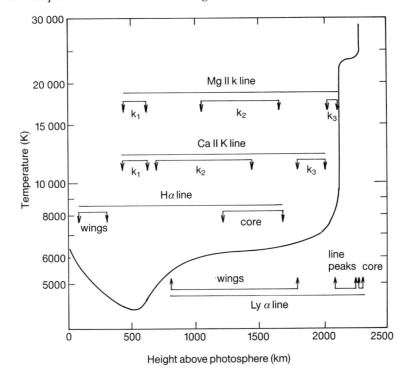

Fig. 4.12 The variation of temperature with height in the solar atmosphere up to
the transition region for an average quiet-sun region. Also indicated are
the height ranges over which the Hα and Ly-α, Ca II H and K, and
Mg II h and k lines are formed. (After Vernazza, Avrett and Loeser (1981))

pared with the radiation observed from the sun, we may obtain a quantity
equal to the number density of emitting atoms (i.e. the number per cubic
metre) times the number density of free electrons times the height range over
which the line is emitted. This quantity can be plotted against temperature for
all the lines for a particular element. For any one element there will be several
points, one for each line emitted by as many ions of the element as possible to
cover as wide a temperature range as possible; the points trace out a curve
that declines with temperature to a minimum then rises again. Placing the
points due to the ions of all elements on this plot gives several such curves, all
with the same shape, but the curves of each element may be slid vertically so
that they overlap to form a single curve. Figure 4.13 shows such a plot for 10
elements that have intense lines in the solar ultraviolet spectrum. The
abundances of these elements, at least relative to one of them (e.g. carbon),
may be determined by the amount by which each curve must be slid to achieve
overlap with the curve for carbon. The abundances of all these elements can
be related to that of hydrogen if the carbon-to-hydrogen abundance ratio is
known.

The profiles of transition-region emission lines give much information about mass motions and possible mechanisms for transporting energy. In recent years, high-quality observations have been made of the C IV lines at 155.0 nm, for example, with the High-Resolution Telescope and Spectrograph (HRTS) built by Guenther Brueckner and colleagues at the US Naval Research Laboratory, and flown on the Space Shuttle as well as several rockets. The instrument is able to observe the line profile with very high spectral resolution with the telescope forming a high-resolution image of part of the sun on the entrance slit of the spectrograph. In this way, details of the

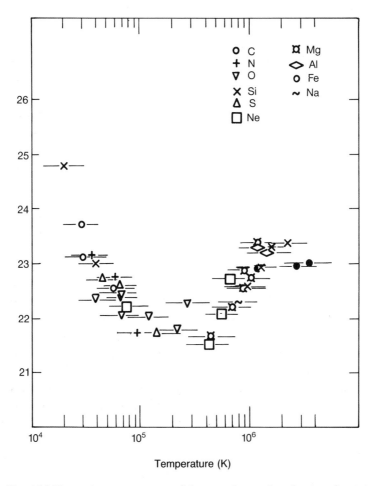

Fig. 4.13 Plot against temperature of the quantity number density of emitting atoms times that of electrons times height range of emission (log scale) for ultraviolet lines emitted by ions of various elements (listed) in the chromosphere, transition region and corona. The abundances of the elements were found by adjusting the points in the plot until a single curve was produced. (After Jordan and Wilson (1971). Reprinted by permission of Kluwer Academic Publishers)

line profile can be matched to particular features like the network structure or sunspots and plages. It has been found, from the Doppler shifts of the line, that both downflows and upflows occur at the network, but with a greater preponderance of downflows. Also, the line is found to be much broader than expected. The C^{3+} ion emitting the lines is expected to exist at a temperature of 110 000 K, as mentioned earlier, and the thermal Doppler line width for this temperature is about 0.012 nm. But the observed widths are more than twice this. The difference has been explained by 'non-thermal' broadening, i.e. Doppler shifts due to the random velocities of a turbulent gas at this temperature, which has led to proposals that these fluctuations arise from wave motions propagated up to the transition region from the photosphere. Much more spectacular are specific events – 'turbulent events' and 'jets' – observed as very large Doppler shifts in portions of transition-region lines like the C IV doublet (Fig. 4.14). Turbulent events occur in tiny (size up to 3000 km) areas over the network and within network cell interiors, last a few minutes and have Doppler shifts to both shorter and longer wavelengths that indicate velocities of up to 250 km/s. Jets are apparently exploding (or at least rapidly expanding) features with velocities of up to 400 km/s, apparently undergoing heating. They have also been directly seen on the sun's limb in ultraviolet spectroheliograms. Their frequency in the HRTS spectra suggests that about 24 per second occur all over the sun. Such a rate could mean that they have significance for the heating of the corona (§5.7).

The upper chromosphere and features associated with solar activity (sunspots and plages) are strong emitters of radio waves at relatively short wavelengths – a few centimetres and less. Astronomy at millimetre and sub-millimetre wavelengths is a relatively new venture, owing to the need for specialized telescope and detector technology and high-altitude sites to avoid atmospheric absorption. The James Clerk Maxwell Telescope in Hawaii has been used for observing the solar chromosphere at such wavelengths. Cen-timetre-wavelength emission is easily detected, but a major limitation has been the very poor resolution of a single telescope compared with an optical telescope (see §10.2). This can be overcome using several linked telescopes in an interferometer array like the Very Large Array in New Mexico with which images of the chromosphere have been acquired with a resolution of 10 arc seconds (7250 km on the sun). The images of the quiet sun at centimetre and millimetre wavelengths with these instruments clearly show the chromos-pheric network, with those network elements that are brightest in Ca K spectroheliograms appearing brightest in the radio. The emission at these wavelengths is optically thick and approximately given by a black-body radi-ation distribution. The observed intensity may be equated to that of an equivalent black body at some temperature – the so-called *brightness temperature*, approximating the kinetic temperature (i.e. that describing particle speeds). The measured intensities indicate brightness temperatures of network elements increasing from 5000 K at a wavelength of about 1 mm to

25 000 K at 6 cm: evidently over this wavelength range the chromosphere is being viewed from its base to just below the transition region.

The most puzzling observations of the chromosphere have been made in the infrared, at a wavelength of 4.7 μm, where there are absorption bands (i.e. numerous spectral lines crowded together to form a continuous absorption feature) due to the molecule carbon monoxide. This appears to be present at altitudes well into the chromosphere, where the temperature has risen to 6000 K or more, yet the molecule can only exist at temperatures of about 4000 K – at higher temperatures it would break up into its constituent carbon

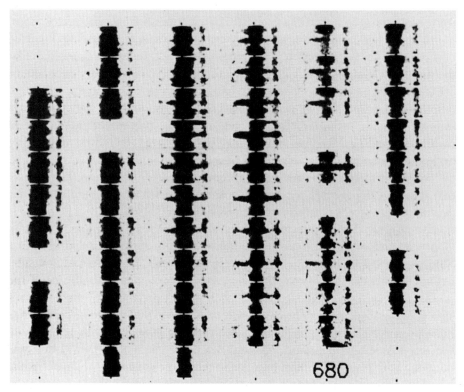

Fig. 4.14 Development of a jet (explosive event) observed with the HRTS instrument on board a rocket on 1 March 1979. Each small image is of the profile of the two C IV lines (154.8 and 155.0 nm) summed, with a portion of the sun, 16 arc seconds long, imaged along the spectrograph slit. The slit was repositioned six times, with 2 arc seconds between each position, to give the six columns of images. The different images in a column were obtained in a time sequence, with 20 seconds separating each image and the earliest times at the top of the figure. The jet appears as a sudden broadening of the line profile, forming a short horizontal streak, with Doppler shifts towards long and short wavelengths corresponding to approaching and receding velocities of up to about 100 km/s. (Courtesy K. Dere *et al.* (1989). Reprinted by permission of Kluwer Academic Publishers)

and oxygen atoms. It seems that the hot chromosphere has pockets of much cooler gas. This has altered our perception of the chromosphere's nature, which we will now consider.

4.5   The nature and heating of the chromosphere

Ultraviolet observations show that there is a general increase of temperature with height above the temperature-minimum region, which may be thought of as constituting the chromosphere proper. This applies to the network elements and the bright cell points within the cell interiors. But at the level of the temperature minimum, these features cover less than 20% of the total area, and only up to 60% even where there are active regions. What about the remaining regions? Though ideas are still vague, it is likely that they are occupied by cool gas – at temperatures of about 4000 K or less – where the carbon monoxide molecules observed in the infrared are able to form. They might almost be called carbon monoxide 'clouds'. Possibly they continually form through an 'overshoot' of the supergranular convection up to the temperature-minimum region, but suffer disruption and break-up as the bright cell points flash up as they periodically do.

In any event, we still have to explain why there is a rise of temperature for the true chromosphere, the parts between the carbon monoxide clouds. This is not expected if radiation were the only form of energy being transported up from the photosphere, since as more and more radiation leaks to space going outwards a steady decrease of temperature should occur. Thus, there must be a non-radiative form of energy present to cause the temperature rise of the chromosphere, as well as the transition region and the corona. Indeed, it is thought that this non-radiative energy is dissipated even at the level of the photosphere.

Before describing what form this non-radiative energy might take, we shall consider the way which the temperature rises in the chromosphere. We use Fig. 4.12, a model calculation for the 'average quiet sun', as a guide. From the base to the top of the photosphere (0 to 500 km altitude), there is a decrease of temperature owing to a decrease in the density of H^- ions, thus reducing the ability of the photospheric gas to absorb energy and so maintain its temperature. However, above 500 km altitude, the transport of non-radiative energy, whatever form it takes, leads to a rise of temperature. This results in an increase in the ionization of hydrogen atoms, so there is a greater number of free electrons and protons. The electrons are available for collisional excitation of certain atoms and ions, which de-excite by emitting line radiation. The emission lines include the strong hydrogen Hα and other Balmer lines, the Ca II H and K lines, and the ultraviolet Mg II h and k lines. The result is that, although energy is still delivered to the middle regions of the chromosphere in some form, the temperature hardly rises at all because that energy is being radiated away; the radiating atoms and ions act as a 'thermostat', and a broad temperature plateau is thereby formed. But there is a limit to this effect, as the supply of neutral hydrogen atoms becomes depleted and further input of energy does not produce such a large number of free electrons. The amount of energy radiated away cannot now compete with the energy that continues to pass into the chromosphere, so that the temperature rises sharply. There is thought to be a second, very narrow, temperature plateau at about 20 000 K due to radiation loss in the form of the intensely bright Lyman-α line emitted by hydrogen, so acting as a second thermostat. Further up, the increase of temperature produces more ionization of hydrogen, and so radiation loss by the Lyman-α line ceases to be significant. The temperature rises extremely sharply beyond this level as now there are no further effective thermostats until the temperature reaches coronal values. In the corona, there are a number of very highly ionized atoms that radiate a lot of energy, much more so than neutral or singly ionized atoms. This would be expected to result in a flattening out of the temperature rise, as in fact happens, with the temperature now 2 000 000 K. Table 4.2 summarizes what the chief radiation energy losses are in the various regions of the solar atmosphere above the photosphere.

The amount of energy lost by radiation above the photosphere can be estimated from observations of the principal emission lines. Over the entire quiet sun, i.e. devoid of sunspot regions, this radiation loss comes to between 1×10^{22} W and 2×10^{22} W. Though this energy loss is small (less than 0.01%

Table 4.2. *Radiation losses in the solar atmosphere*

Atmospheric region	Temperature (K)	Height above photosphere[a] (km)	Main contributors to radiation
Temp. minimum	4 400	500	
Low to mid-chromosphere	6 000	1 000–2 000	Hα, Ca II H and K, Mg II h and k lines
Upper chromosphere	20 000	2 200[b]	Hydrogen Lyman-α
Transition region	100 000	2 500	UV lines (e.g. C IV, 155 nm)
Corona	2×10^6	>5 000	Soft X-rays

Notes:
[a] Defined to be optical depth unity at 500 nm, solar disc centre.
[b] Note spicules occur up to 9000 km height.

Sources: Data from Athay (1981) and Vernazza, Avrett and Loeser (1981).

of the photosphere's radiative energy output), it must be balanced by some energy input, i.e. a form of non-radiative energy supplied from the photosphere or interior. Much effort has been taken into finding what form this takes (we deal with the energy supplying the corona in Chapter 5).

Wave motions generated in the photosphere are a promising candidate for heating the chromosphere. Sound (or pressure) waves are, as we saw in Chapter 2, very significant in the solar interior where they travel around trapped inside cavities, giving rise to the five-minute oscillations. Some years ago, the five-minute oscillations were considered as a heating mechanism for the chromosphere, but it was later found that only those waves with much shorter periods than five minutes could propagate through the temperature-minimum region, assuming vertical wave propagation, and so be available for heating. A sound wave by itself does not heat the gas it passes through, but merely causes the gas particles to oscillate. But a sound wave travelling outwards from the photosphere soon encounters gas with much reduced density and because of this is altered. The wave initially has a simple sine-wave form in which regions of compressed and rarefied gas travel at equal speeds, as indicated in Fig. 4.15. As the wave reaches lower-density gas, the regions of compressed gas travel faster than those of rarefied gas, so the form is more and more distorted until a shock 'front' is formed. The passage of such a front through the atmosphere does give rise to heating. This mechanism is thought to be responsible for the heating of the lower chromosphere.

Above the low chromosphere, the rôle of the magnetic field at the network

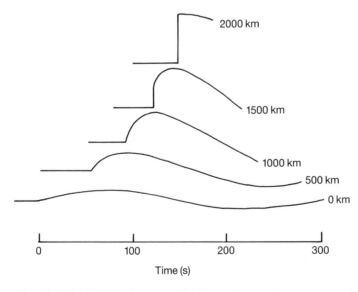

Fig. 4.15 The initially sine-wave-like form of a sound wave generated at the base of the solar atmosphere (bottom curve) is altered to one with sharp leading edge or shock front (top curve) as the wave progresses further up (heights indicated by each curve in the figure). (After Bird (1964))

is thought to be important. Magnetic field lines in an ionized gas or plasma are subject to wave motions called magnetohydrodynamic (MHD) waves. A particular type is known as Alfvén waves. These arise when a field line is displaced sideways and released; a tension set up in the field lines tends to restore them to their original shape, but in doing so sets up an oscillation, the Alfvén wave, propagating at the Alfvén speed. One can imagine the field lines behaving like tensed elastic bands. If the ionized gas has perfect electrical conductivity, the oscillations in the field lines cause the plasma to oscillate with them because the field lines are 'frozen in'. An Alfvén wave would under these circumstances continue indefinitely. But the wave would give up some of its energy if the plasma were not perfectly conducting, i.e. were slightly resistive, and this energy could be available for heating the gas it passes through. It was once thought that, if Alfvén waves were generated below the photosphere, they would be damped as they passed through the photosphere and low chromosphere, and therefore would not reach the upper chromosphere. However, this is not so if they propagated up via the tiny filigree regions (§3.1). Actually, a wave motion that is directed at some angle to a set of magnetic field lines will generate three sorts of MHD wave models. One of these is a pure Alfvén wave, and the other two are known as slow and fast MHD waves which combine the characteristics of sound and Alfvén waves. Thus, the heating of the chromosphere above the lowest level may well in part be by the dissipation of Alfvén or other MHD waves, produced by disturbances deeper down. Relevant in this respect are the observations of non-thermal broadening in the C IV transition-region line (§4.4) which may indicate heating at the transition region by waves.

The nature of the transition region has been debated over the years. We have so far described the solar atmosphere above the photosphere with reference to Fig. 4.2: the chromospheric network structures are connected to the high-temperature corona along magnetic field lines, with a steady rise of temperature from 6000 K (lower chromosphere) to 2 000 000 K in the corona, and the transition region is a thin region where the temperature rises from 20 000 K to the coronal 2 000 000 K. The energy of the transition region seems to be mostly supplied by thermal conduction and the downflow of hot gas from the hot corona *above*, i.e. dissipation of wave energy from the photosphere *below* is not so significant. Recently, this interpretation of the observations has been questioned. Limb measurements of transition-region ultraviolet lines point to the emission coming from tiny (less than 100 km across) structures not resolvable by any instrument so far used. U. Feldman considers that much of the observed ultraviolet line emission is in fact due to these 'unresolved fine structures', as he has called them, and only a small part to the 'true' transition region, i.e. the interface between chromosphere and corona.

Chapter 5

The solar corona

The solar corona is an extremely hot, tenuous part of the solar atmosphere appearing in white light as streamers, plumes and other structures extending out from the chromosphere. The temperature is around 2 000 000 K – higher in regions associated with sunspots – which means that the largely hydrogen gas is a fully ionized plasma giving rise to X-rays, extreme ultraviolet (wavelengths less than about 100 nm) and radio (metre-wave) radiation. Thus, although the radiation from the corona is but a tiny fraction of that from the photosphere, it covers a far wider wavelength range. In this chapter we will review the observational evidence about the corona's nature, the coronal 'holes' and their links with the solar wind, magnetic fields, prominences and the still controversial questions concerning how the corona is heated to such high temperatures.

5.1   The white-light corona

The white-light corona may be observed during a total eclipse or with coronagraphs, instruments producing an artificial eclipse with an occulting disc to blot out the photospheric light (§§1.5, 10.1). Coronagraphs are sited at mountain-top observatories to ensure the most transparent skies or, in the past few years, on satellites. They have allowed the gradual changes of coronal forms to be almost continuously recorded, while extremely rapid changes – coronal 'transients' – are now recognized: these will be described in §6.4. The gradual changes are related to the sunspot cycle, the corona having long symmetrical streamers almost aligned with the equator and polar plumes at minimum, and many streamers and lobe-like structures at maximum. A crude measure of coronal shape – the 'ellipticity' – was used until a few years ago, defined by the difference in equatorial and polar diameters divided by the equatorial diameter at a particular light level. This ranges from a value close to 0 at sunspot maximum (i.e. corona roughly circular) to about 0.2 at minimum (i.e. a flattened ellipse).

The corona's white-light radiation is simply scattered photospheric light, and hence its colour is almost identical to the photosphere. Both free electrons and dust grains in the corona do the scattering, and so give rise to two main components of the coronal emission. The electron-scattered component, known as the K corona (standing for the German *Kontinuierlich*), dominates from near the photosphere out to about two solar radii from the sun's centre, or about 700 000 km above the photosphere. Its spectrum is a featureless continuum like that of the photosphere but without the Fraunhofer lines. The dust-scattered component has a spectrum that resembles the photospheric *with* the Fraunhofer lines; it is known as the F (for Fraunhofer) corona. Both the K and F components decrease in intensity with increasing distance from the sun, but beyond $2\frac{1}{2}$ solar radii from the sun's centre the F component is more intense than the K. Figure 5.1 shows how the 'surface' brightness (i.e. brightness per unit area of the sky) of the F and K components varies with distance from the centre of the sun. The surface brightness scale in the figure is logarithmic and expressed in terms of the photospheric surface brightness. The distances in the figure are measured from the sun's centre ('heliocentric' distances) and are in units of the solar radius, sometimes written as $1 R_\odot$, equal to 696 000 km; they extend to the distance of the earth's orbit (149 600 000 km or $215 R_\odot$). The K corona's surface brightness at relatively small distances is plotted for sunspot maximum and minimum; that at sunspot maximum is greater by about 50%. The coronal surface brightness, as this figure shows, is at most about a millionth of that of the photosphere, and not too different from that of the full moon. It is comparable in intensity to the sky surface brightness near the sun for ideal atmospheric conditions at sea level, so illustrating the difficulty of trying to observe the white-light corona outside total eclipses.

The absence of Fraunhofer lines in the K corona's spectrum is due to the extreme speeds of the coronal electrons off which photospheric light is scattered. At a temperature of 2 000 000 K, the electrons move with an average speed of 10 000 km/s – some 3% of the speed of light – and so the Fraunhofer lines of the photospheric spectrum are Doppler broadened to extremely wide profiles – 40 nm wide in the case of the Hα line, for example – and so the lines are no longer perceptible as such.

The F corona, on the other hand, is due to sunlight scattered off dust particles in interplanetary space. These particles are tiny solid grains, a few micrometres in diameter, forming a disc-like structure round the sun roughly in the ecliptic plane. The disc structure has a hole in it centred on the sun with a diameter of about four solar radii (about 3 000 000 km). The F corona extends so far out that it can be seen against the night sky from the earth as a

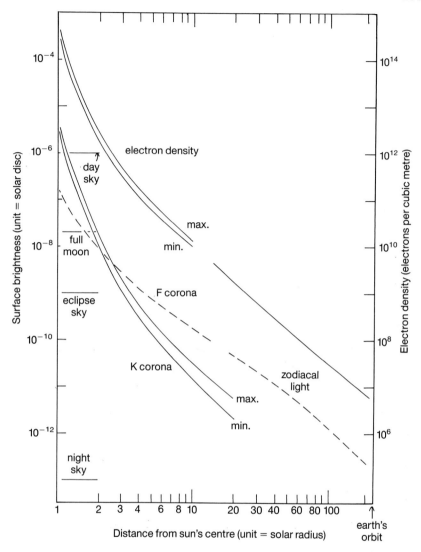

Fig. 5.1 Variation of coronal surface brightness and electron density with distance from the sun's centre (units of a solar radius, or 696 000 km). The surface brightness scale (units of the mean solar disc) is the left vertical one, the electron density (in electrons per cubic metre) the right – each is logarithmic. The K corona surface brightness is shown for solar maximum and minimum (equatorial values for the latter). The F corona values (dashed line) are continuous with the zodiacal light. For comparison, the surface brightness for the full moon and clear sky for day and night and during a total solar eclipse are indicated. Electron densities are plotted separately for solar maximum and minimum for small distances. (Data from Allen (1973) and Blackwell *et al.* (1967))

faint cone of light along the ecliptic, stretching up from the horizon. This *zodiacal light* is especially conspicuous at tropical latitudes but can also be seen elsewhere. Its emission has been detected from spacecraft out to 3 astronomical units. A faint patch of light, visible only under the clearest conditions exactly opposite the sun's direction, known as the *Gegenschein*, also arises from sunlight scattered by dust in the disc-shaped cloud beyond the earth's orbit. Compared with the electrons giving rise to the K corona, the dust grains that form the F corona move more slowly, and their motions are not random as with the electrons but 'orbital', like that of the planets round the sun (in the case of the earth, the orbital speed is about 30 km/s). Thus, the Fraunhofer lines remain visible in the F coronal spectrum, unlike the K.

There is another distinguishing feature. Scattering, like any kind of reflection, produces polarization (see §3.6 and Fig. 3.12). The detection of polarized light is made by special analysing filters such as Polaroid®, widely used as material for sunglasses. Such filters are like a very fine grid of parallel bars which transmit polarized light only if the plane of polarization is parallel to the grid bars. This direction is called the transmission plane of the analysing filter. Thus, unpolarized light incident on a piece of analysing filter emerges as plane-polarized because all light waves having planes of vibration inclined to the analysing filter's transmission plane are stopped and only those with planes parallel to it can pass through. The white-light radiation of the corona is found to be highly polarized. The coronal light observed during a total eclipse or with a coronagraph is unattenuated when the analysing filter's transmission plane is tangential to the sun's limb, but greatly attenuated when perpendicular to the limb. The four photographs of the corona during a total eclipse in 1970 (Fig. 5.2) were taken through an analysing filter orientated at various angles, and shows how coronal streamers disappear when the analysing filter's transmission plane is parallel to their direction. The amount or *degree* of polarization of coronal radiation is expressed by $(T-R)/(T+R)$, where T is the intensity of coronal light with the polarization plane tangential to the limb, and R that with the polarization plane perpendicular to the limb (i.e. in a radial direction). The degree of polarization would thus be 0 for unpolarized light ($T=R$), and 1 for plane-polarized light ($R=0$). It is found that the degree of polarization varies from about 0.2 for coronal light near the sun's limb to about 0.6 at one or two solar radii beyond the limb.

The corona's polarization is due to the fact that the light scattered by the coronal electrons is incident from a particular direction, viz. the sun itself. Figure 5.3 shows light rays from the sun and a point in the corona (P) as seen by an observer on the earth (the double-headed arrow convention of Fig. 3.12 is used in this figure). Unpolarized white light (indicated by the crossed arrows) is emitted by the sun (S) in all directions, and an electron at point P scatters this unpolarized light. For point P sufficiently distant, the sun can be considered a point source. The observer on earth (O) is so distant that the angle between S, P and O is almost a right angle. (Other points along the line

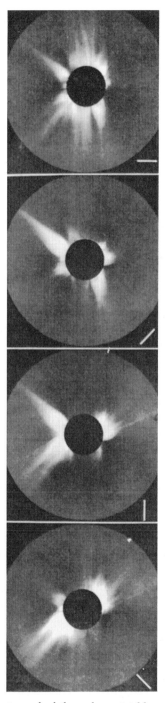

Fig. 5.2 The corona photographed through a rotatable analysing filter during the total solar eclipse of 7 March 1970. The orientation of the analysing filter's transmission plane is indicated by the short white line in each of the four images. Coronal streamers disappear when they lie along this orientation. (Courtesy K. Saito, Tokyo Astronomical Observatory and *Sky and Telescope*)

$P'P''$ also give rise to scattering, but as their radial distance from the sun is larger, the density of electrons, and consequently the amount of scattering, is smaller.) The observer on earth thus sees radiation with electric-field vibrations given by the projection of the crossed double-headed arrows against the sky; this radiation is thus plane-polarized, with the electric-field vibrations in a plane parallel to the solar limb, as indicated. The observer viewing the corona through an analysing filter sees the corona's light reduced by a large amount with the analysing filter's transmission plane orientated along the sun's radius and hardly reduced at all with the transmission plane parallel to the sun's limb. Actually, the degree of polarization is smaller than unity, considerably so near the sun, partly because the sun is not actually a point source and partly because points such as P' and P'' in Fig. 5.3 do make some contribution to scattered light, producing radiation with plane of polarization in a radial direction.

As mentioned, the K (electron-scattered) corona predominates for comparatively small distances from the photosphere. Now, the greater the density of electrons, the greater the K corona's surface brightness should be since there are more electrons to scatter light. This offers a means of determining the density of electrons. In scattering the photospheric emission, each electron acts as a tiny mirror, whose area, or 'cross section', is 10^{-28} square metres. We imagine a line of sight passing through the corona. It encounters N such electrons, say, the cumulative effect of whose scattering gives the observed surface brightness. That is, N times 10^{-28} must give the K corona's surface brightness in terms of the photosphere's brightness. According to Fig. 5.1, the maximum surface brightness is 10^{-6} of that of the photosphere, so

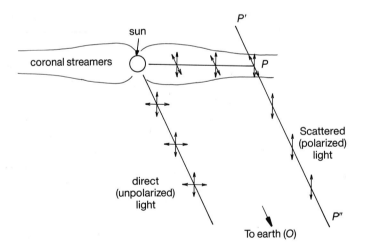

Fig. 5.3 While direct light from the sun is unpolarized (as indicated by the crossed arrows), white-light emission from coronal streamers is polarized with the electric-field vibrations parallel to the solar limb.

that N is about $10^{-6}/10^{-28}=10^{22}$ electrons. To get the number of electrons per cubic metre, we assume that our line of sight crosses a typical coronal streamer which averages about 100 000 km in extent, giving the density of electrons as 10^{14} per cubic metre.

Deriving more accurate electron densities is far from easy, especially for large distances from the photosphere, and other methods are also used, such as the occultation of cosmic radio sources by the corona and, near the earth, direct measurements with spacecraft. Figure 5.1 shows how electron density (obtained from various measurements) varies with distance from the sun. For small distances the density is plotted for both solar maximum and minimum conditions.

5.2   The hot corona

The large visible extent of the corona during eclipses is an important clue to its extremely high temperature. As most of this emission is due to the K corona, the large extent is really another way of saying the electron density decreases relatively slowly with distance. We can find a rough value of temperature, assumed constant, using a pressure scale height argument like that in §3.4, i.e. assume a hydrostatic equilibrium. We take temperature to be approximately constant, and use the fact that pressure is proportional to temperature times density. A scale height can thus be found from the density plot of Fig. 5.1, i.e. the height range over which density decreases from one value to 37% of that value; it is about 60 000 km. Equating this to 0.03 times temperature, as in §3.4, gives the temperature as 2 000 000 K.

The identification of the coronal 'green' (wavelength 530.3 nm) and 'red' (637.5 nm) lines with lines observed in the spectra of spark discharges in the laboratory by Edlén and others some years ago (§1.5) was a strong indication of the corona's high temperature. The green and red lines are so-called forbidden lines in the spectra of Fe XIV (i.e. due to 13-times ionized iron, Fe^{13+}) and Fe X (nine-times ionized iron, Fe^{9+}) respectively. 'Forbidden' means that the electron transitions involved are highly improbable by certain quantum-mechanical selection rules. We will illustrate the excitation of these and other forbidden lines with the specific case of the Fe XIV green line.

The ion Fe^{13+} is an iron atom with 13 of the usual 26 electrons removed, so with 13 remaining. This configuration is similar to that of aluminium atoms in their neutral (or 'un-ionized') state, and the pattern of the spectral lines in each shows resemblances. Indeed, the spectra of all ions with 13 electrons, such as singly ionized silicon, twice ionized phosphorus, etc., show such correspondences, and form what is called an 'iso-electronic sequence': in this case, the spectra are said to belong to the 'aluminium' sequence. In each spectrum there are two strong, or resonance, lines due to transitions of the outermost electron from a common upper energy state with $n=4$ (we will refer to it as the '4d' state) to either of two closely spaced 3p energy states, separated by an amount of energy corresponding to the electron spin energy (see §3.5). This is shown by an 'energy level' (sometimes called Grotrian) diagram, in which the 4d (A) and 3p (B and C) levels are represented as horizontal lines and transitions between them as arrows (Fig. 5.4). Figure 5.4 is drawn for the case of Fe XIV. Edlén had wavelength measurements of the resonance lines of neutral aluminium (396.1 nm and 394.4 nm) and other members of the iso-electronic sequence up to scandium (Sc IX), which gave him the energy 'splitting' of the two 3p levels (B and C). By a process of extrapolating, this splitting could be deduced for Fe XIV. It was found that the energy difference for Fe XIV corresponded to the green forbidden line, viz. 530.3 nm. The two Fe XIV resonance lines, incidentally, fall in the X-ray region, at 5.96 nm and 5.90 nm.

The improbability of a transition occurring between the 3p states B to C is increasingly evident going up the iso-electronic sequence. A neutral aluminium atom spends only about 10^{-8} seconds in state B before a transition to state C occurs, but for iron it would in theory spend almost a day in state B before this transition occurs. In a laboratory plasma like a spark discharge, the density of electrons is great enough that an Fe^{13+} ion in state B would be de-excited by electron collisions to state C or further excited to state A or some other state; the forbidden line, then, would never be observed. The fact that it is observed in the coronal spectrum implies not only an extremely high temperature but also an extremely low density otherwise de-excitation would occur by electron collisions. This is indeed the case, the coronal electron density of 10^{14} electrons per cubic metre being many millions of times less than that of a spark discharge.

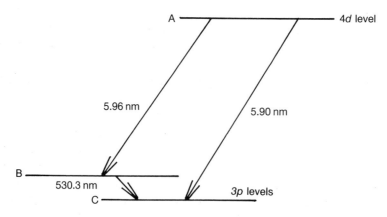

Fig. 5.4 Energy level, or Grotrian, diagram for 13-times ionized iron, with transitions giving rise to the Fe XIV lines indicated.

Table 5.1. *Coronal forbidden lines*

Wavelength (nm)	Emitting ion	Notes
332.8	Ca XII	
338.8	Fe XIII	
360.1	Ni XVI	
398.7	Fe XI	
408.6	Ca XIII	
423.1	Ni XII	
511.6	Ni XIII	
530.3	Fe XIV	'Green' line
569.4	Ca XV	'Yellow' line
637.5	Fe X	'Red' line
670.2	Ni XV	
706.0	Fe XV	
789.1	Fe XI	
802.4	Ni XV	

The other emission lines in the corona's visible spectrum can all be ascribed to similar forbidden transitions in ions of elements such as calcium and nickel, all of them very highly stripped. Table 5.1 is a selection of the more intense lines in the visible and also the near-ultraviolet spectrum, giving wavelengths and emitting ions.

The red Fe X and green Fe XIV line emission can be observed over much of the corona which, as previously mentioned, generally has a temperature of about 2 000 000 K. But over sunspot regions the coronal temperatures is enhanced, and might typically be around 4 000 000 K. It is at these locations that the normally faint Ca XV 'yellow' line (wavelength 569.4 nm) appears intense, showing that this is a 'hotter' line than the red or green lines. The intensities of these coronal lines, then, clearly depend on temperature. Temperature first of all determines how much of a particular ion is present relative to all the other ions of an element, and secondly causes the excitation of the ion from its ground state (level C in Fig. 5.4) to the upper state (level B) from which the reverse de-excitation process occurs resulting in the emission of the line photons.

We will deal first with the question of ionization in the corona. If a portion of the corona is in a 'steady state', with no change in its temperature with time, we speak of the gas as having an ionization 'equilibrium'; this means that, although many atomic processes are occurring all the time resulting in the creation or destruction of a particular ion, the number density of that ion (i.e. number per cubic metre) remains the same. To take the example of the Fe^{13+} ion which emits the Fe XIV green line, the atomic processes that result in the further ionization to the Fe^{14+} ion are exactly balanced by processes that result in the recombination of Fe^{14+} ions to form Fe^{13+} ions. We now describe the processes involved in the corona.

Ionization generally proceeds by collision of the ion with free electrons, these being extremely abundant. It requires a relatively large amount of energy to ionize, e.g., the Fe^{13+} ion to form Fe^{14+} – about 28 times the energy to ionize a hydrogen atom – but at the extremely high coronal temperatures at least some electrons have the required energy. The process of 'collisional ionization' is illustrated in Fig. 5.5a: an incoming free electron (e_1) removes one of the ion's bound electrons, e_2 (usually the outermost one) and the ion is left in the next higher 'stage', i.e. with its charge increased by one unit. Note that ionization by photons – 'photoionization' – is in principle possible but there are too few high-energy photons to make this process important for the corona.

Recombination can occur by two main processes. In 'radiative recombination', illustrated in Fig. 5.5b, a free electron is captured by the ion in one of its available orbits, but in general the electron has more than the required energy, so the excess must be removed, and appears as a photon (hence the name 'radiative' recombination). The second process, 'dielectronic recombination' (Fig. 5.5c), is more involved, but actually occurs more frequently in the corona (as was first shown by Alan Burgess in 1964). A free electron e_1 is captured by the ion which results in the 'double' excitation of the ion – two electrons, one of them the originally free one (e_1), are placed in excited orbits.

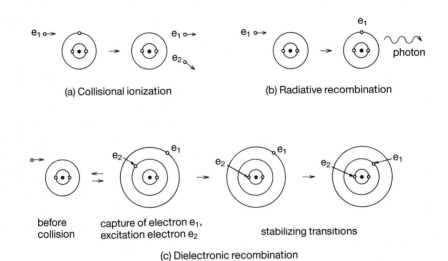

       (a) Collisional ionization   (b) Radiative recombination

before   capture of electron e_1,
collision   excitation electron e_2 stabilizing transitions

                  (c) Dielectronic recombination

Fig. 5.5 Pictorial representations of the processes of (a) collisional ionization; (b) radiative recombination; (c) dielectronic recombination. The electrons are indicated by the open circles, the atomic nuclei the filled circles, and electron orbits round the nucleus the large circles.

The electron e_1 has to have precisely the required energy for this to happen, i.e. without any excess energy. The highly unstable doubly excited configuration may relax by the ion's undergoing the exact reverse process (hence the reversed arrows in Fig. 5.5c), in which case no recombination occurs, or by the ion's undergoing a step-by-step stabilization, with first one electron (e.g. e_2) then the other (e_1) falling back to vacancies in inner orbits, so that recombination is eventually accomplished.

Ionization equilibrium, then, is a balance of collisional ionization with the sum of radiative and dielectronic recombination. All these processes depend on temperature (and only negligibly on density), and may be calculated or sometimes measured in the laboratory. This allows the proportion of ions of an element to be calculated for various temperatures. Figures 5.6a and 5.6b show respectively the distribution of ions of iron and oxygen with temperature (horizontal scale, given as the logarithm of temperature). The vertical scale gives the fraction of all the ions of the element (iron or oxygen) that is in the form of a particular ion. Thus, at a temperature of exactly $1\,000\,000\,\text{K}$ (logarithm=6.0), the ion Fe^{8+} predominates over other iron ions, and the O^{6+} over other oxygen ions. There are some ions that exist over much larger temperature ranges than others: these include Fe^{16+} which has an electron configuration like that of neutral neon, and O^{6+} which is like neutral helium. Now, both neon and helium have filled electron orbits (see §3.5) – $n=3$ in the case of neon, and $n=2$ in the case of helium – and it is therefore relatively difficult to remove any one of these electrons. The same applies to neon-like and helium-like ions such as Fe^{16+} and O^{6+}, and so they can continue to exist up to comparatively high temperatures before collisional ionization eventually reduces their proportions. Hence their curves in Fig. 5.6 cover wide temperature ranges.

Excitation of a coronal forbidden line – taking the ion from its ground state to an upper state – can proceed by either the absorption of a photon ('photoexcitation') or collisions with a free electron ('collisional excitation'). These two processes compete for the excitation of the Fe XIV green line; close to the sun, where there is an abundance of photons, photoexcitation is often more important, but for sufficiently dense coronal regions (more than 10^{15} electrons per cubic metre) collisional excitation is more important since the number of collisions depends on the number density of electrons. The relative importance of the two processes can sometimes be observed to change with distance from the sun, particularly in a regular coronal structure like a long streamer, with photoexcitation predominating near the sun, collisional excitation further out.

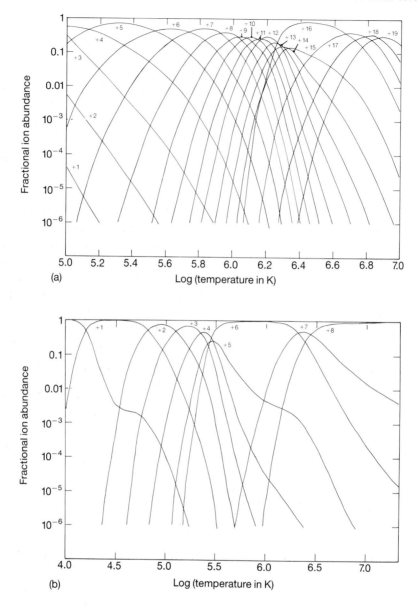

Fig. 5.6 Ionization equilibrium ratios for (a) iron, (b) oxygen ions. Both the vertical (fractional ion abundance) and horizontal (temperature) scales are expressed logarithmically. (Arnaud and Rothenflug (1985))

5.3   Coronal radiation in non-visible wavelengths

The corona's visible-wavelength radiation can only be observed off the sun's limb during eclipses or with coronagraphs; it is far too weak to be seen on the disc against the background of the intense photosphere. However, the situation is quite different in X-rays, or at extreme ultraviolet or radio wavelengths. Such radiation is emitted strongly by the corona but not at all by the photosphere. Consequently, we can observe the corona on the sun's disc as well as off the limb.

The corona's X-ray emission is in the form of both emission lines and a continuum. The emission lines arise from highly stripped ions similar to those emitting the visible-wavelength forbidden lines. The two resonance lines in the Fe XIV spectrum, at 5.90 and 5.96 nm (§5.2, Fig. 5.4), are both emitted by the corona, since the temperature is high enough to excite Fe^{13+} ions from the $3p$ (C) ground state to the $4d$ (A) state. Other resonance or strongly 'permitted' lines also occur in the X-ray spectrum due to ions that are abundant in the corona. The graphs of Fig. 5.6a and 5.6b show which ions these are for the case of iron and oxygen. Those parts of the corona over sunspot regions, where the temperature is enhanced, show X-ray emission from several ions, all of which are abundant in the temperature range 2 000 000 K to 4 000 000 K approximately (6.3 to 6.6 on the log scale). For iron, these include the neon-like Fe^{16+} ion especially, so lines of Fe XVII at 1.5 and 1.7 nm feature very strongly, as shown in the spectrum of Fig. 5.7. Helium-like ions of light elements are also represented. Resonance lines of helium-like carbon (C V), oxygen (O VII) and neon (Ne IX) occur at 4.03, 2.16 and 1.345 nm respectively (the Ne IX line features in Fig. 5.7). Similarly, there are lines of hydrogen-like ions, equivalent to the familiar Lyman lines of hydrogen itself at ultraviolet wavelengths. The 'Lyman-α' lines of hydrogen-like carbon (C VI) and oxygen (O VIII) are at 3.37 nm and 1.90 nm respectively (the O VIII line appears in Fig. 5.7).

In the wavelength range between X-rays and the ultraviolet – the 'extreme ultraviolet' range, which may be rather arbitrarily defined to be from 10 nm to 100 nm – are several intense lines due to lithium-like ions. Lithium has three electrons, with the outer in the $n=2$ orbit. Excitation may take it from a $2s$ sub-orbit to a $2p$, and de-excitation back to $2s$ results in two closely spaced lines forming a doublet, separated in energy by the electron spin energy (§3.5). The lithium-like carbon (C IV) lines at 155 nm, emitted at transition-region temperatures, have already been mentioned (4.4). The corresponding doublet is emitted by oxygen (O VI) in the coronal spectrum at 103.2 and 103.8 nm, by magnesium (Mg X) at 61.0 and 62.5 nm, and by silicon (Si XII) at 49.9 and 52.1 nm. The coronal image of Fig. 4.10 was taken in the Mg X 62.5-nm line.

Some strong lines in the region 17.1–21.1 nm are due to $3p$–$3d$ transitions in a range of iron ions, from Fe X to Fe XIV, and were first observed in early rocket spectra. Much interest surrounded them some years ago as many of the lines were also seen in the spectra of plasmas formed in the first-generation fusion machines, notably the Zeta machine at Harwell Laboratory. Temperatures of the order of 1 000 000 K were being obtained in these devices, with densities that were not too different from those in the corona. The excitation conditions of these artificial plasmas and the corona were thus comparable, hence the similarity of their spectra.

Figure 5.8 shows four of a remarkable sequence of slitless ultraviolet spectrograms taken during the total eclipse of 7 March 1970, the same eclipse that features in Figs. 4.3 and 5.2. They were taken at moments around and during the totality by an ultraviolet spectrograph with automatic film loading on board a rocket launched into the moon's shadow. As the spectrograph was slitless, each spectral line appears as an image of either the chromosphere, transition region or corona, depending on its emitting temperature. Thus, the image corresponding to the C IV doublet at 155 nm (1548 Å on the figure) is a thin ring with irregularities, this being an image of the transition region with prominences around the sun's limb, while the Fe XII image at 217 nm (2170 Å) shows the diffuse structures characteristic of the corona, this ion being abundant at about 1.6×10^6 K. The image in the hydrogen Lyman-α

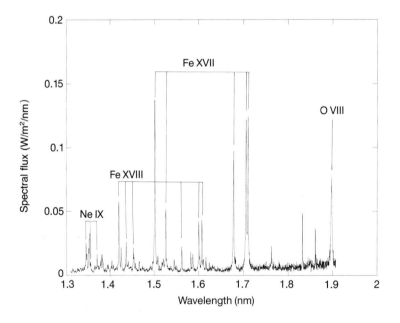

Fig. 5.7 X-ray spectrum between 1.3 and 1.9 nm of an active region taken with the X-ray polychromator (Bragg crystal spectrometer) on the *Solar Maximum Mission* spacecraft. The ions responsible for the strongest lines are identified.

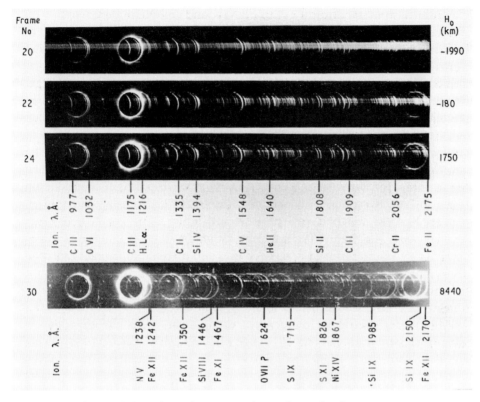

Fig. 5.8 Four slitless ultraviolet spectra taken with a rocket-borne spectrometer during the total eclipse of 7 March 1970. Frames 20, 22 and 24 were taken at the start of totality, and frame 30 within totality. Principal spectral lines giving rise to the images of the chromosphere and corona are identified, with wavelengths in ångströms (1 Å=0.1 nm). The image in hydrogen Lyman-α (121.6 nm, about a quarter of the way from the left) shows strong emission from the corona, despite the fact that at coronal temperatures most hydrogen atoms exist in an ionized form. (Gabriel *et al.* (1971). Reprinted with permission from *The Astrophysical Journal*)

line at 121.6 nm (1216 Å) very clearly shows the corona. This was surprising, as it was not expected that appreciable amounts of hydrogen could exist in a neutral form in the extremely hot corona. In fact, at 2 000 000 K or so, about one in a million hydrogen atoms do remain neutral. This small proportion is more than offset by the overwhelming abundance of hydrogen compared with any other element. A millionth part of the total number of hydrogen atoms is comparable, for example, to the total number of argon or calcium atoms, as can be seen from Table 3.3. Excitation of the single electron in these neutral hydrogen atoms, from the $n=1$ to 2 orbits, occurs both by electron collisions and by the Lyman-α photons radiated by neutral hydrogen in the chromosphere.

As well as spectral lines in extreme ultraviolet and X-ray wavelengths, the corona emits continuous radiation. Continua similar to the chromospheric Lyman continuum are formed in the X-ray region by recombination of electrons with the bare nuclei of carbon and oxygen. In addition, continua are formed by free electrons passing near protons. The negatively charged electron experiences the positive electrostatic field of the ion, and, if not actually captured by the ion, travels in a curved path (geometrically, a hyperbola) round the ion. The electron may suffer some 'braking' in its trajectory, resulting in a change of its path from one hyperbola to another, and the emission of a photon. The photon is unquantized, so a continuum is formed, which peaks, for coronal gas at $2\,000\,000$ K, at wavelengths of around 4 nm, in the soft X-ray region. This continuum is known as Bremsstrahlung (German for braking radiation) or free–free emission.

Solar radio emission with wavelengths longer than 0.1 m arises almost exclusively from the corona. Unlike X-rays and extreme ultraviolet radiation, also emitted by the corona, the earth's atmosphere is transparent to this radiation, so it may be observed from the ground. However, resolving coronal features or even the solar disc (half a degree in diameter) is a formidable task unless interferometric techniques are used. Solar radio emission may be broadly classified as follows: a background emission which is very nearly constant, showing a slight variation with the eleven-year sunspot cycle; a slowly varying (or S) component which changes with a roughly 27-day periodicity, i.e. with the sun's synodic rotation period; and a 'burst' component consisting of very short-lived enhancements. The S component is due to active regions as they pass across the sun, and the burst component to flares: we will deal with these in Chapter 6 (§§6.2, 6.3). The background radiation is Bremsstrahlung emitted by the quiet sun. Its spectrum consists of a broad continuum extending in wavelength from a few centimetres to a few metres. Unlike coronal soft X-ray and ultraviolet emission, radio emission with wavelengths up to 1.5 m or so is optically thick, and its intensity may be expressed in terms of the equivalent black-body or brightness temperature (§4.4), an approximation to the kinetic temperature of the emitting region. Brightness temperature is often used to specify a radio intensity even if the observed emission is optically thin and is far from being like a black body.

Radio waves in the corona only propagate if they have a frequency exceeding a certain critical value – the *plasma frequency*, related to the density of the coronal gas. It is equal to $9\sqrt{N}$ Hz, where N is the number density of electrons (in electrons per cubic metre). Now, on a large scale, a fully ionized plasma like the corona is electrostatically neutral – equal numbers of negatively charged electrons and positively charged ions – but if there is some disturbance on a local scale such that there is an excess or deficiency of electrons, the plasma tends to restore its neutrality by either repelling the electrons from the volume or attracting them; the electrons 'overshoot', so that the reverse process occurs, resulting in oscillations having a frequency equal to the plasma frequency. Any electromagnetic wave passing through the plasma with frequency less than the plasma frequency is suppressed. The consequence is that any observed low-frequency radiation such as radio waves with wavelengths longer than 1 m (e.g. 2 m, corresponding to 150 MHz) must come from relatively high in the corona, where the electron density is low, about 3×10^{14} per cubic metre, since a plasma frequency of 150 MHz corresponds to this density. The brightness temperature is measured to be about 1 000 000 K, equal to the gas kinetic temperature. High-frequency radiation comes from low in the atmosphere; thus, centimetre-wave emission largely originates where the plasma frequency equals 3 GHz (equivalent to a wavelength of 1 cm), which corresponds to the chromosphere; again, the brightness temperature is, as expected, about 10 000 K, roughly equal to the chromosphere's kinetic temperature.

Figure 5.9 shows the distribution of radio intensity (expressed as brightness temperature) across the sun's disc. It illustrates some of the above points. Thus, the disc brightness temperature decreases from about 1 000 000 K at wavelengths of 1.5 m to 10 000 K at centimetre wavelengths, the emission arising from the corona to the chromosphere over this range. The extensive emission off the limb at the longer wavelengths also shows that it is coronal. There is a slight decrease of the disc brightness temperature for wavelengths longer than 1.5 m because the emission is no longer optically thick, and the brightness temperature becomes less than the kinetic temperature of the emitting gas. There is a predicted limb brightening at wavelengths between a

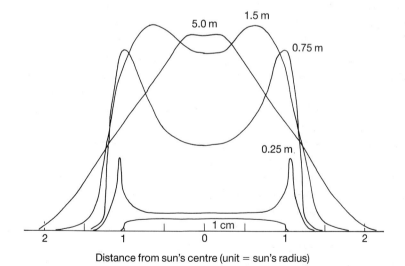

Fig. 5.9 Radio intensity distribution across the disc of the quiet sun at various wavelengths. (After Kundu (1965))

few centimetres and 1.5 m, arising from the increase of temperature with height, but contrary to expectation, a limb darkening occurs at very long wavelengths. This is due to the fact that rays which appear to originate at the limb in fact have suffered appreciable refraction, starting their trajectories at much higher altitudes where the optical thickness is very small: the optical depth of a ray at sun centre is in consequence larger than at the limb, giving a limb darkening.

5.4   The form of the corona: coronal holes

Individual features of the white-light corona have been recorded from the 1940s with coronagraphs, most notably those at Pic du Midi and the Sacramento Peak Observatory in New Mexico, and it was recognized from these observations that coronal structures were generally either 'closed', resembling arches, or 'open', i.e. directed outwards but not necessarily radially. A distinction was made between large, closed forms, 'arches', and smaller ones, 'loops', with arches being fainter and more stable, loops brighter and often associated with solar flares.

The first soft X-ray images of the corona, obtained with telescopes aboard stabilized, sun-pointed rockets, showed the same features that had been seen on the limb, but with the advantage that structures on the solar disc as well could be observed. The corona was shown to be composed *entirely* of loops or arches with their footpoints attached to the photosphere or chromosphere. As had been noted from coronagraph images, there were both larger arches, stretching to higher altitudes and appearing relatively faint, and smaller, brighter loops. The large structures are often hundreds of thousands of kilometres long, i.e. large fractions of a solar diameter, and connect active regions, sometimes across the solar equator. The small, brighter loops are connected with sunspot groups, and are around 10 000 km long; they will be dealt with in §6.2.

The total eclipse on 7 March 1970, mentioned several times previously, provided an opportunity to study the relation of the white-light structures seen at the solar limb during totality with those in X-rays, these being observed with a rocket-borne X-ray telescope, built by the American Science and Engineering Corporation, launched at around the time of the eclipse. Figure 5.10a is a superposition of a photograph of the white-light corona at totality with one of the soft X-ray corona. The X-ray brightenings on or just within the solar limb form the bases or 'footpoints' of white-light structures that take the form of streamers or rays. There are a few white-light streamers without an X-ray counterpart, but this is probably because their X-ray foot-points are located on the invisible hemisphere. The coronal counterparts of active regions are revealed as having extended white-light streamers that are closed or loop-like but with tops that tail off to a point.

(b)

(a)

Fig. 5.10 (a) The corona in white-light at the time of the total eclipse of 7 March 1970 superimposed on an X-ray image of the corona taken with a rocket-borne X-ray telescope shortly afterwards (the X-ray image is within the moon's disc of the white-light image). The white-light streamers nearly all have bases that are marked by X-ray bright patches. (VanSpeybroeck *et al.* (1970); © Macmillan Journals Ltd) (b) The calculated (potential) coronal magnetic field on the same occasion. (Newkirk (1971))

X-ray coronal features can be compared with the appearance of photospheric magnetograms. Both the larger and small-scale, sunspot-related X-ray loops connect regions of opposite magnetic polarity. The heights of the loops are generally related to the separation of the regions of opposite polarity – small loops being associated with compact sunspot groups and giant arches reaching to great altitudes connecting widely separated regions of opposite polarity. A detailed study of the extremely high-resolution X-ray image shown in Fig. 5.11 shows an intimate relation to features in the corresponding photospheric magnetogram and also the Hα chromosphere. It is in fact clear that coronal X-ray loops can be identified with magnetic 'flux tubes', i.e. tubes defined by the magnetic field geometry, that mark the extension of the photospheric magnetic field into the sun's outer atmosphere. The X-ray emission arises from the fact that the magnetic flux tubes are filled with very hot plasma. We will deal with this in more detail in §5.5.

A notable discovery from the early X-ray photographs of the corona was that of numerous bright, very confined areas that have become known as *X-ray bright points*. Unlike sunspot-related coronal regions, which occur in a belt roughly 40° either side of the equator, bright points occur over the entire sun, though they are most numerous at low latitudes. Some 40 may be visible at any one time on the visible solar hemisphere. Their lifetimes are short, some only a few hours, and over the course of a day more than a thousand may appear. There is a range of sizes, with tiny ones (less than about 7000 km) the most common and with the larger ones probably no different from small sunspot groups; an average size cannot be properly defined. Bright points are, like all other coronal features, closely related to the photospheric magnetic field, and in fact are associated with the ephemeral bipolar regions mentioned in §3.8. Again, the separation of the areas of opposite polarity are related to the X-ray bright point's size. An X-ray bright point evolves with time by spreading out, with a corresponding increasing separation and weakening of the bipolar region.

A striking feature of X-ray photographs is certain sharply defined areas of very low or even zero X-ray emission. They became known as coronal *holes* as it appeared as if there were a void in the corona at these locations. A coronal hole is always present at each of the solar poles, while there are often others at

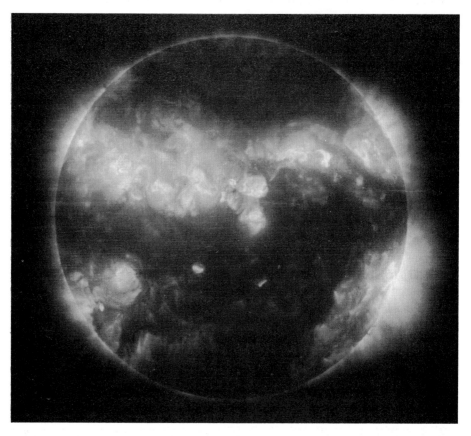

Fig. 5.11 X-ray image taken on 11 September 1989 with the Normal Incidence X-ray Telescope (NIXT) on board a NASA sounding rocket. The image, which has 0.75 arc sec resolution, was taken at the wavelength of an Fe XVI line at 6.35 nm. The instrument was built by the Smithsonian Astrophysical Observatory and IBM Thomas J. Watson Research Center, principal investigator Leon Golub. (Courtesy L. Golub, IBM Research and Smithsonian Astrophysical Observatory)

much lower latitudes, occasionally straddling the equator. Figure 5.12, photographed by the American Science and Engineering Corporation's telescope aboard the *Skylab* spacecraft, shows a particularly extensive example in 1973. Coronal holes were discovered as long ago as 1957 by M. Waldmeier using coronal maps constructed from green-line observations of the corona at the limb; he referred to them as *Löcher* (German for holes). Observations from some of the NASA *Orbiting Solar Observatory* (*OSO*) spacecraft, launched in the 1960s and early 1970s, also indicated their presence, e.g. from areas of low intensity in the light of the coronal extreme ultraviolet MgX line at 62.5 nm. They could also be picked out when on the limb: an example is the one on the sun's south-west limb in Fig. 5.10a, noticeable by both the absence of white-light streamers and a decrease in X-ray emission.

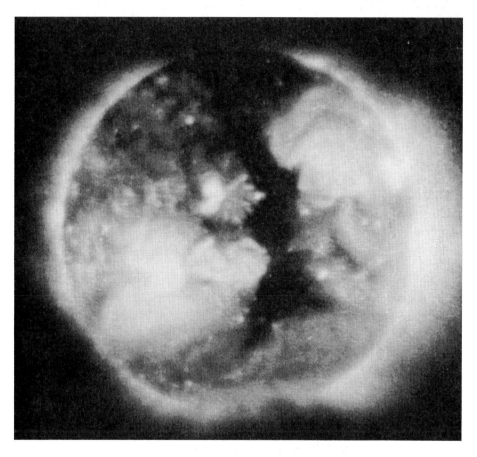

Fig. 5.12 A huge coronal hole stretching across the visible disc of the sun as photographed by the American Science and Engineering Corporation's X-ray telescope aboard *Skylab* on 1 June 1973. (Courtesy L. Golub and American Science & Engineering, Inc., Cambridge, Massachusetts)

There is a pattern to the appearance of low-latitude holes with the solar cycle: they are most evident in the declining years of sunspot activity, a little before solar minimum. The *Skylab* mission, operating between May 1973 and February 1974, two years before the 1976 minimum, was thus well timed to observe them, and the observations taken by the various instruments, as well as those on the ground, were the subject of one of three *Skylab* Workshop conferences. Some of the properties of coronal holes indicated here are findings of this Workshop.

Over the *Skylab* period, up to 20% of the visible solar hemisphere was occupied by coronal holes, with the polar holes making up a large fraction of this. Individual low-latitude holes varied greatly in size, with the largest having areas up to 10% of the visible hemisphere. The six low-latitude holes seen had a strong association with large 'unipolar' magnetic field regions seen in photospheric magnetograms, with the hole boundaries marked by magnetic inversion lines (§3.8). The most extensive holes did not seem to show differential rotation as do photospheric features.

The low-latitude holes had a definite evolutionary sequence. Each started as a small, isolated feature, with its area growing with time. On attaining a fairly large size, there was a coalescence with one of the polar holes, the one having the same magnetic polarity as the low-latitude hole. Eventually, the hole would diminish and disappear. This sequence, at least in 1973, took about 6–10 synodic rotations. The coronal hole shown in Fig. 5.12, the first to be recognized during the *Skylab* period, was at the time of the photograph a coalescence of a low-latitude hole with one at the north pole, so that its full extent was from the north pole to well south of the equator. There was a pattern to the occurrence of such holes during the *Skylab* period, with holes separated by between 90° and 120° around the solar equator. When the *Skylab* sequence of observations ceased in early 1974, there were two very large holes situated on opposite sides of the sun, one consisting of a low-latitude extension of the north polar hole, the other a low-latitude extension of the south polar hole. The entire corona had a very simple structure, in the form of a sheet between the two holes that surrounded the sun. When seen edge-on at the limb, the sheet appeared as two long streamers extending radially out on opposite sides of the sun, inclined at a small angle with the solar equator. This is very typical of the solar-minimum corona seen during total eclipses, the two streamers connected to the sun by an arch-like structure below which is a dark cavity. 'Helmet' streamers is a name often given to such an appearance.

There were relatively few X-ray images of the solar corona around the time of the succeeding sunspot minimum, in 1986, though a fine series of white-light coronal images was made by the High Altitude Observatory's white-light coronagraph on the *Solar Maximum Mission* (*SMM*) spacecraft. These showed two bright symmetrical streamers which, in 1984, were tilted by 30° to the solar equator, as in the previous solar minimum. In time, the inclination slowly decreased, reaching 5° to 10° by early 1986. Figure 5.13 is an image from the *SMM* instrument obtained in 1985.

Although coronal holes are most evident in soft X-ray images, they are prominent in spectroheliograms formed in the light of coronal extreme ultraviolet lines, e.g. the Mg IX line at 36.8 nm and the Mg X line at 62.5 nm (Table 4.1). They can also be made out in ultraviolet lines emitted by the transition region and chromosphere, though the effects are subtle and sometimes ambiguous. The general appearance of the sun in such lines is dominated by the presence of the chromospheric network, and it is found that the network is rather less clear within coronal holes. The intensity of transition-region lines on the limb shows a gentler decline outwards in the region of coronal holes, indicating a less sharp temperature gradient passing through the transition region. Within the polar holes, faint polar plumes are evident in Mg IX spectroheliograms, similar to white-light features often seen during

Fig. 5.13 The solar corona seen in the form of two bright streamers by *Solar Maximum Mission* on 11 March 1985, near the time of solar minimum. (Courtesy A. Hundhausen and C/P investigators on the *Solar Maximum Mission* spacecraft)

eclipses. Also within the polar holes, very large spicule-like features, known as 'macrospicules', have been observed in He II 30.4-nm line spectroheliograms: they extend to 40 000 km above the limb, compared with up to 9000 km for Hα spicules. The helium lines generally are unambiguous indicators of coronal holes, though formed in the chromosphere and transition region. Use has been made of the near-infrared line of neutral helium, at 1083.0 nm, which can be observed from the ground. Spectroheliograms in this line show active regions as *dark* areas while coronal holes show up as areas rather lighter than the background.

Radio observations have also shown the presence of coronal holes. At short wavelengths, around 2 cm or less, where radiation is largely from the transition region or lower, the emission in coronal holes is, rather surprisingly, marginally greater than elsewhere, but this is probably explained by the increased thickness of the transition region indicated by ultraviolet observations. At metre wavelengths, the holes appear as areas of reduced emission with lower brightness temperatures (about 750 000 K) than the surrounding quiet corona.

Coronal holes are not merely a curiosity of X-ray or other images of the corona, but have great significance for the connection between the sun and interplanetary space. They are the source of high-speed particle streams, and are to be identified with the mysterious M-regions that Bartels had hypothesized to explain recurring geomagnetic disturbances (§1.9). We shall deal with this in more detail in Chapter 7.

Table 5.2 summarizes the characteristics of the various coronal features mentioned in this section: large-scale coronal features, sunspot-associated (active) regions, bright points and coronal holes.

Table 5.2. *Typical properties of coronal features*

Feature	Typical dimension (in 10^3 km)	Temperature (10^6 K)	Particle density (per cubic metre)
Coronal hole:			
low-latitude	300–600	1–1.5	4×10^{14}
polar	700–900		
Large-scale structure	>100	1.6	1.6×10^{15}
Active regions	10	2–4	$1-7 \times 10^{15}$
X-ray bright point	5–20	2.5	1.4×10^{16}

5.5 Coronal magnetic fields

The various features in the X-ray or white-light corona strongly suggest that they arise because of a magnetic field pervading the corona, an extension of the photospheric field. This has already been indicated (§5.4) in the case of the loops and arches which are identifiable with magnetic flux tubes. Here we will explore how this comes about.

The high degree of ionization of the corona ensures that its electrical conductivity is extremely high, which in turn implies that any magnetic field will have its field lines frozen into the plasma (§§4.5, 2.4). On a microscopic scale, this means that the motions of the constituent electrons and ions are largely determined by the field. An electron or ion moving with a velocity perpendicular to a field line experiences a magnetic force which acts perpendicularly to the field and to the particle's instantaneous velocity, with the result that the particle performs a circular motion. A particle with initial velocity not perpendicular to the field describes a helical motion along the field line instead of a circle. The radius of the gyrating motion ('gyro-radius') and the frequency of gyrations depends on the field strength and the mass of the particle: in general, electrons perform spirals with small radius very rapidly and the much heavier ions perform spirals with large radius much more slowly (see Fig. 5.14). For a 1-mT field that is typical of much of the corona, and for electron and ion velocities typical of a temperature of 2 000 000 K, the electron's gyro-radius is about 30 mm and its gyro-frequency 30 MHz; for a proton, the gyro-radius is 1.3 m and the gyro-frequency 15 kHz. Either way, the radii of gyrations are minute compared with the lengths of coronal loops (at least many thousands of kilometres) along which the particles travel.

From time to time, an electron or ion travelling along a particular field line of a loop will 'collide' with another particle: a collision in this context means that the two particles approach each other so closely that they come under the influence of each other's electrostatic charge. The spiral path of the electron or ion along the field line is thereby interrupted, but resumed a few gyro-radii away along a new field line. Because the corona is so tenuous, such collisions are infrequent, and so the electron or ion may travel many thousands of kilometres between successive collisions. Thus, its motion is very closely 'controlled' by the magnetic field.

It needs only a very small field to control the particles of a coronal plasma, this control being determined by the ratio of gas pressure to what is known as magnetic pressure. If this ratio (sometimes called β) is less than one, the plasma is controlled by the field to a high degree. Despite the high coronal temperature, the gas pressure is only about 0.03 pascals near the base of the corona, and a field strength of only 1 mT or less is needed to achieve a value for β of less than one.

What happens to an electron travelling along the field lines when it arrives at one of the loop's footpoints? If there were no collisions, the spiral path it takes eventually reverses on itself: the footpoint acts as a 'mirror' for the electron. Actually, because there is a greater density of other particles at the footpoints, collisions become more frequent, so that a particle that has travelled relatively unimpeded along the top part of the loop may impart much of its kinetic energy to the surrounding particles it collides with near the footpoints. This is a microscopic description of the conduction of heat energy: energy in the hot, upper parts of an X-ray loop is transported by electrons (ions play only a minor rôle) to the footpoints where it is dissipated in the denser, cooler chromospheric layers. The X-ray loops, then, cool by loss of energy along field lines to the chromosphere. By contrast, heat conductivity perpendicular to the field lines is strongly inhibited, electrons being constrained to travel in their helical paths along the field.

Although heat conduction is a cooling mechanism for X-ray loops, the conducted energy is not lost from the sun as a whole, since it is simply distributed to a different part of the atmosphere. But there is another energy loss mechanism for the loop – the very fact we observe it, particularly in X-rays, means that energy is being radiated away, often in amounts comparable to conductive loss. And in this case the energy is truly lost from the sun in the sense that it passes into space and never returns.

The magnetic field, then, though not particularly strong, plays an important rôle for the corona. The large-scale arches and the active region loops evident in X-ray photographs are actually magnetic-field structures that are 'filled' with plasma, i.e. the ions and electrons forming the plasma spiral along the field lines, emitting X-rays and other radiation as they do so. The coronal holes, by contrast, have magnetic field lines that are open to interplanetary space, so that plasma is not confined as with loops but is free to travel out from the sun.

The strength of the coronal magnetic field can only be approximately measured. The Zeeman effect cannot be used since the splitting of the Zeeman line components is too small to be measurable for coronal spectral

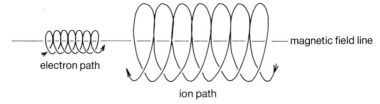

Fig. 5.14 The helical path taken by an electron and an ion round magnetic field lines.

lines. Other methods have been used, but with limited success. The magnetic field affects the polarization properties of some spectral lines and radio emission, and attempts have been made to measure fields, particularly in prominences, using this fact.

Some more or less successful efforts have been made to calculate the coronal field using the observed photospheric field combined with certain assumptions. A widely used method, due to H. U. Schmidt, is to assume that the field in the corona is entirely determined from the distribution of field strength in the photosphere; the coronal field is then known as a *potential* field. This implicitly means that there are no electric currents flowing in the corona, since a current produces its own magnetic field. The problem is rather like that of finding the electric field close to a surface which has a distribution of electrostatic charge. For the solar case, the surface is the photosphere, and it is assumed to have a distribution of 'magnetic' charges as indicated by photospheric magnetograms. Beyond a certain radial distance from the sun, the field lines are assumed to be dragged out radially by a steady expansion of the outer corona (the solar wind). 'Synoptic' maps, showing coronal magnetic field as individual lines of force, have been constructed (e.g. by a group at the High Altitude Observatory) using whole-disc magnetograms for successive days of an entire solar rotation. Figure 5.10b shows such a map made from magnetograms around the time of the total eclipse of 7 March 1970 (Fig. 5.10a). This map indicates the associations of X-ray loops with regions of closed magnetic field lines while the coronal hole on the south-west limb (referred to earlier) corresponds to a region of field lines open to interplanetary space. Some degree of success for these calculations is indicated, then, by the relation of observed coronal features with the calculated field-line patterns.

The typical solar-minimum corona, as indicated in §5.4, consists of two giant coronal holes at each pole with a sheet-like structure between which gives rise to the two symmetrical helmet streamers when seen edge-on. A magnetic-field configuration that would give rise to this appearance has been calculated (see Fig. 5.15) on the assumption that the sun's general magnetic field is in the form of a dipole (i.e. similar to that produced by a bar magnet) but stretched out by the solar wind. The coronal plasma is free to flow along the open field lines to interplanetary space. For the equatorial region, the field configuration is arch-like close to the sun but at larger distances it is stretched out to a sheet which contains a surface current flowing round the sun. This is therefore a 'non-potential' magnetic field configuration. We shall return to this model of the coronal field in §7.2 in explaining high-speed particle streams.

5.6   Quiescent prominences

Solar prominences are tongues of material emitting a visible spectrum typical of the chromosphere (e.g. emission lines of hydrogen, helium and various metals), so having a temperature of about 10 000 K, but are suspended in the corona and so are surrounded by gas of about 2 000 000 K. They are conspicuous features at the time of total eclipses when, like the chromosphere, they appear red because of the bright Hα line. Prominences often have a beautiful, intricate and rapidly changing structure. Figure 5.16 is an image in Hα of a prominence above the solar limb. This is the most familiar appearance of prominences, but they are also conspicuous features of Hα spectroheliograms, where they show up as dark *filaments*: Figs. 4.5 and 5.17 show examples.

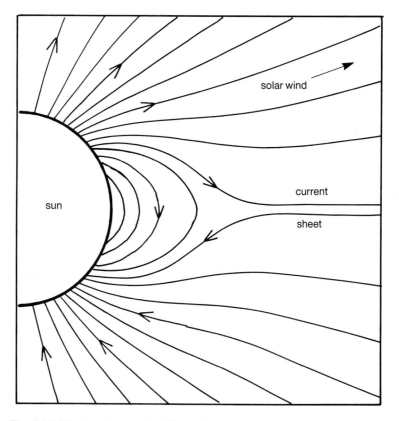

Fig. 5.15 Calculated magnetic field configuration at the time of solar minimum: a
          dipolar field stretched out by the solar wind to form a helmet structure
          near the equator and a current sheet. (After Pneuman and Kopp (1971).
          Reprinted by permission of Kluwer Academic Publishers)

Fig. 5.16 Quiescent hedgerow prominence in Hα. (Courtesy Big Bear Observatory)

Different authorities have different classification schemes for prominences. In this section, we will deal with prominences that are associated with quiet regions of the sun – 'quiescent' prominences. Other types are associated with sunspot (or active) regions and flares; these are described in §§6.2 and 6.3.

The quiescent prominences are the longest-lasting of all prominences. They appear over most of the sun – from the relatively low latitudes of sunspots to very high latitudes. The most well developed may last several months, i.e. several solar rotations, though more typically the lifetime is about one month. The prominence of Fig. 5.16 and the filament of Fig. 5.17 are both known as 'hedgerows', having extensive structure well above the limb but attached to the sun's surface at only a few points; they are the most common type. Quiescent prominences and filaments are roughly 100 000 to 600 000 km long and 5000 to 10 000 km thick, with heights ranging up to 50 000 km above the photosphere.

Quiescent prominences are found in particular latitude belts on the sun. Some develop from the prominences associated with active regions, and are found to the poleward side of sunspot groups north and south of the equator. There they form two 'royal zones' which steadily move with the spots towards the equator as the solar cycle progresses. On the disc, it is found that the filaments of this group lie along magnetic inversion lines that separate large areas of weak magnetic field with opposite polarity, the vestiges of old active regions. A second group of quiescent prominences exist at much higher latitudes, often forming a 'polar crown'. These prominences appear a few years after sunspot maximum, and then show a poleward migration, reaching either pole at about the time of the succeeding maximum. Most likely, the prominences of the lower-latitude first group migrate polewards to form the prominences of the polar second group.

The changes that quiescent prominences generally show are usually gradual, though individual features may move – either up or down – with velocities of a few kilometres per second. Sometimes a rotation is apparent. The most spectacular motion occurs at the end of a quiescent prominence's life, when the prominence suddenly starts to ascend, attaining a velocity of up to a few hundred km/s, and finally disappearing; the phenomenon is called a prominence *eruption*. On the disc, the eruption of a quiescent filament occurs by the sudden disappearance (sometimes known as a *disparition brusque*), since

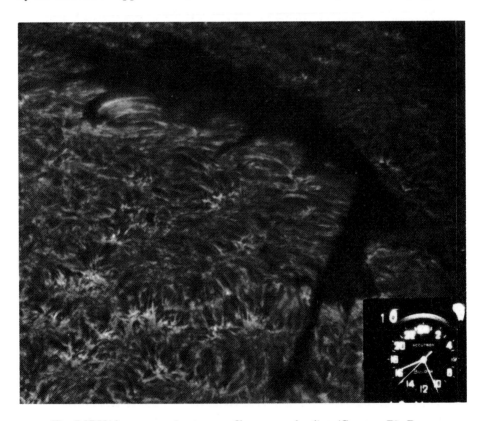

Fig. 5.17 Hedgerow prominence as a filament on the disc. (Courtesy Big Bear Observatory)

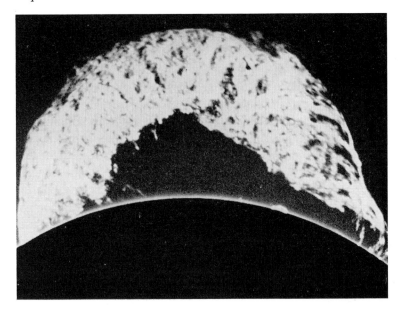

Fig. 5.18 Large eruptive prominence ('Anteater') in Hα on 4 June 1946. (Courtesy High Altitude Observatory)

the Hα line emitted by the prominence is Doppler-shifted to the violet out of the narrow spectral band of the filtergram or spectroheliogram. Figure 5.18 shows the giant 'Anteater', the most well-known example of a prominence eruption, on 4 June 1946. As is typical, the prominence was quiescent and in this case had persisted for several months. Its eruption occurred in little over an hour, a sequence of photographs showing its rapid upward motion. A fine thread-like structure is visible which appeared like an unwinding helix as the prominence ascended. Most prominence eruptions are preceded by an increase of activity in the form of random velocities in the prominence, and a gradual ascending motion. An eruption occurs within a day or so of the height attaining about 50 000 km. As we shall see in §6.2, the more active sunspot-related prominences also show eruptions but they are much faster and are related to the subsequent occurrence of flares. The eruption of quiescent filaments is accompanied by less spectacular phenomena, viz. a brightening of the underlying chromosphere and a slow increase and decrease ('gradual rise and fall') of soft X-ray emission, sometimes lasting several hours. Many of the large-scale ejections of coronal mass that are observed (these will be described in §6.4) are followed in their paths outwards from the sun by the remnants of quiescent prominence eruptions.

Usually a prominence erupts quite spontaneously, i.e. no other event on the sun seems to be responsible. Sometimes a large flare (not necessarily near by) sets off an eruption by a 'Moreton wave', which has the appearance of a shock front travelling out from the flare across the chromosphere and lower corona. If directed at the filament, the filament either undergoes the sudden disappearance just described or oscillates up and down, resulting in it continually disappearing and re-appearing as its Doppler-shifted Hα line passes into and out of the range of the filter through which it is observed. This has been called a 'winking' filament phenomenon.

As for so many objects in astronomy, the spectra of prominences hold the key to understanding the physical conditions, e.g. temperatures, pressures, etc. The visible spectra of all prominences are dominated by the hydrogen Balmer lines. The Ca II H and K lines are also strong. There are also many weaker lines due to neutral helium and metals, such as iron, titanium and chromium, in a singly ionized state. The appearance of such spectra gave rise at one time to the name 'metallic'; they were distinct from the spectra of more active prominences which do not show these metal lines. The difference between the two groups is in fact not one of the presence or absence of metals, but rather of temperature, the quiescent prominence spectra showing relatively narrow lines whose widths, if due solely to thermal Doppler broadening, imply temperatures of about 5000–6000 K. The more active prominences are much hotter (more than 10 000 K), and at such temperatures the metal lines are no longer so prominent.

As prominences are situated in a corona, it might be expected that there would be a 'transition region' separating the cool prominence gas from the hot coronal gas. The existence of a region of enhanced temperatures has been confirmed by ultraviolet observations. Figure 5.19 shows a large eruptive prominence associated with a coronal mass ejection in the light of the ionized-helium line at 30.4 nm, the equivalent of Lyman-α in this hydrogen-like ion, and emitted at about 50 000 K. Such a temperature is far higher than that deduced from visible-wavelength spectra. The image is from the US Naval Research Laboratory's slitless spectrograph on the *Skylab* mission. The emission from the sun itself is from the chromosphere (near this image are other, coronal, images in the light of much hotter lines). There are also indications of a transition region from radio observations. Filament emission between centimetre and decimetre (tens of centimetres) wavelengths is optically thick, so that the measured brightness temperature is approximately the gas kinetic temperature. Over this wavelength range, the brightness temperature increases from 15 000 to 60 000 K. Quite possibly, then, a transition sheath completely surrounds the filament, over the thickness of which the temperature increases from a few thousand K to the coronal 2 000 000 K.

The dark appearance of Hα filaments against the chromosphere must mean that the source function of the prominence is less than that of the chromosphere. Now, in the chromosphere the Hα line source function is controlled by radiation processes: collisional processes do not contribute significantly, while scattering is a neutral process, neither adding nor sub-

Fig. 5.19 Large eruptive prominence on 9 August 1973 photographed from *Skylab* in the light of the ionized helium (He II) 30.4-nm line. (Courtesy K. G. Widing, US Naval Research Laboratory)

tracting line photons (see §4.2). For a quiescent prominence, the particle densities, measured from collision broadening of Balmer lines, are about 10^{17} per cubic metre. This very low density means that collisional excitation is still very small. Scattering, however, has a different effect from that in the chromosphere. Photospheric radiation is incident on the prominence from below, but atoms in the prominence scatter this radiation in all directions, up and down, with the result that the photospheric radiation is weakened. A quiescent filament thus appears dark. The hot, active prominences, on the other hand, may be sufficiently dense that collisional excitation of $H\alpha$ is important, and they may appear bright on the disc.

Prominences are denser than the surrounding corona and so there must be a mechanism which holds them up against gravity (although it has been pointed out that the sun apparently has the opposite problem of holding prominences down, such is their eruptive tendency). Several magnetic models have been proposed. One, due to R. Kippenhahn and A. Schlüter in 1957, has a magnetic-field configuration like that of a loop arcade, with the prominence material lying across the arcade, supported by deformed magnetic field

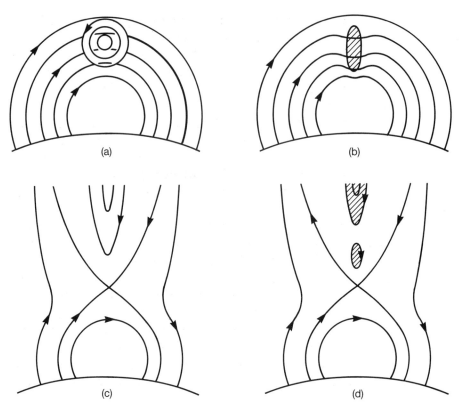

Fig. 5.20 Means of supporting a prominence in the model of Kippenhahn and
Schlüter (a and b). An electric current flows perpendicularly to the
magnetic field lines of an arch-like structure, the associated field lines of
which add vectorially to the field of the arch to give a 'sagging' loop
geometry. In the model of Kuperus and Raadu (c and d), material
condenses in a current sheet, with isolated knots forming which sink down
until supported by the field below.

lines which act like elastic bands supporting a heavier weight (Fig. 5.20b). To
be more precise, the prominence carries an electric current whose flow is
perpendicular to the magnetic field lines of the loop; this current has an
associated field-line pattern consisting of circles centred on the prominence,
shown end-on (Fig. 5.20a). These reinforce the field below the prominence,
but oppose the field above, so giving the 'sagging' loop pattern of Fig. 5.20b.
The force of gravity, then, is opposed by a magnetic force.

Magnetic reconnection, i.e. the reconnection of oppositely directed field
lines, to be described in the context of solar flares in § 6.5, has been used in a
more recent model for support of prominences. A flat, current-carrying sheet
is supposed to lie vertically above a loop arcade (seen end-on in Fig. 5.20c). In
the sheet, opposite field lines reconnect, so forming islands that contain cool

prominence material; these sink down until supported by the horizontal field below (Fig. 5.20d), so explaining the falling knots of Hα material often seen in prominences. More elaborate models by Eric Priest and colleagues at St Andrews University explain the formation and support by a thermal instability at the tops of several loops in an arcade structure, into which material from the chromosphere is 'siphoned'.

An objection to these and similar models is that, although quiescent prominences do indeed lie along the boundary between areas of opposite magnetic polarity, the field lines do not apparently run perpendicular to the prominence's length but almost parallel to it. The magnetic field geometry could in fact be quite complex, e.g. helical, as is indicated by the fine structure in the Anteater prominence (Fig. 5.18).

Matching theoretical models to observations in a much more detailed way is clearly desirable, but our measurements of prominence magnetic fields is extremely rudimentary. Field strengths have been measured by the Zeeman effect within prominences (they are of the order of 1 mT for quiescent prominences), but the fields around prominences have not been measured. Field directions have been deduced on the assumption that they are identical with the alignment of chromospheric and prominence structures, and a rough idea may be derived from calculations using the photospheric field. But until we can measure coronal field strengths and directions, there is little hope of knowing the detailed means of prominence support or how the field is disturbed during eruptions.

5.7 Heating of the corona

A fundamental, and it must be admitted to a large extent unanswered, question of solar physics is why the corona is so hot. As with chromospheric heating (§4.5), we know that non-radiative energy must be supplied to the corona and that the origins of this energy ultimately lie at the level of the photosphere or below. Again, the amount of energy supplied is modest compared with the vast output of radiative energy from the photosphere, but none the less its existence has to be explained, and what form it takes has been the focus of great debate and controversy for many years. A question related to coronal heating is the acceleration of the solar wind, this being the corona in a steady expanding state. We will outline here some of the theories for heating mechanisms that have been considered, discussing first the energy requirements for these mechanisms.

The problem is rather different from the chromosphere in several respects. Energy is supplied to the chromosphere and most is simply radiated away. Energy supplied to the corona is disposed of in several ways. For the closed loops that make up the 'quiet' corona, the energy is partly radiated (nearly all in the form of X-rays), but some is returned downwards to the transition region. This occurs by two means. Firstly, conduction takes heat from the hot corona to the cooler transition region and lower layers, the abundant supply of free electrons doing the conducting, which proceeds down along the legs of each coronal loop. The amount of conducted energy depends on the temperature of the gas in each loop and the temperature gradient (rate of change of temperature with distance) at the loop footpoints; both are very high, so conduction operates very efficiently. Secondly, energy is conveyed by a downflow of the hot coronal gas, again along the legs of coronal loops, the evidence for which is the Doppler shifts of transition-region lines like the ultraviolet C IV lines at 155 nm (see §4.4). The energy conveyed consists of the thermal energy of the flowing gas plus its flow energy, the sum being called 'enthalpy'. The enthalpy, conductive and radiative energy losses are all thought to be comparable to each other (though enthalpy is difficult to estimate accurately), and the total energy amounts to 300 W per square metre. If the entire sun were covered with quiet corona (using the fact that the solar surface area is 6×10^{18} square metres), the total energy lost by the corona would be 2×10^{21} W.

We know that a sizeable fraction of the corona is often in the form of coronal holes. The energy losses for these regions of open magnetic fields are rather different from those for the closed loops of the quiet corona. By their very nature, coronal holes are known to lose very little energy by radiation; they appear almost black in X-ray photographs, at least compared with surrounding coronal loops. Conduction, too, plays a considerably reduced rôle since both the temperature and temperature gradient are smaller than for loops. On the other hand, enthalpy is still important, and in addition there are energy losses through the expansion of coronal gas as it flows out via the open field lines to form high-speed streams in the solar wind. The total energy loss comes out to be more than that for closed loops: between 1000 and 3000 W/m^2, the range reflecting uncertainties in the enthalpy estimates.

Theories for the energy output that balances these losses at present centre on two main ideas: the conversion to heat energy by the dissipation of the energy in electric currents or magnetohydromagnetic (MHD) waves. Sound waves progressing into shock waves are not a likely mechanism, as there is probably no energy available after they have heated the lower chromosphere for heating the atmosphere higher up.

Coronal heating by electric current dissipation is equivalent to the reconnection of magnetic field lines, e.g. in narrow sheets. This was mentioned as a prominence support mechanism in §5.6 and will be described further as a flare mechanism in §6.5. It relies on the coronal gas having a certain amount of electrical resistivity; when electrons carrying a current encounter resistance, energy is taken from the current and turned into heat, just as in a domestic electric appliance. As we saw earlier (§5.5), very often a 'potential',

i.e. current-free, calculation of coronal magnetic field based on photospheric magnetograms agrees with visible coronal forms such as white-light or X-ray loops and coronal holes. This would seem to imply an absence of currents in the corona on such occasions, but in fact they may be present with energy sufficient to heat the corona but with associated magnetic field that alters the potential field configuration to an insignificant extent.

The dissipation of the current may be in the form of small impulses or it may be continuous. As evidence of the former, observations of the sun with a very sensitive hard X-ray detector in 1981 showed the existence of frequent very small flares, each lasting only a few seconds. There is some uncertainty about how much energy each small flare releases, but it is not out of the question that enough is released over the surface of the sun to explain the heating of the quiet corona. E. N. Parker has considered that the corona is heated by countless releases of small amounts of energy which he calls 'nano-flares'.

There is no direct evidence of a continuous dissipation of currents, but it has been suggested that it occurs in slowly evolving coronal active regions, i.e. above sunspot groups, with the currents being maintained by the twisting of magnetic flux tubes. The recent discovery of vortices of photospheric granules (§3.1) may be a manifestation of such twisting at the photospheric level.

Coronal heating and the acceleration of the solar wind by the dissipation of MHD waves has been studied for many years. As mentioned earlier (§4.5), such waves may be important in heating the upper chromosphere. It is possible that Alfvén waves are generated below the photosphere, pass up to the upper chromosphere and corona via the filigree regions of high magnetic field, and through conversion to fast and slow MHD waves, are dissipated. Waves with very long periods (of an hour or more) might be responsible for the heating of the largest (and coolest) coronal loops, while smaller, hotter coronal loops may be heated by the dissipation of fast-mode MHD waves, since they can be refracted into regions of higher densities which the smaller loops have. It has been calculated that MHD waves with periods of about five minutes could be dissipated in the surfaces of loops and so lead to a heating of the material within the loops.

There is a possibility that the jets seen in the transition-region C IV lines at 155 nm (§4.4) are important for heating the corona. Their velocities (400 km/s) are much larger than the local Alfvén speed of about 60 km/s, suggesting the formation of MHD shocks which could in turn heat the corona. The observed rate of such jets implies that very roughly $60\,W/m^2$ of energy could be released, a substantial fraction of the heating requirements for the quiet corona.

Chapter 6

The active sun

To a very good approximation, the sun is a vast thermodynamic machine in which nuclear reactions at its core generate prodigious amounts of energy which are transferred through its interior to the surface from where most is radiated to interplanetary space and beyond. The temperature increase with height through the thin atmosphere shows that this picture is not fully accurate: a non-radiative form of energy must pass up from below and be released in the atmosphere. Also, the sun's magnetic field gives rise to much complexity in the quiet sun. More particularly, it is the origin of solar activity. In this chapter we will describe the many forms this activity takes – sunspots, active regions, flares, ejections of coronal material – and what ideas are being used to explain them.

6.1   Sunspots

Sunspots are regions in the solar photosphere which have reduced temperatures compared with their surroundings, making them appear dark when viewed in white light, and which occur in relatively low latitudes, between about 40° north and south. Their lifetimes have an enormous range, from less than an hour to half a year. The sizes of individual spots similarly vary from about the resolution limits of the best telescopes under good seeing conditions – about 300 km – to around 100 000 km, i.e. many times the earth's diameter (12 740 km). Even so, no more than about 1% of the sun's visible hemisphere is covered by spots at any given time. There is a strong tendency for spots to occur in pairs or groups which may be very complex in the largest cases; thus spots are not randomly distributed over the sun, even over the latitude zone in which they occur. When small, a sunspot is little more than a sharply defined dot. The photospheric granulation is visible right up to the boundary of such a small spot, and may even intrude into the spot edges giving rise to a notched appearance. The spot itself is much darker than the dark lanes separating granules. More developed spots have a dark interior, the *umbra*, surrounded, or at least partly so, by a lighter area known as the *penumbra*. The penumbra consists of fine, light filaments which are approximately radial for spots with circular outline, but non-radial and irregular for complex sunspots. The penumbral filaments are lighter than the umbra but less bright than the photosphere, with decreasing intensity going towards the umbra. Figure 6.1 shows a large, almost regular spot showing radial penumbral filaments, and Fig. 6.2 a large irregular spot with a large penumbra consisting of non-radial filaments and containing many umbrae. Figure 6.3 shows a very extensive, irregular group of spots and the neighbouring granulation.

All sunspots start their existence as *pores*. Opinions differ as to their definition, but most observers call pores those features that are small dark areas but no darker than the integranular lanes, lasting up to an hour, frequently much less. A spot develops from a pore when the pore shows a sudden darkening.

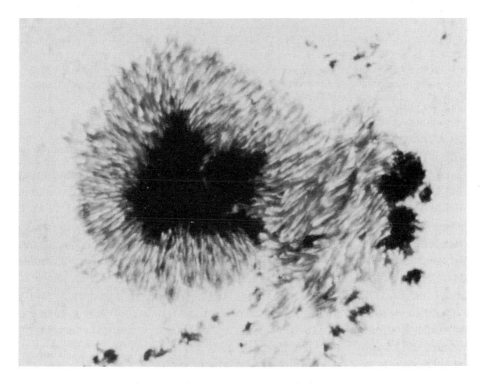

Fig. 6.1 Large sunspot showing radial penumbral filaments and grains within filaments. Photograph taken on 5 July 1970 with 10-nm-wide filter centred on wavelength 528.0 nm. (Courtesy R. Müller, Observatoire de Pic du Midi. Reprinted by permission of Kluwer Academic Publishers)

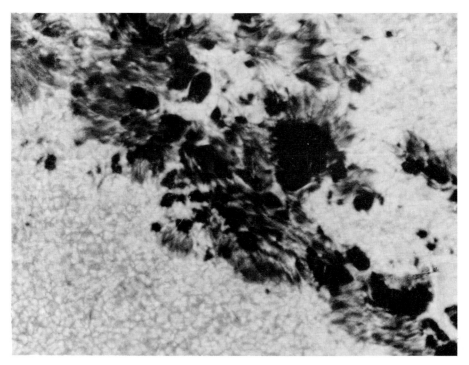

Fig. 6.2 Extensive sunspot group showing umbrae and non-radial penumbral
      filaments surrounded by photospheric granulation. (Courtesy National Solar
      Observatory/Sacramento Peak)

Most spots are extremely small, penumbra-less features that are born and die
within only a few hours. Larger spots that survive this initial phase occur in
pairs that may evolve into groups. Frequent photographs of a developing
region taken as a 'patrol' sequence show that a pair of spots is formed that is
highly inclined to a parallel of latitude, the leading or more westerly spot
being nearer the solar equator than the following. For pairs that go on to
become groups, there is a rapid growth in area and separation between the
two spots in the first one or two days, with the growth, including penumbral
development, continuing until maximum area and complexity is attained,
about ten days after the first appearance. The leading spot is generally nearly
circular in outline, while often several spots make up the following part of the
group and other small spots appear between the two. The initial separating
motion of the spots actually consists of 'proper motions' (i.e. motions apart
from those due to solar rotation) of both the leading and following spots: the
leading spot drifts westwards at first but eastwards and slightly poleward later,
while the following spot drifts slightly eastwards and poleward before coming
to a stop. Because of these motions, the inclination to a line of latitude
decreases with time from its initial high value. The decay of the group occurs

by the gradual disappearance of the following spot (often by subdivision into smaller spots) and the small intermediate spots, but with the leading spot, still quite circular and retaining its penumbra, only gradually decreasing in size; it may survive for several weeks after the disappearance of the rest of the group.

Sunspot groups have an infinite variety of forms and sizes, ranging from single small spots to giant groups with complex structure and containing dozens of spots. None the less, it has been found possible to define broad categories, the most widely used scheme until recently being devised by M. Waldmeier, known as the 'Zurich' classification scheme. This has been replaced by a scheme due to Patrick S. McIntosh at the US National Oceanic and Atmospheric Administration's Space Environment Laboratory in Boulder, Colorado, and is used in the NOAA bulletin of solar activity *Solar–Geophysical Data* and other publications. By it, spots are classified according to three descriptive codes which are summarized here. Figure 6.4 illustrates particular codes with the spots or spot groups they refer to.

The first descriptive code is a modification of the Zurich scheme, with seven broad categories. It is illustrated by the first column of sunspot photographs in Fig. 6.4. (In the following, 'unipolar' refers to a single spot, 'bipolar' to two or more spots, implying, as we shall see shortly, magnetic polarities.)

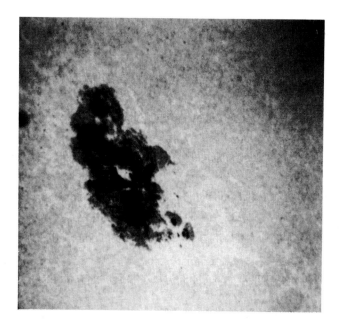

Fig. 6.3 Extensive irregular group of spots, associated with one of the most flare-prolific active regions ever seen. Photograph taken on 12 March 1989. (Courtesy US Air Force/Air Weather Service, Holloman Solar Observatory)

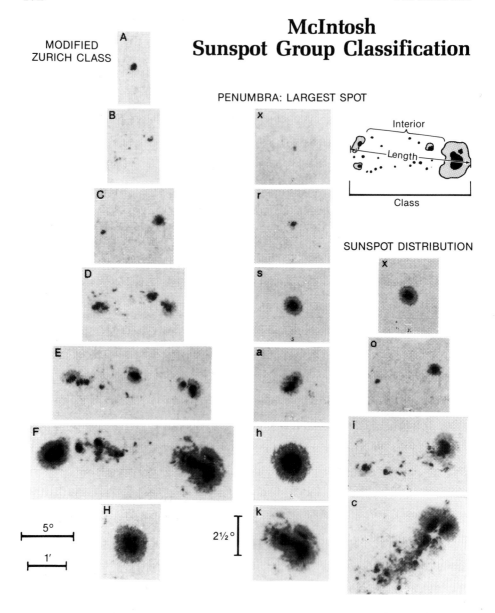

Fig. 6.4 The McIntosh sunspot classification scheme. Three letters describe in turn the class of sunspot group (single, pair or complex group), the penumbra of the largest spot in the group, and the spot distribution. (Courtesy P. S. McIntosh, NOAA (1990))

    A: Unipolar group with no penumbra, at start or end of spot group's life

    B: Bipolar group without penumbra on any spots

    C: Bipolar group with penumbra on one end of group, usually surrounding largest of leader umbrae

    D: Bipolar group with penumbrae on spots at both ends of group, and with longitudinal extent less than $10°$ ($120\,000$ km)

    E: Bipolar group with penumbrae on spots at both ends of group, with length between $10°$ and $15°$ ($120\,000$ and $180\,000$ km)

    F: Bipolar group with penumbra on spots at both ends of group, and length more than $15°$ ($180\,000$ km)

    H: Unipolar group with penumbra. Principal spot is usually the remnant leader spot of pre-existing bipolar group

The second code (illustrated by the middle column of Fig. 6.4) describes the penumbra of the largest spot of the group.

    x: No penumbra (class A or B)

    r: Rudimentary penumbra partly surrounds largest spot, either forming or decaying

    s: Small, symmetric penumbra, elliptical or circular. There is either a single umbra or compact cluster of umbrae, mimicking the penumbral symmetry. N–S size smaller than $2\frac{1}{2}°$ ($30\,000$ km)

    a: Small, asymmetric penumbra, irregular in outline. N–S size smaller than $2\frac{1}{2}°$

    h: Large, symmetric penumbra, N–S size larger than $2\frac{1}{2}°$

    k: Large, asymmetric penumbra, N–S size larger than $2\frac{1}{2}°$

The third and final code (final column of Fig. 6.4) describes the compactness of the spots in the intermediate part of a group.

    x: Assigned to (but undefined for) unipolar groups (class A and H)

    o: Open: few, if any, spots between leader and follower

    i: Intermediate: numerous spots between leader and follower, all without mature penumbra

    c: Compact: many large spots between leader and follower, with at least one having mature penumbra. An extreme case (bottom photograph of final column and Fig. 6.3) is when the entire spot group is within one penumbral area

In the McIntosh scheme, up to 60 classes of spots are covered (not all combinations of descriptive letter codes are allowed). A particular group will go through a number of categories in its lifetime (e.g. A*xx*–B*xo*–A*xx*). A statistical analysis of patrol photographs of spots indicates that 'groups' of type A*xx* (unipolar spots without penumbra) and B*xo* (simple bipolar pair without penumbra) are the most numerous (20% and 15% of total). The scheme has proved more useful than the older Zurich classification in distinguishing groups of spots likely to produce flares; in particular the category F*kc* (large bipolar group with many intermediate spots in a compact arrangement) is most prone to give large flares.

As was discovered by Hale in 1908 (§1.7), sunspots have strong magnetic fields, indicated by the Zeeman splitting of certain Fraunhofer lines. 'Bipolar' as used above strictly refers to the fact that opposite magnetic polarities are both present, 'unipolar' to only one being present. When complex, the leading spot of a group may have one polarity, the following spots the reverse, with frequently a mixture of both polarities for the intermediate spots. The degree of magnetic complexity is specified by a magnetic classification – the so-called Mount Wilson scheme – as follows.

> α: Unipolar: one or more spots with the same polarity (this being the same as the polarity of leading spots of pairs in the same hemisphere). Large areas of opposite-polarity field with no associated spot may occur near by
>
> β: Bipolar: either a single spot pair with leader and follower spots having opposite polarities or more complex groups with one polarity at one end, the reverse at the other
>
> γ: Complex: spot groups in which individual spots have polarities distributed in a much more irregular way than with bipolar groups. Occurs less frequently

Very large, flare-productive groups often have γ classification, while most spots or groups have either α or β classification. Some groups are classed $\beta\gamma$, indicating polarity distribution intermediate between β and γ.

Hale's further discovery that the leading and following spots of pairs or groups reverse their polarities at the beginning of each sunspot cycle has been demonstrated for every cycle since 1913. Although there is a general migration of spots towards the equator as the eleven-year cycle proceeds, spot groups hardly ever straddle the equator itself. On one occasion when this did occur, for a $\beta\gamma$ group in May 1921, the parts of the group north of the equator had polarities like those of other groups in the northern hemisphere, but the reverse was the case for parts south of the equator.

Improved techniques for measuring fields have enabled the determination of field strengths in various parts of individual spots. The field is greatest at the centres of umbrae (strength of about 0.2–0.4 T), and directed perpendicularly to the solar surface. It decreases towards the edge of the spot (strength=0.08 T) with increasing inclination to the vertical, and seems to extend for a short distance beyond the spot edge. High field strengths persist above sunspots, a value of 0.1 T being measured in the transition region by the Ultraviolet Spectrometer Polarimeter experiment on the *Solar Maximum Mission* spacecraft. Pores have smaller field strengths – generally around 0.15 T – than spots. The evolution in time of magnetic field and associated spots is closely related. Very often, a spot with a particular polarity may grow by the coalescing of small magnetic elements of the same polarity, while decay proceeds by the breaking up of the field into outward-streaming magnetic elements. High-resolution magnetograms have revealed the existence of numerous tiny magnetic 'knots' – regions where the field is enhanced to about 0.14 T – surrounding large sunspots yet different from pores in that they correspond to no particular white-light feature other than a dark, inter-granular channel, though very often they are associated with bright features in spectroheliograms at the wavelength of a magnesium line at 517.3 nm. Often, too, a large spot is surrounded by small 'satellite' magnetic features, some but not all marked by a small spot, which have the opposite polarity to the large spot.

A sufficiently complex group may have umbrae of opposite magnetic polarity within the same penumbra, a so-called δ configuration. A very high value for the magnetic field gradient, i.e. change of field strength per unit distance, may occur for such configurations, and under these circumstances they may be the site of very energetic flares.

Rudolf Wolf's systematization of sunspot number – the relative sunspot number – is still in use, though the Zurich sunspot programme is now carried out by the Sunspot Index Data Centre in Brussels. Wolf's procedure was to define a relative sunspot number, R, that could be compared for all observers, enabling values of R to be obtained from years past and better establishing its cyclical nature. Numbers of groups (g) visible as well as of individual spots (f) are included in Wolf's formula, with a factor (k) to represent the efficiency of the observer and telescope:

$$R=k(10g+f)$$

An international sunspot number is now in use, obtained by a network of more than 25 observing stations, and with the observatory at Locarno, Switzerland, as the reference station to preserve continuity with Wolf's Zurich series. They appear in the *Quarterly Bulletin of Solar Activity*, published by the International Astronomical Union.

Analyses of relative sunspot number confirm the uniformity of the sunspot number cycle, though the periodicity is by no means exact: the average length between sunspot maxima is 11.1 years, but varies in individual cases between 8 and 15 years. The rise from minimum to maximum (average 4.8 years) takes an appreciably shorter time than the fall (average 6.2 years), giving the typical cycle an asymmetric form. This asymmetry is much more marked, on average, for cycles with large maxima, those with small maxima having almost equal rise and fall times. Some success has been claimed for methods of predicting the maximum spot number from the rate of rise of R shortly after minimum: a sharp increase is generally correlated with a large maximum which is earlier than normal. There is a very wide range in the amplitudes of individual cycles; the maxima of 1805 and 1816, for example, had Rs of 49, a quarter of the amplitudes of the huge recent maxima of 1957, 1979 and 1989. Smoothed monthly sunspot numbers from 1749 to 1988 are shown in Fig. 6.5. A serial numbering scheme for sunspot cycles instituted by Waldmeier is still widely used, and will be referred to here. On this scheme, cycle number 1 is that which reached its maximum in 1761. The recent maxima in 1957, 1968, 1979 and 1989 are cycles 19, 20, 21 and 22 respectively.

Sunspot numbers are a convenient way of characterizing the sunspot cycle and are readily obtained observationally. The sunspot measurements of the Royal Observatory, Greenwich, and other observatories were described in §1.6. In general, they show the same eleven-year cyclical pattern as the number measurements (see Fig. 1.6), though in detail the relation between spot area and number R is not so close. A proportionality does exist between monthly averages of area and R: the spot area in millionths of a solar hemisphere (1 millionth = 3 040 000 square kilometres) being approximately $17R$. Individual spot groups may attain areas of several thousand millionths, though the more typical small pairs have areas of only one or two hundred millionths. A spot is visible to the naked eye if its area exceeds about 500 millionths. There is a very rough correlation of spot group lifetime and maximum area attained.

A curious phenomenon is the marked asymmetry of sunspot occurrence in both latitute and longitude. For whole cycles at a time, one hemisphere may have many more spots than the other. In longitude, there is marked evidence for the occurrence of active sections of the sun in which sunspot groups continually tend to be born.

Much detail within spots has been recorded with high-resolution photographs. As well as the radial and non-radial penumbral filaments mentioned earlier, there is also a short-lived rudimentary form that seems to take on a fuzzy appearance but which in fact consists of elongated photospheric granules separated by unusually dark intergranular lanes. The very best sunspot photographs show the light penumbral filaments to consist of fine grains that are very bright – almost as bright as the surrounding photosphere. Individual grains last about an hour, over which time they drift from the spot's outer edge to the umbra.

Despite their appearance in Figs. 6.1 to 6.3, even umbrae are not uniformly dark but contain isolated light 'dots', revealed in photographs of umbrae that have been carefully over-exposed. These umbral dots have been compared to photospheric granules but are smaller (300 km instead of 1100 km typical

MONTHLY MEAN SUNSPOT NUMBERS
January 1749 − December 1988

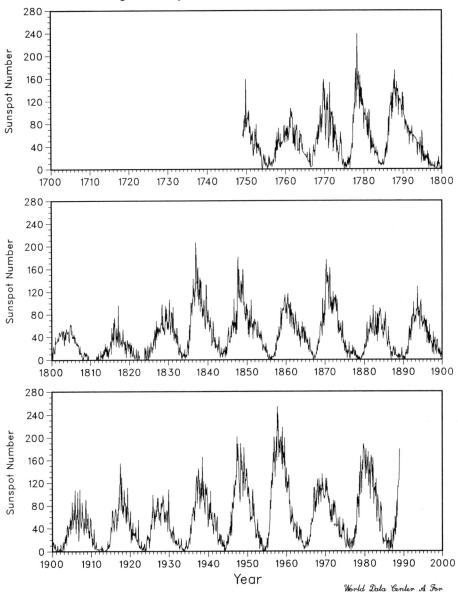

Fig. 6.5 Smoothed monthly sunspot numbers from January 1749 to December 1988. (Courtesy World Data Center A for Solar Terrestrial Physics, NOAA)

diameter) and rather more long-lived (25 minutes instead of 18 minutes on average). It appears that umbral dots are the penumbral grains that have just entered the umbra.

Invariably surrounding sunspot groups and visible in white light are the bright patches known as the photospheric *faculae*. They are best seen near the limb. They are closely associated with the chromospheric network and with bright patches in the chromosphere. The brightness of the photospheric faculae arises from their being a little hotter – by about 300 K – than their surroundings. High-resolution photographs show them to be composed of chains or clusters of tiny bright points – 'facular bright points' – very similar to the network bright points of quiet-sun areas (§3.1) in brightness, size (0.2 arc seconds – 150 km – or less) and lifetimes (about 20 minutes). Like the network bright points also, facular bright points are associated with small regions of high magnetic field (about 0.1 T) near the chromospheric network elements and with chromospheric bright patches.

Faculae are often connected with sunspot structure. A large sunspot very often has a temporary ridge of bright material – a *light-bridge* – crossing it. The growth of such structures normally proceeds by extension from one edge of a spot's umbra right across to the other, sometimes taking several days to do this. At the limb, they appear particularly bright; and it is then that a connection of the light-bridge with the faculae surrounding the sunspot group is often revealed.

Light-bridges are associated with a dividing process that large spots sometimes undergo shortly before their dissolution or, conversely, with the collision or merging of two spots. In the first case, spots are seen to split into two with the bright, streamer-like material of the light-bridge turning into ordinary photosphere. As an illustration of the second case, the interpenetration of one spot into a larger has been known: in a period of 12 hours, a smaller spot moved over a distance of 10 000 km to within the umbra of a larger, still retaining its identity by a surrounding circular light-bridge.

The occurrence of the Wilson effect (§1.2) – the displacement of the umbra towards the sun's centre as a spot approaches the limb – would suggest that sunspots are shallow depressions. But unequivocal examples of the Wilson effect in sequences of sunspot photographs are very few. Sometimes spots showing the Wilson effect on rotating round at the east limb do not do so when approaching the west, while others quite definitely show no Wilson effect even when very close to the limb. Doubts about the alleged observations of the effect in the past have been expressed on the grounds that few spots retain a circular profile throughout their passage across the disc (some 13 days): some evolution is bound to occur, and this sometimes gives rise to the appearance of a Wilson effect, though none is present.

Sunspots have a vertical structure, or more properly have an atmosphere that extends into the corona. Both the umbrae and penumbrae are visible in

Hα spectroheliograms, and in addition there is an impressive structure of dark, almost radial filaments that extend out by up to about 10 000 km beyond the normal penumbra of mature sunspots: this is the 'superpenumbra' (Fig. 6.6). Although long-lived as a whole, individual filaments making it up last only around 20 minutes. Further up in the solar atmosphere, ultraviolet and X-ray photographs (e.g. from *Skylab*) have revealed that coronal X-ray emission is reduced above large spots while ultraviolet transition-region emission is enhanced. Apparently the normal coronal heating processes are inhibited above a sunspot so that temperatures of the order of a few hundred thousand K, rather than more than a million K, are attained. Very strong emission in centimetric radio wavelengths is associated with the chromosphere above sunspots, the emission arising from non-thermal processes (see §6.2).

Sunspots are the sites of a number of remarkable dynamic phenomena, many of which have only recently been recognized. Chief among these is the famous Evershed velocity flow in spot penumbrae. This was first discovered spectroscopically as a horizontal flow of material proceeding radially outwards from a spot's umbra and penumbra and sometimes a little beyond. The outflow beyond the spot is visible as a so-called 'moat' containing outward-

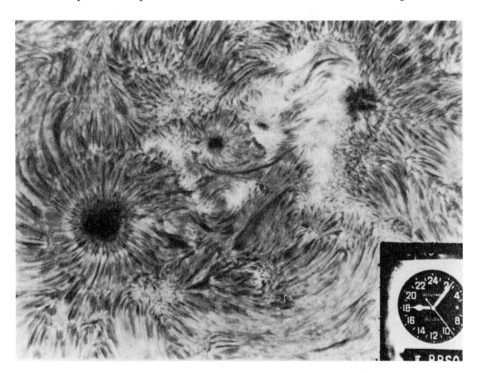

Fig. 6.6 Active region including large sunspot in Hα light. The spot has a superpenumbra surrounding the white-light penumbra. Dark fibrils and plage occur elsewhere in the active region. (Courtesy Big Bear Observatory)

moving facular points. The maximum velocity is around 2 km/s, attained within the spot penumbra for large spots, nearer the edge or beyond it for small spots. The flow is best observed for spots near the limb, for there the horizontal flow is most nearly directed towards and away from the observer on opposite sides of the spot. For chromospheric lines like Hα and calcium K, there is an 'inverse', i.e. inward, Evershed flow. In this case, the flow is mostly along the dark radial filaments making up the superpenumbra. The velocity is much larger – about 20 km/s – than the photospheric flow, and is directed slightly downwards towards the umbra rather than exactly horizontally. The inverse Evershed effect is seen for ultraviolet lines, such as the C IV lines at 155 nm. These show that there is not only an inflow at high altitudes but a downflow just above the sunspot umbra. Figure 6.7 shows a section through a sunspot atmosphere with the regions giving rise to various emission lines and the associated flows shown schematically.

As well as the continuous Evershed flow in spots, there are also oscillatory and wave phenomena which are present at the photospheric level but are more readily detectable in the chromosphere, i.e. can be seen in the light of the Hα or calcium K lines. Running penumbral waves, for example, are visible in Hα (particularly the line wings) as light and dark bands propagated alternately from the edge of a spot umbra. They move outwards, generally

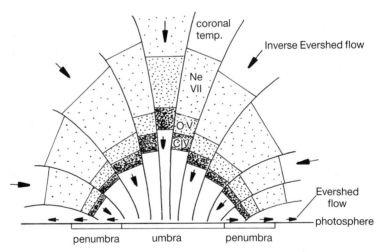

Fig. 6.7 Schematic section through a sunspot from the photosphere to the corona. Magnetic field lines converge on the spot, becoming increasingly inclined to the vertical going away from the spot. Within individual flux tubes defined by the field lines, C IV emission occurs relatively low down, the higher-temperature O V and Ne VII emission further up, while the X-ray emission from coronal temperatures occurs highest of all, though it may be almost absent for flux tubes emerging from the spot centre. Evershed and inverse Evershed flows are indicated by the short arrows. (After Nicolas *et al.* (1981))

from not more than a small part of the umbra's circumference, at a speed of about 20 km/s, fading to invisibility just before reaching the penumbral edge. The oscillatory motions present in spots show up as Doppler shifts in Fraunhofer lines. There are marked photospheric oscillations within the umbra with periods of 3 and 5 minutes, the later being related to the five-minute oscillations of the undisturbed photosphere (§2.3). Longer-period oscillations are present in the spot penumbra. In the Hα line, oscillations occur in small 'cells' with a size of a few thousand kilometres within the umbra. They have a period of 2 to 3 minutes, and are sustained for a few cycles. Apparently related to these oscillations, but better seen in the chromospheric calcium K line rather than Hα, are the umbral flashes, appearing as small, bright points at the umbra's centre, moving outwards, brightening at first then fading as they do so in their two-to-three-minute lifetime.

The effective temperature (§3.4) may be measured for a sunspot umbra as for the photosphere by equating the radiative flux to σT_{eff}^4, where σ is the Stefan–Boltzmann constant. It may be roughly estimated from the measured flux ratio – 0.27 – of umbral to normal photospheric radiation; as T_{eff} for the photosphere is 5778 K, T_{eff} is equal to $(0.27)^{\frac{1}{4}} \times 5778$ or about 4165 K. Similarly, the penumbra is estimated to have an effective temperature of 5500 K.

The low umbral temperature allows the formation of molecules. Several are observed, many not familiar in a terrestrial environment; examples include magnesium hydride (MgH), diatomic carbon (C_2) and hydrogen monoxide (OH), all having constituent atoms of abundant elements. One molecule that ought to be abundant in sunspots but whose spectrum is not easily observed is diatomic hydrogen, H_2. In fact, it was only definitely discovered a decade ago from studies, by Carole Jordan and collaborators, of ultraviolet spectrograms from the HRTS instrument mentioned in §4.4. A number of emission lines near the hydrogen Lyman-α line, visible only at the location of a sunspot and a pore, were identified as components of a complex band of lines that the H_2 molecule normally emits in this region. The lines are excited by fluorescence, in which H_2 molecules are raised from one energy state to another by the absorption of Lyman-α line photons.

6.2 Active regions

A sunspot group is but a part of a much more comprehensive entity known as an active region or, in older literature, centre of activity. Active regions extend several tens of thousands of kilometres into the solar atmosphere above sunspots, and have a distinctive appearance in the chromosphere, transition region and corona. Compared with quiet-sun areas, active regions have enhanced emission over a broad spectral range, from soft X-rays to decimetric radio waves. In 1953, the French solar astronomer L. d'Azumbuja defined an active region as the sum total of all visible phenomena occurring when sunspots appear. It was recognized in time that sunspots were not in fact a necessary part of an active region; many small, short-lived active regions exist without ever producing spots, while active regions formerly having spots persist for some time after their spots have disappeared. In a sense, the coronal X-ray bright points (§5.4) are small-scale active regions, there being a continuous spectrum of sizes. Both X-ray bright points and the larger active regions (with or without sunspots) have a basically bipolar magnetic field structure, simple in the case of bright points but with varying degrees of complexity for active regions. Hence, to generalize d'Azumbuja's earlier definition, an active region comprises *all* phenomena associated with the emergence of magnetic field at the photosphere from deeper layers: the scale of the emerging field determines how large the region will be. This definition then includes X-ray bright points as well as extended, weak-field remnants of active regions. However, the internationally accepted serial numbering scheme for active regions, used by NOAA, assigns a number to a region only if there are sunspots present or if it gives rise to a flare.

The chromospheric part of active regions is visible, as with quiet-sun regions, in spectroheliograms taken at the wavelengths of strong Fraunhofer lines, notably the Hα and ionized calcium H and K lines. For many years, these were our only means at imaging active regions in the solar atmosphere above the photosphere. Observations are now carried out in some other chromospherically formed lines in the visible and near-infrared range (an example of the latter being the neutral helium line at 1083.0 nm), and in the ultraviolet, e.g. the hydrogen Lyman-α (121.5 nm) and ionized magnesium h and k lines at 280 nm. Centimetric radio maps of active regions also show their chromospheric structure. Active regions extend into the corona, as was first indicated by slow variations in decimetric radio emission and coronal 'green line' enhancements, and was later established from images made in coronal ultraviolet lines, soft X-rays and radio. Thus, images in the C IV (155.0 nm), Ne VII (46.5 nm) and Fe XV (28.4 nm) lines show structures ranging in temperature from 100 000 K to 2 000 000 K, while those in soft X-rays indicate higher temperatures still, up to about 4 000 000 K, which are present in young, rapidly developing active regions.

In Hα and K line spectroheliograms (see Fig. 4.7), an active region consists of bright patches having a granular structure which are called *plages* (see §4.3) or sometimes chromospheric faculae, the latter name aptly indicating a general correlation with the photospheric faculae seen near sunspots close to the limb. The K line plages are brighter compared with surrounding areas and rather larger than corresponding features in Hα. In Hα particularly, a mass of *fibrils*, or dark elongated features, pervades the area. They are much longer and more conspicuous than the quiet-sun fine dark mottles: typical dimensions might be 20 000 km long and 1500 km diameter. They are often arranged in a striking radial or spiral pattern centred on a sunspot, extending beyond the superpenumbra filaments (Fig. 6.6). They are low-lying features as they are not visible at the limb. Winding through the region is a dark filament (appearing as a prominence on the limb), often with one end close to a large, usually the leading, sunspot. This *sunspot* or active-region filament (or prominence) lasts for a few days, i.e. much less than quiescent filaments (§5.6). Sunspot filaments are also characterized by rapid motions along their length, with material flowing into the sunspot. At high resolution, nearby fibrils may be seen to run parallel for much of their lengths to the filament, but diverging in direction at their ends.

As with the photospheric features (spots, pores, facular points), there is a close relationship of all chromospheric structures with the photospheric magnetic field. The fibrils connect areas of opposite magnetic polarity, and it is believed that they are aligned with the local magnetic field. Sunspot filaments, as with their quiescent counterparts, separate areas of opposite magnetic polarity, following the magnetic inversion line. However, also like quiescent filaments, the field direction (as indicated by neighbouring fibrils) is almost parallel to the magnetic inversion line. The bright Hα and calcium K line plages are generally areas of enhanced field strength.

An active region evolves with time through changes in the chromospheric features already mentioned and in its associated magnetic field. An active region first appears in calcium K spectroheliograms as brightenings in the network structures, often at or near the site of a previous active region. The brightenings increase in intensity, the network cells becoming filled in with very bright plage which by now is elongated in an east–west direction. Any sunspots that form do so within a few days or so of the active region's first appearance in Hα or the K line and small flares also start to occur. An 'emerging flux region' is the name often given to regions that develop into fully fledged active regions, 'flux' referring to magnetic field flux. In magnetograms, an active region is born when a bipolar region appears, small at first, with the areas of opposite polarity slightly inclined to a parallel of latitude, the leading part being equatorward, as with the first sunspots that appear. Also like the spots, the areas of opposite polarity initially undergo rapid separation and become less inclined to the equator. New magnetic flux emerges on the inside of the active region, at the boundary between the two polarities, while the major sunspots are near the outward edges of the separating areas of magnetic polarity. The newly emerged flux shows up in Hα spectroheliograms as dark, low-lying filaments known as arch filament systems, the loops joining opposite polarity directly across the magnetic inversion line. At their footpoints are small, very bright points that show up prominently in filtergrams made in the wings of the Hα line – around 0.1 nm away from the line core: some very good examples appear in Fig. 4.7. They are known as 'bombs' (christened thus by F. Ellerman in 1917) or 'moustaches', the latter name deriving from the extraordinary appearance of their spectrum, which shows bright emission extending out to 1 nm or more either side of the dark core of the Hα and other Balmer lines. They last from about 10 to 40 minutes, rapidly developing and rapidly fading, are very small (most beyond telescopic resolution) and for large active regions very numerous (100 or more). They are the sites of small-scale Hα ejections called surges (larger surges are flare-associated). The Ellerman bombs and small-scale surging characterize a developing, complex active region.

Maximum development for a large active region is attained when the Hα and K line plage is very bright with frequent flaring, the associated sunspots have maximum size and number, and the magnetic fields have a considerable degree of complexity, with patches of positive and negative polarity intermingled. Decay occurs by expansion and fading of the plage, diminishing size of spots and simplification of magnetic field distribution. The plage usually persists for a few weeks after its initial appearance, and long outlasts the associated spots. It ends up as a large, faint structure, 100 000 km or so in diameter. In magnetograms this structure shows up as a large area of weak bipolar field, the following area gradually migrating poleward. The last survivor is the active-region filament which by this time has also migrated polewards and is classifiable as a 'quiescent' filament.

The radio emission of active regions, particularly over the wavelength range from about 1 cm to 2 m, gives rise to the slowly varying or S component, one that rises and falls with the 27-day synodic rotation period. Although active regions could not be resolved on the sun when the S component was first discovered in the 1940s, it was realized from the correlation of its intensity with sunspot area and the fact it decreased when a sunspot group was occulted by the moon during an eclipse that it must have its origin in active regions. In fact, chromospheric plages, the atmosphere above sunspots and coronal loops all contribute to the S-component emission.

The plage-associated radio emission is weaker than the spot emission, but covers a larger area, approximately equal to the chromospheric plage, and is unpolarized. It is attributed to Bremsstrahlung, as with the quiet-sun radiation (§5.3). The sunspot-associated emission has a brightness temperature of a few million K and shows circular polarization. High-resolution maps made with large interferometer arrays show that the emission arises from the chromosphere above the spot penumbra. It is not due to Bremsstrahlung, but is associated with the spot's magnetic field. In §5.5 we discussed the way electrons spiral round magnetic field lines. Because the electron undergoes an acceleration in the spiralling motion (a 'centripetal' acceleration), it emits radiation at the gyro-frequency, equal to $28 \times B$ GHz (where B is the field in tesla), the radiation being circularly polarized when viewed along the field line. The emission occurs only if the electron does not collide too often with surrounding particles so that the spiralling motion is sustained. The spectrum of this radiation should in principle be like an emission line, with frequency equal to the gyro-frequency, as well as several multiples, or 'harmonics', of this frequency; but because in the solar atmosphere the field varies from place to place, a rather broad continuum is formed. Sunspot emission is ascribed to this 'gyro-magnetic' radiation. The measured brightness temperature of this radiation is not related to (and in fact much exceeds) the gas kinetic temperature. The coronal loops of the active region give rise to rather strong emission with brightness temperatures of about 1 000 000 K. It covers the whole plage (not just the spots) and is due to both Bremsstrahlung and gyro-magnetic radiation.

Soft X-ray images of the sun have provided great insight into the form of the coronal part of active regions. Figure 6.8 shows images of the sun in soft X-rays (1.50 nm wavelength) and ultraviolet radiation (138.8 nm wavelength). The X-ray active regions extend into the corona, as can be seen when they appear as in the calcium K line, viz. as bright plage regions at the bases of the coronal X-ray regions. High-resolution images from X-ray telescopes on *Skylab* show that the X-ray structures are in the form of loops, with footpoints in regions of opposite magnetic polarity. These images also show that the intensity of the loops decreases sharply with the active region's age.

Extreme ultraviolet images from the slitless spectrograph on *Skylab*, men-

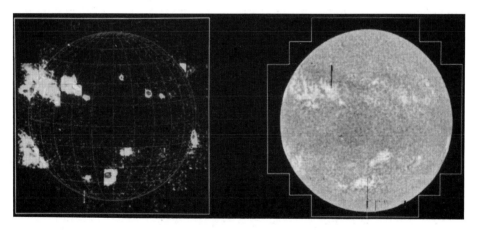

Fig. 6.8 The sun in soft X-rays (left) and ultraviolet radiation (right), showing the correspondence of the chromospheric parts of active regions seen in ultraviolet with the coronal in X-rays. The wavelength of the ultraviolet emission is 138.8 nm (part of the solar continuous spectrum), the X-rays the wavelength of a strong Fe XVII line at 1.50 nm. The data are from the X-ray Polychromator and Ultraviolet Spectrometer/Polarimeter aboard the *Solar Maximum Mission* spacecraft which performed a series of pointings across the sun's disc so that a whole-disc image could be obtained. (Courtesy of J. Gurman and other *Solar Maximum Mission* investigators)

tioned in §5.6, show the predominantly loop geometry of the coronal active region and give information about the temperature structure of individual loops. Figure 6.9 shows images of an active region in the Ne VII (46.5 nm), Mg IX (36.8 nm) and Fe XV (28.4 nm) lines, formed at temperatures of 500 000 K, 1 000 000 K and 2 000 000 K respectively, together with a photospheric magnetogram. There is a gradation from the short, spiky structures in Ne VII to diffuse, generally complete loop structures in Fe XV. The Ne VII spikes appear to be the legs of loops, implying cooler temperatures there. The hotter, Fe XV loops range from low-lying loop arcades to more diffuse loops that reach to greater heights. The low-lying loops connect the penumbra of the leading spot to regions of opposite magnetic polarity marked by chromospheric plage rather than spots. As mentioned in §6.1, there is an absence of hot, coronal material in loops connecting directly to a spot umbra, though material at transition-region temperatures is present. The higher, more diffuse loops connect regions of opposite polarity on the farthest edges of active regions, and show a resemblance to the patterns of magnetic field lines obtained from potential field calculations.

The appearance of the soft X-ray loops forming the coronal active region

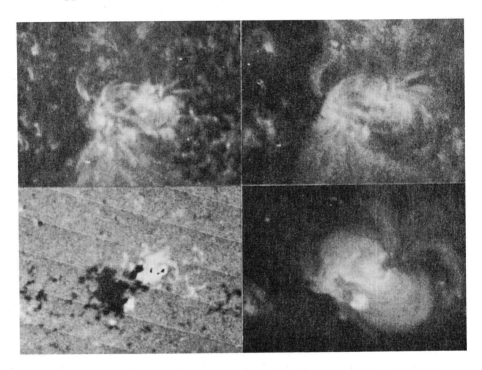

Fig. 6.9 Images of an active region in ultraviolet light – emission lines of Ne VII, 46.5 nm (top left), Mg IX, 36.8 nm (top right) and Fe XV, 28.4 nm (bottom right) – and a Kitt Peak magnetogram (bottom left) in November 1973. (Courtesy N. Sheeley (1980))

varies markedly with its age. A young, vigorous region is very bright and compact in X-rays, with X-ray line emission indicating temperatures as high as $5\,000\,000$ K. A more evolved region is fainter and more diffuse, and has temperatures of more typically $3\,000\,000$ K. Those that have reached a stage when the spots have disappeared are cooler still. The decrease in X-ray brightness of particular loops is largely attributable to decrease of density with age rather than temperature, the intensity depending on the square of density. Values of around $1-7\times10^{15}$ per cubic metre are measured, appreciably higher than quiet-coronal structures (see Table 5.2).

6.3   Solar flares

The most striking instance of solar activity is the solar flare, a sudden release of energy appearing as electromagnetic radiation over an extremely wide range and as mass, particle, wave and shock-wave motions. Flares invariably occur in active regions, being commonest and largest when the active region is in a rapidly developing stage; however, they may also occur when the region is decaying and has lost all its sunspots – seemingly the region's last gasp before dying. Numerous small flares may occur when the region is in its initial stages, when an arch filament system is present but the sunspots have yet to form. Because of the association with active regions, the frequency of flares follows the eleven-year sunspot cycle, but with some deviations: there is often a burst of flare activity late in the cycle, as with the large flares in August 1972, half-way down the decline of cycle 20.

The various phenomena accompanying flares may last for as little as a few minutes, but for giant flares it may be a day or so before everything returns to the pre-flare state (unless other flares occur in the same active region in the meantime). As much as 10^{25} J are radiated in large flares, with about a quarter of this amount appearing in visible wavelengths. At least as much energy may appear as mass motions. Such a flare might typically last an hour and cover an area of 3×10^{9} square kilometres, or 0.1% of the sun's visible hemisphere. Thus, the flare's visible radiation is emitted at the sun at the rate of about 2×10^{11} W per square kilometre. Now, the energy normally emitted by the photosphere amounts to 6×10^{13} W per square kilometre. Hence the visible radiation of flares is normally quite insignificant compared with the sun's normal radiative output, and only very rarely are flares visible at all in white light: Carrington and Hodgson's 1859 observation (§1.7) was the first indication of their existence. But it is not the amount of energy in a flare that is of interest so much as the suddenness of its release. Few astronomical phenomena – X-ray emission associated with certain binary-star systems, cosmic gamma-ray bursts, and flares on other stars are among them – are as rapid as solar flares.

The extreme diversity of flare-associated phenomena and the continuing search for a physical mechanism by which energy is so suddenly released have made the study of flares a particularly burgeoning one in solar physics. An extra impetus in recent times is the recognition that flare-like phenomena with even greater energy output occur on some stars. In the following, we will describe in turn the flare in visible wavelengths; the high-energy (X-rays, gamma rays, etc.) flare; an outline interpretation of these observations; and the emission of flare-accelerated particles.

When observed on the disc in Hα, the very first stage of a flare is a darkening and a rising motion in the active-region filament running along the magnetic inversion line. A few minutes later, the flare proper begins with two bright areas either side of the magnetic inversion line which, for flares that develop to some size, are in the form of 'ribbons'. These bright areas show a rapid expansion, often accompanied by a strong increase of brightness, called the 'explosive' or 'flash' phase (some observers reserve the term flash phase for the moment when, as happens in all flares, there is a rapid increase in the width of the Hα line). At the explosive phase, the active-region filament is ejected from the sun altogether. The two ribbons (in the case of large flares) then start to separate with speeds of several km/s at first, declining later on. The ribbons do not always expand uniformly, for the presence of sunspots or large concentrations of magnetic field seem to inhibit their passage. For large flares, dark loops – *loop prominences* – start to form across the location of the now-invisible dark filament, connecting the inner edges of the two bright ribbons. The ribbons and loops eventually fade from view after a few hours.

On the limb, a flare in Hα appears as a bright point and develops into a brilliant mound of emission. At flare onset, there may be temporary ejections of material in the form of surges or sprays. *Surges* have ascending, then descending motions along an almost straight path, with ascent velocities of between 50 and 200 km/s, reaching heights of about 100 000 km (Fig. 6.10). There is a tendency for surges to recur at the same location. *Sprays* are explosive events in which fragments of Hα-emitting material are ejected at velocities of up to 2000 km/s – higher than the solar escape velocity, 618 km/s (Fig. 6.11). At late stages of a large flare on the limb, the loop prominences mentioned earlier may be visible, first as a small bright mound, then as small bright loops, successive ones forming above each other, apparently condensing out of the surrounding hot corona, the lower ones disappearing by the draining of material along the loop edges (Fig. 6.12). The loop shape is sometimes only apparent by the path that bright blobs – *coronal rain* – take as they fall back towards the solar surface. As with the bright ribbon separation in disc flares, the loop height continues to increase with time but at a rate that steadily decreases. For very intense limb flares, the loops may persist for almost a day, by which time they attain a height of 100 000 km above the limb.

Hα flares have been systematically recorded for many years at observatories round the world, either visually with observers using spectrohelioscopes or on film, with the sun photographed several times a minute. Many such patrols operated during the International Geophysical Year between 1957 and 1958, around the maximum of cycle 19, when links between solar and terrestrial phenomena were specifically studied. Flare reports are received world-wide at the World Data Center A for Solar Terrestrial Physics at NOAA in

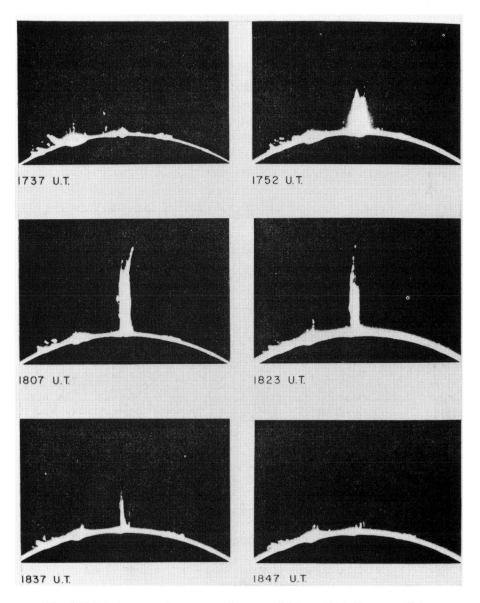

Fig. 6.10 Development of surge prominence on 12 June 1946. (Courtesy High
   Altitude Observatory)

Boulder, Colorado, and are collated and published in the NOAA bulletin *Solar–Geophysical Data*. A simple scheme for classifying Hα flare importance is used by NOAA and extensively elsewhere, which is based on both flare area (corrected for foreshortening) and brightness as measured in Hα. This is given in Table 6.1a. On this scheme, the smallest and faintest flares are classed as Sf, and the largest and brightest 4b. As well as flare importance, times (start, maximum brightness, end) and positions on the disc (latitude and longitude measured east or west from the central meridian), are given in *Solar–Geophysical Data*.

The visible spectra of flares have a remarkable appearance, with many of the strong Fraunhofer lines 'reversed' from absorption to emission lines. In large flares, the central intensity of the Hα line may be more than twice the nearby continuum and with wings extending out up to 1 nm either side of the centre (Fig. 6.13); sometimes there is a central depression ('self-reversal') in the emission and a slight enhancement of the red side of the line profile (a 'red asymmetry'). Many of the lines in the hydrogen Balmer series may appear in emission – up to the H22 line (i.e. due to transitions between the $n=22$ and $n=2$ levels of hydrogen) in disc flares, and H30 in limb flares. So, too, are numerous lines due to neutral and singly ionized metals – the Ca II H and K

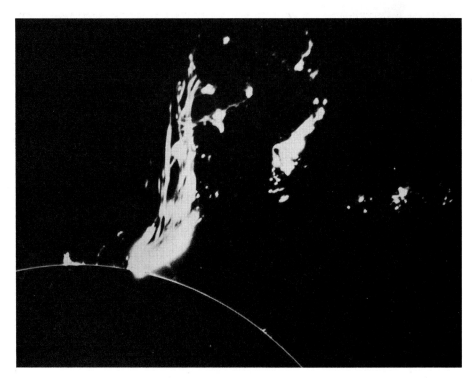

Fig. 6.11 Large spray prominence on 3 January 1969. (Courtesy Institute for Astronomy, University of Hawaii)

Table 6.1. *Classification schemes for flares*

(a) *Hα flares*

Importance	Flare area (square degrees)	Flare area (millions square km)	Flare brilliance[a]
S (subflare)	Less than 2	Less than 300	
1	2.1– 5.1	300– 750	
2	5.2–12.4	750–1850	f, n, b
3	12.5–24.7	1850–3650	
4	More than 24.7	More than 3650	

Note:
[a] f=faint; n=normal; b=bright

lines particularly, but also lines of magnesium, aluminium, iron and titanium. Neutral helium is represented by the 587.6-nm line and, in the near-infrared, by the 1083.0-nm line. These lines are either in absorption or (for larger flares) in emission.

Fig. 6.12 Flare loop prominence in Hα (above) and the Fe XIV coronal ('green') line. (Courtesy Institute for Astronomy, University of Hawaii)

There are some notable differences in the visible spectra of disc and limb flares which point to different excitation conditions with height. Most of the Hα emission seen in disc flares comes from lower down, whereas in limb flares the emission is in the corona. Higher temperatures, then, are expected for limb flares, as are indeed observed: a good indicator is the He II line at 468.6 nm, requiring a temperature of about 25 000 K for its excitation, which is sometimes present in limb flares. There are also differences in the profiles of the metal lines, which are broader in limb than in disc flares. This is attributed to non-thermal motions occurring at greater heights.

The wide profiles of the hydrogen Balmer lines in flare spectra are due to collisional broadening. The electron density can be estimated from high members of the Balmer series, as the broadening increases with density. Electron densities of up to 4×10^{19} per cubic metre are measured at the maximum of some disc flares, with apparently smaller values before and after. Lower electron densities (10^{18} per cubic metre) are measured for the higher limb flares.

Occasionally, for limb flares, the Balmer continuum (edge at 364.6 nm) is present. This is normally the only continuum in visible spectra, but as already indicated 'white-light' flares have been observed, showing a continuum that covers the entire visible range and into the ultraviolet and infrared. Up to the peak of cycle 22, only about 60 such flares had been recorded since Carrington and Hodgson's 1859 flare, and only a few of these captured on film. Figure 6.14 is a white-light image of the great flare on 7 August 1972,

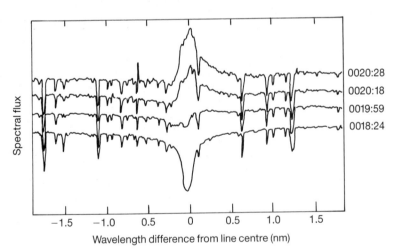

Fig. 6.13 Successive Hα line profiles at a location in a flare on 20 June 1982 (times in hours, minutes and seconds indicated to the right of each curve). The line initially has an absorption profile but develops a strong emission core, with a slight red asymmetry. The wavelength scale is measured from the rest position of the centre of the Hα line. (After Ichimoto and Kurokawa (1984))

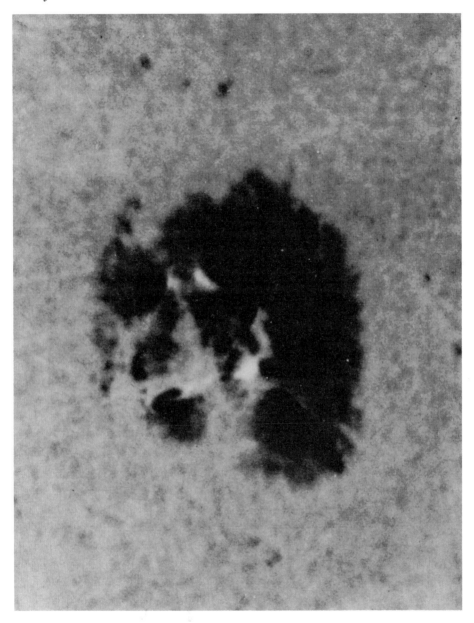

Fig. 6.14 White-light flare on 7 August 1972 at 15:20 UT. (Courtesy National Solar Observatory/Sacramento Peak)

showing patches of emission near a large spot. These patches usually occur in pairs, disposed on either side of the magnetic inversion line in a region of large magnetic gradient, and are short-lived, reaching maximum brilliance before the Hα emission. The spectra of white-light flares generally show the

Balmer continuum, though it definitely seems to be absent for a few flares. Thus, two types of such flare may exist: those whose spectra show the Balmer continuum, for which the continuum arises in the chromosphere, and those that do not, for which the continuum arises near the solar temperature minimum and is radiated by the H^- ion. The emission lines normally seen in intense flares without white-light continuum are present, often very intense and with broader profiles. Most likely white-light flares do not form a special category, but are simply very intense – possibly all flares have a continuum that is generally too weak to be seen.

The tendency of an active region to flare depends on the region's magnetic complexity, those of class $\beta\gamma$ or γ on the Mount Wilson magnetic scheme being especially prone to flaring. Corresponding sunspot types are also known to be associated with frequent flaring; these include compact, complex groups, with many intermediate spots (class Fkc on the McIntosh scheme). Detailed studies of the photospheric magnetic field have shown the circumstances under which flaring occurs. There is, for example, a relation of flaring with how field lines (as indicated by the $H\alpha$ fibrils) are 'sheared' across the magnetic inversion line, i.e. the angle they make with this line. The more sheared they are, i.e. the more nearly parallel to the inversion line, the greater is the observed flare frequency. The changes in the shear angles appear to be the result of proper motions of sunspots, either changes of position ('translations') or spot rotations. The emergence of new magnetic flux may lead to areas of strong field gradient. A particular instance is the δ magnetic configuration (§6.1), in which an area of one polarity is embedded in the penumbra of a spot of the opposite polarity. Emerging new flux may also give rise to a pronounced 'kink' in the magnetic inversion line or to satellite sunspots. All these areas are likely sites for subsequent flaring.

A remarkable property of some flares is the ability to recur with the same geometry and time development within an active region. It might be thought that a flare would significantly alter the magnetic configuration of an active reigon, so that subsequent flares would have differing forms, but this does not seem to be so. Among the many remarkable instances of these so-called 'homologous' flares is a series of five such flares in a δ region within 13 hours in May 1980.

The detailed relation of flare $H\alpha$-emitting structures and the magnetic field measured with vector magnetographs have recently been studied. Richard Canfield and colleagues at the University of Hawaii have mapped both the magnetic field and the $H\alpha$ line profile over the area of several flares. Because the complete magnetic field can be measured with their instrument, not just the longitudinal component, associated currents can be derived and mapped. Figure 6.15 shows the longitudinal magnetic field as well as areas of high line-of-sight current density for a flare in 1989: the latter are shown as black or white depending on the current's direction. $H\alpha$ line profiles were measured over the same flare area using an imaging spectrograph. The Hα line shows the usual wide emission profile for most of the flare area, but in certain locations the profile has a self-reversal, these being thought to be the sites of particularly high energy release. In the flare of Fig. 6.15, these locations occur at the edges of the areas of major line-of-sight current density.

As interesting and spectacular as visible-wavelength observations of flares are, much valuable extra information has been obtained from complementary

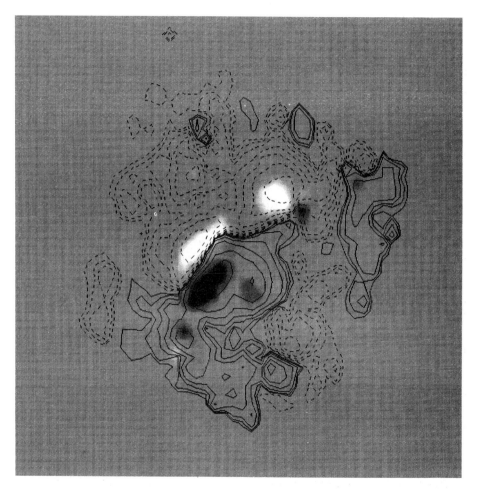

Fig. 6.15 Longitudinal field (shown by the contour lines, full and dashed lines indicating opposite polarity) and vertical component of current density (shown by black and white areas), derived from measurements of the complete magnetic field, in an active region on 20 October 1989. Regions of non-thermal electron precipitation, deduced from observations of the Hα line profile, are located between the two largest areas of vertical current density, at their edges. (Courtesy R. C. Canfield *et al.* (1991))

observations in the radio, X-ray and other wavelength regions. Such observations have indicated that the flare energy release is sited in the corona, and that the chromospheric (e.g. Hα) flare occurs as a consequence of this release. Many of the important discoveries of flares have been made very recently, and observations obtained with spacecraft and with radio interferometers.

Perhaps the most revealing of the recent flare observations have been made in the hard X-ray range. 'Hard' X-rays in this context means wavelengths less than about 0.06 nm, though usually photon energies are specified rather than wavelengths. Such energies are normally expressed in units of electron volts (1 eV=1.6×10^{-19} J) or kilo-electron volts (keV); a wavelength of 0.06 nm corresponds to 20 keV. Many instruments sensitive to hard X-rays do not form images at all, as it is very difficult to do so, but simply collect photons incident from all over the sun's visible hemisphere, so that the data consist of 'light curves' or 'time profiles' of hard X-ray emission. Such observations and others in the radio and ultraviolet regions have led to the recognition of an *impulsive phase* near the start of most flares. This impulsive phase is generally missing in the time profiles of soft X-rays (wavelengths more than about 0.06 nm, photon energies less than 20 keV). Figure 6.16 shows typical time profiles for hard and soft X-rays during a large – actually a double – flare on 5 November 1980 as observed by instruments (HXIS, the Hard X-ray Imaging Spectrometer, and HXRBS, the Hard X-ray Burst Spectrometer) on the *Solar Maximum Mission* (SMM) spacecraft. The differences between the two are striking, the softer energies showing two enhancements having gradual variations and the harder energies showing several short-lived bursts at the onset of each soft X-ray enhancement. These bursts can be resolved into many extremely short pulses, with fluctuations as short as 0.02 seconds for some flares. The rapid pulses also occur at centimetre radio and ultraviolet wavelengths. Collectively, they constitute the impulsive phase. Over these widely different spectral regions, the emission is simultaneous to a very high degree – often less than a second. Over the period 1980–89, covering the decline of sunspot cycle 21 and the rise of cycle 22, some 12 776 flares were observed by the HXRBS instrument aboard *SMM*, and the vast majority have the characteristics of the flare of Fig. 6.16.

The spectra in hard X-rays show a featureless continuum that declines rapidly with increasing energy. Very often the spectral flux (photons/m^2/s/keV) has a roughly 'power-law' dependence on energy, i.e. is proportional to $E^{-\alpha}$, where E is the photon energy (say, in keV) and the index α is a positive number; thus, on a graph which has logarithmic y (intensity) and x (energy) axes, the spectrum is a straight line sloping downwards from right to left (see Fig. 6.17). During the flare impulsive stage, the spectrum is steeply falling (α large, e.g. between about 5 and 10) at first, flattens at the maximum (α reduced, e.g. to around 3), then steepens once more during the decline. (The spectra of the flare illustrated in Fig. 6.17 are unusual in that they do not steepen during the decline.) The spectrum is, then, 'harder' at the maximum, i.e. is relatively intense at very high energies. The cause of the emission is very likely to be Bremsstrahlung, or free–free transitions of energetic electrons in the electrostatic field of protons. Bremsstrahlung accounts, at least partly, for the soft X-ray continuum of the quiet corona and active regions, and is emitted by a plasma at a temperature of a few million degrees Kelvin. However, the hard X-ray Bremsstrahlung of flares requires much more energetic conditions. The continuum is known to extend to hundreds of kilo-electron volts, so that the kinetic energy of the electrons moving past protons in the emission process must have at least this amount of energy. If these electrons have a distribution of energies like those in a hot gas – a so-called thermal or Maxwellian distribution – the temperature must be enormous, 10^8 or 10^9 K. This has led to the idea that the electrons are not thermal at all, but are accelerated by some process in the flare to energies that extend to very high values and with the numbers of electrons in particular energy ranges proportional to energy raised to a power or index, like the emitted spectrum (the index of the electron energy distribution is in fact $\alpha - 1$ if the emitted spectrum has index α).

Such 'non-thermal' electrons are met with in many astrophysical and laboratory conditions. If flare electrons are accelerated in this way, they would quickly escape from the acceleration site (an electron with energy of 100 keV,

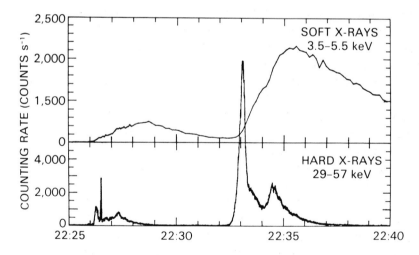

Fig. 6.16 Time profiles in hard and soft X-rays for a flare on 5 November 1980. Times are indicated in hours and minutes and the X-ray flux by the photon counting rate in the HXIS (upper) and HXRBS (lower) instruments on *Solar Maximum Mission*. (Courtesy C. de Jager, B. Dennis and HXIS and HXRBS investigators on the *Solar Maximum Mission* spacecraft; from Dennis (1985))

e.g., having a velocity of 0.55×the velocity of light). Figure 6.18 shows schematically how the electrons relate to an assumed magnetic-field configuration of the flare, taken to be a loop connecting opposite-polarity regions. The acceleration site could be anywhere along the loop length, e.g. near its apex, so that the electrons travel down both legs of the loop, quickly encountering much denser gas just above the transition region. There they are stopped by collisions with the particles of the dense gas, with a small proportion (one in 10^5 or so) emitting hard X-rays. The dense gas forms what is called a 'thick target'. The spectral changes – harder at event maximum – are ascribed to changes in the energy distribution of the electrons as they are accelerated.

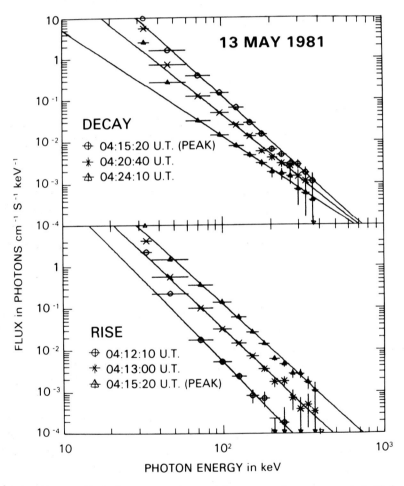

Fig. 6.17 Hard X-ray flare spectra (spectral flux versus photon energy in keV) for the rise and decay phases of a flare on 13 May 1981. (Courtesy B. Dennis and HXRBS investigators on *Solar Maximum Mission* spacecraft; from Dennis (1985))

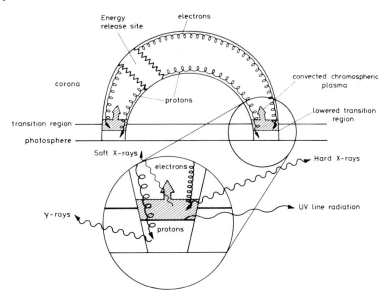

Fig. 6.18 Single flare loop shown schematically to illustrate the acceleration of electrons and protons from the flare energy release, chromospheric evaporation at loop footpoints, and regions of the emission of hard X-rays, gamma rays and ultraviolet lines. (After Dennis and Schwartz (1989))

An alternative to this 'non-thermal' picture of flare hard X-ray emission has been proposed. In this, the electrons are accelerated at some point in the loop, but rather than escaping from this point they collide with each other to form a thermal gas with temperature of 10^8 K or higher. This gas rapidly expands to fill the loop, previously supposed empty, in just a few seconds, the electrons dissipating their energy in collisions with the denser gas of the loop footpoints.

Despite the difficulties of forming images in hard X-rays, there have been some recent advances, and the results give tantalizing insight into the nature of hard X-ray flares but have not led to unequivocal interpretation. A case in point is the 5 November 1980 flare seen by the HXIS instrument on *SMM* (the time profiles for this flare are illustrated in Fig. 6.16). Figure 6.19 shows the flare at the time of the main impulsive stage, 22:33 Universal Time (UT) in intermediate energies (16–30 keV). There is a general area of emission, about 2 arc minutes (90 000 km) long, with three bright points, two on one side of the magnetic inversion line, one on the other. This was thought at the time to be strong evidence for the non-thermal, thick-target picture, these bright points being identified with the footpoints of magnetic loops crossing over the magnetic inversion line (as in Fig. 6.18) where energetic electrons encounter denser gas. However, it is now recognized that the image is not inconsistent with the thermal picture either, so direct imaging has not so far been able to distinguish between the thermal and non-thermal interpretations. Some evidence for the height structure of impulsive events has come from the observation of flares from two widely separated spacecraft, the *International Cometary Explorer* and the *Pioneer Venus Orbiter*, so giving a 'stereoscopic' view. The impulsive emission from three of the five flares observed occurred at low altitudes (less than 2500 km), implying that the thick-target picture of hard X-ray bursts is more correct, but this was not so for the other two.

Some hard X-ray events have been noted that occur just after the impulsive event and are unusually long-lasting, of the order of several minutes. Although not many such events have been observed, their character is so distinctive that there is little doubt that they form a category of their own. Their spectrum starts 'soft' (α large), becomes steadily 'harder' (α small) at the maximum, and still harder beyond the maximum. Some spectra from an event of this type (on 13 May 1981) are illustrated in Fig. 6.17. Such events lack the impulsive emission of most hard X-ray bursts. There is very often a simultaneous radio burst with similarly gradual variations. A telling observation was made by an imaging hard X-ray instrument on the Japanese *Hinotori* satellite of the 13 May 1981 flare. The hard X-ray emission occurred at an

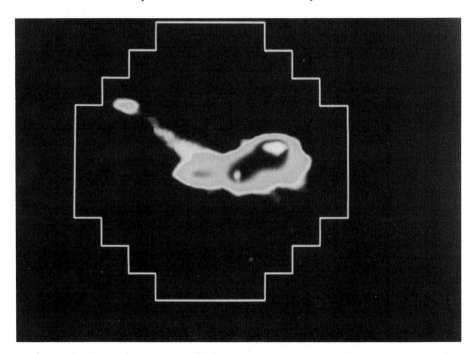

Fig. 6.19 Image of flare on 5 November 1980 at 22:33 UT, the time of the main impulsive stage (see Fig. 6.16). The image was taken in intermediate-energy (16–30 keV) X-rays. (Courtesy C. de Jager and HXIS investigators on *Solar Maximum Mission* spacecraft)

altitude of 40 000 km above the corresponding Hα flare, and radio emission at 35 GHz seems to have been coincident with it. Other observations show that the hard X-ray source is very large, perhaps tens of thousands of kilometres across. The impression is that the emission is Bremsstrahlung due to energetic electrons as with impulsive bursts but that these electrons remain trapped in a bubble-like magnetic field configuration (referred to as a 'plasmoid') rather than encountering denser gas at lower altitudes.

Flare soft X-ray emission appears to arise from a very hot plasma, with temperatures of up to about 20×10^6 K . As indicated earlier, the variations are gradual, not impulsive, and show a rise and decay similar to those of the Hα intensity. The time of maximum soft X-ray intensity follows that of the hard X-ray impulsive stage by an interval that depends on the X-ray wavelength – typically one or two minutes for 0.2 nm emission, more than ten minutes for 2 nm emission. A gradual soft X-ray rise is sometimes observed before the impulsive stage, especially for larger flares. This may last a few minutes or sometimes much longer, and appears to be due to the heating and intensification of a loop, not necessarily the one in which the major energy release takes place.

Photographs of soft X-ray flares indicate a variety of sizes, from single, compact loops about 1400 km (2 arc seconds) long, to arcades containing many loops. The loops are seemingly magnetic flux tubes connecting areas of opposite polarities. The footpoints of loops in an arcade are marked by the ribbons of the Hα flare.

Soft X-ray emission consists of both continuum and lines. The continuum is mostly free–free and free–bound emission, the latter arising from recombinations of electrons on to highly stripped ions of elements such as carbon. As the temperature of the emitting plasma is so high, atoms of elements like hydrogen and helium exist in a completely ionized form, while those of other elements are generally stripped of all their $n=2$ and 3 orbiting electrons, leaving only two or one of the remaining $n=1$ electrons (i.e. the ions are in the helium-like or hydrogen-like stages). As with the general corona, excitation of the remaining bound electrons occurs by collisions with free electrons, with photons emitted by the downward transition. Thus, resonance lines of hydrogen-like and helium-like ions of abundant elements (carbon, oxygen, neon, magnesium, silicon, sulphur, argon, calcium and iron) feature prominently.

Although the emitting plasma has temperatures of up to 20×10^6 K, much lower temperatures also exist. This is indicated by the presence of lower-temperature lines in flare spectra. Thus, lines due to a range of iron ions are very prominent at wavelengths of less than 2 nm, generally ranging from the neon-like stage Fe^{16+}, with its complete complement of electrons in the $n=1$ and 2 orbits, to the helium-like Fe^{24+}, the $n=2$ to 1 transition of which gives rise to a strong line feature at 0.19 nm. Their presence shows the existence of temperatures from as low as 2×10^6 K to 20×10^6 K. Plasmas with such different temperatures can only co-exist if they are in separate magnetic flux tubes, so the soft X-ray flare should be thought of as several different loops with the observed temperature range.

A balloon-borne X-ray instrument flown in 1980 detected a continuum between soft and hard X-ray energies during a flare that indicated a temperature of 35×10^6 K. This is much more than the maximum temperature normally observed in a soft X-ray flare, and appears to be due to a 'superhot' component. The existence of this component has been confirmed by the presence of the Lyman-α lines of hydrogen-like iron (Fe^{25+}), at 0.178 nm, during very energetic flares, indicating similar temperatures.

A striking phenomenon is observed at the flare impulsive stage – the broadening of soft X-ray line profiles. This has been particularly observed for high-temperature lines such as those due to helium-like iron and calcium, but it is known to occur for lower-temperature ions also. The broadening is symmetric for limb flares, but for disc flares the line profiles show asymmetric broadening, with excess emission on the short-wavelength side. Interpreted as a Doppler effect, this line shift indicates the motion of hot plasma towards the observer at velocities of about 400 km/s, exceptionally even higher. The motion has been ascribed to the convection of gas at the footpoints of a loop or loops in the model indicated by Fig. 6.18. Particles accelerated at the energy release site are guided down the loop legs and reach the dense gas at the loop footpoints. This gas is unable to radiate away the dumped energy of the energetic particles fast enough, so converts some of the energy into convective motion, the now-heated gas moving upwards into the loop. An observer viewing the flare on the disc sees the plasma moving with approaching velocity, so the spectrum has a shift towards shorter wavelengths. The broadening of X-ray spectral lines has commonly been attributed to turbulence in the convecting gas, but since the gas is heated to a fully ionized plasma, it is likely to be guided by magnetic field lines without the formation of turbulent eddies; more likely, the line broadening is due to motion of the plasma along the curved path of the loop, the approaching line-of-sight velocity varying depending on how far the plasma is up the loop.

The description just given is the 'chromospheric evaporation' picture that has been widely advanced in recent years. Some tentative evidence for its correctness is the fact that, at the flare impulsive stage, when X-ray lines show a Doppler shift to shorter wavelengths, the Hα line profile has a red asymmetry (Fig. 6.13), i.e. shows a Doppler shift to longer wavelengths. This has been interpreted by the presence of a downward-moving condensation in the chromosphere at the site where the energetic electrons dump their energy; calculations suggest that very roughly the momentum of the upward-moving hot gas equals that of the downward-moving chromospheric gas, so obeying Newton's third law of motion. However, there is not a unanimous acceptance of the chromospheric evaporation idea, some claiming it to be too simplistic, and the debate about interpretations of observations continues.

Table 6.1. *Classification schemes for flares*

(b) *X-ray flares*

Importance	Peak flux in 0.1–0.8 nm range (watts per m^2)^a
B	10^{-7}
C	10^{-6}
M	10^{-5}
X	10^{-4}

Note:
^{a} Flare X-ray importance is indicated by a number following one of the above letters, the number being a multiplier: e.g. C3.2 stands for peak flux of $3.2 \times 10^{-6}\,\mathrm{W/m^2}$.

Classification of flares has already been described by their extent and brightness in Hα, but a flare's importance can also be described in terms of the maximum soft X-ray emission from a flare. There is, as might be expected, a general correlation between the two, but not a detailed one, so that flare importance in soft X-rays provides complementary information. A scheme has been adopted that relies on data from monitoring satellites called the *Geostationary Operational Environmental Satellites* (*GOES*) that are in geostationary orbits above the earth's western hemisphere. The scale is described in Table 6.1b. Data from *GOES* in the form of lists of X-ray flares, including their times and peak emission, are made available by NOAA in *Solar–Geophysical Data*.

Radio emission from flares has been detected over a large wavelength range, extending from the so-called microwave region (wavelengths less than 10 cm, frequencies greater than 3 GHz) to the kilometre region (wavelengths up to 10 km, frequency down to 30 kHz), the latter being observed from spacecraft. There is a great complexity of phenomena, though there are important links with emission observed at other wavelengths. Thus, the microwave emission very closely follows that in hard X-rays, showing the same fine time structure, perhaps with a very slight lag of a second or so. High-resolution images of radio flares obtained with the Very Large Array interferometer show a pair of small patches of emission either side of the magnetic inversion line at very short wavelengths (2 cm) but a single more extensive area joining these patches at longer (6 cm) wavelengths. The shorter-wavelength emission seems to come from the footpoints of a loop structure, the longer-wavelength from the whole length of the loop.

It is thought that the electrons emitting hard X-rays by impact with the denser gas at loop footpoints give rise to the microwave emission by a process that is in principle similar to gyro-magnetic radiation that occurs in active regions and sunspots (6.2), i.e. the electrons emit radio-wave radiation as they gyrate about the magnetic fields, probably near the top of the loop (see Fig. 6.18). However, the emitting electrons in the case of flares have velocities which are large fractions of the speed of light (are 'relativistic'), which appreciably modifies the emission process. Firstly, the radiation is highly directional, visible only when the electron velocity is almost exactly directed at an observer. Secondly, the emission occurs not only at the gyro-frequency and the first two or three harmonics but at a great number of closely spaced harmonics – typically 10 to 50 – which are broadened into a continuum. This 'synchrotron' emission is believed to be due to emitting electrons with energies of 100 keV to 1 MeV which probably belong to the same population that gives rise to the hard X-ray burst.

Following the impulsive radio bursts is often a gradual enhancement called a 'post-burst increase'. This appears to be the equivalent of the soft X-ray thermal flare, and the emission mechanism is Bremsstrahlung.

At wavelengths longer than the microwave region, there is a whole range of phenomena associated with flares and also with non-flaring active regions. There is a classification system which can best be described with a *dynamic spectrum*, i.e. a graph of radio frequency (*y*-axis, on a logarithmic scale) and time (*x*-axis); Fig. 6.20 shows one schematically. In this plot, the microwave burst appears as an area in the lower left-hand corner. At metre wavelengths, a broad continuum called a type I burst very often appears, persisting for several hours after a flare, with fine structure superimposed on it in the form of short bursts. Type I 'storms' are similar but last for several days and are associated with non-flaring active regions having large sunspots. The exact details of the radiating mechanism are not known, but it is believed that such bursts arise from plasma waves that couple to electromagnetic waves at radio frequencies, with the emission occurring at the local plasma frequency.

Type III (or 'fast drift') bursts are almost as common as type I bursts but are very different in nature: they generally appear as short-lived bursts at high frequency, drifting extremely rapidly to low frequencies, so tracing a nearly vertical line across a dynamic spectrum. They occur in groups at the start of a flare or more often without flare association. Some type III bursts show 'reverse' drifts, i.e. from low to high frequencies, while occasionally a burst shows first a forward drift, then a reverse drift, forming a U-shape in a dynamic spectrum. Some bursts have been traced to extremely low frequencies (30 kHz) with receivers aboard the *Radio Astronomy Explorer* spacecraft. The radiation of type III bursts arises from streams of fast electrons, which for the majority of bursts travel out from the sun; for the reverse-drift bursts, the electrons travel inwards, and for U-bursts they travel first outwards, then inwards, following a closed structure like a magnetic loop. The faster electrons in the stream move ahead of the slower-moving ones, and a plasma instability is thereby caused resulting in the formation of plasma

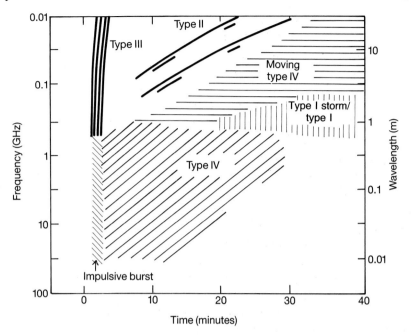

Fig. 6.20 Schematic dynamic radio spectrum showing classification of radio bursts from centimetre to decametre wavelengths. The characteristics of the types of bursts are described in the text.

waves. These are partly converted to radio waves which are emitted at the local plasma frequency and occasionally at twice this frequency (the 'second harmonic').

Powerful flares generate shock waves that move out from the sun and in doing so give rise to so-called type II (or 'slow drift') radio bursts, which appear as emission at the plasma frequency and second harmonic, with a drift from high to low frequencies, much slower than for type III bursts. There are strong associations with gamma-ray flares and those producing large-scale mass ejections. Late in the burst there may be a series of radio pulsations with periods of a second or so, thought to be due to particles trapped in a volume at the top of an active-region magnetic loop across which the shock wave passes.

Finally, a continuum often accompanies type II bursts but at lower frequencies, lasting an hour or so: this is a type IV burst. The emission seems to be due to particles trapped in large loops. A rarer event is the 'moving' type IV burst, the motion of such bursts having been recorded with imaging radio telescopes ('radioheliographs'); Fig. 6.21 shows an example taken with the Culgoora radioheliograph in Australia. They may be due to synchrotron-emitting electrons trapped in a plasmoid, rather like the gradual hard X-ray events mentioned earlier.

Ultraviolet emission during flares shows both impulsive and gradual

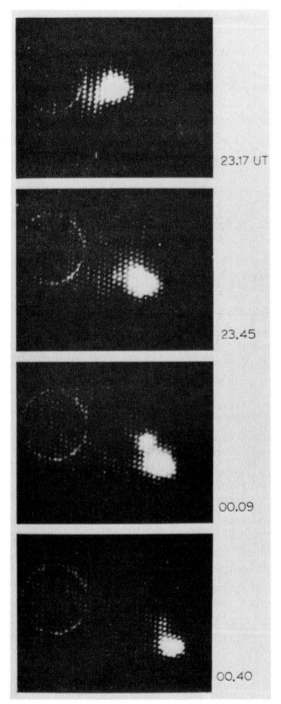

Fig. 6.21 Images (radioheliograms) at 80 MHz (3.75 m) of a moving type IV burst on 1 March 1969 obtained with the CSIRO Culgoora radioheliograph. (Courtesy A. C. Riddle (1970). Reprinted by permission of Kluwer Academic Publishers)

increases. It is observed both directly with space-borne instruments and also by increased ionization in the earth's ionosphere, which affects the transmission of radio signals. The variations are impulsive for emission lines formed in the transition region and the continuum around 150 nm, but are not so for chromospheric and coronal emission lines. The impulsive emission consists of spikes which occur almost simultaneously with (though not always with the same relative intensity as) those in hard X-rays: Fig. 6.22 shows the time profiles during a flare for an oxygen (O V) transition-region emission line at 137.1 nm and hard X-rays. Images in the ultraviolet show that there are often many emission points before the flare occurs, some of which subsequently brighten during the flare, with different bright points giving rise to particular spikes in the time profile. In at least some cases, the bright points are identifiable with loop footpoints. There is a general decrease of ultraviolet emission for flares near the limb.

The impulsive ultraviolet emission can be explained by the model flare loop of Fig. 6.18. As fast particles encounter the loop footpoint regions, they heat the gas there which leads to a lowering of the transition region where the density is higher. The intensity of transition-region ultraviolet lines is propor-

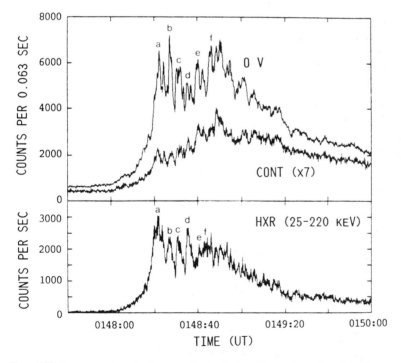

Fig. 6.22 Time profiles of a flare on 24 April 1985 in an O V ultraviolet line at 137.1 nm, the ultraviolet continuum at 138.8 nm, and hard X-rays (HXR), showing the coincidence of the fine time structure. (Cheng *et al.* (1988). Reprinted with permission from *The Astrophysical Journal*)

tional to the square of the density of the emitting gas, so an increase of the intensity occurs each time a pulse of fast electrons reaches a loop footpoint. The decrease of ultraviolet line intensities near the limb is, in this picture, due to obscuration of the emission by chromospheric structures.

Gamma rays, seen in exceptionally intense flares, show the presence of nuclear reactions in solar flares, and have been observed in great detail with the Gamma Ray Spectrometer on the *SMM*. Figure 6.23 shows a spectrum during a large flare from the *SMM* instrument. A broad continuum is present, with intensity falling with increasing energy over the range 0.4 MeV to about 8 MeV. It is partly due to Bremsstrahlung (as in hard X-ray flare emission), with the emitting electrons at relativistic energies (greater than about 1 MeV) and partly due to line emission that is highly broadened by the Doppler effect. Sometimes the continuum extends to energies above 10 MeV: the origin is then partly Bremsstrahlung and partly a nuclear process – the creation by fast-proton collisions of extremely short-lived particles called pions which decay to form gamma rays. Nearly all the flares showing this extremely high-energy continuum are near the limb. This has been explained by the fact that the electrons emitting the Bremsstrahlung must be directed almost exactly towards the observer for the radiation to be seen; for a limb flare, the

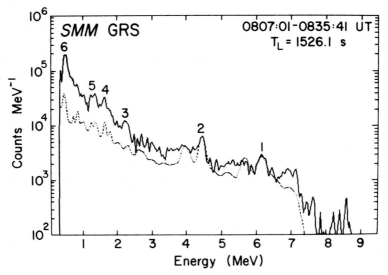

Fig. 6.23 Gamma-ray spectrum for a limb flare on 27 April 1981, integrated over a period of 29 minutes. The gamma-ray emission lines that are numbered are identified as follows: 1, 2, 4, 5=radiative de-excitation of nuclear excited states in oxygen, carbon, neon and magnesium isotopes; 3=(mostly) the 2.2 MeV deuteron-formation line; 6=(mostly) the electron–positron annihilation line at 0.511 MeV. (From Chupp (1984); courtesy T. Vestrand and GRS investigators on the *Solar Maximum Mission* spacecraft)

electrons mirroring at the loop footpoints temporarily move parallel to the sun's surface, and then the radiation is directed towards the observer.

There are several gamma-ray lines present in Fig. 6.23, among which are those at 0.511, 4.4 and 6.1 MeV (lines no. 6, 2 and 1 respectively in the figure). Normally, a line at 2.2 MeV (line no. 3) is the strongest for flares near disc centre, but for this limb flare it is weak. It is formed when neutrons are captured by protons to form deuterons (nuclei consisting of a proton and neutron). The neutrons are themselves produced when protons, accelerated at the flare energy release site, travel towards the flare loop footpoints where, unlike electrons, they penetrate right down to the photosphere, and collide with target protons and other particles of the dense gas there. Initially, the neutrons have extremely high energies, but collisions may slow them down sufficiently for capture by protons to occur, when the 2.2-MeV gamma rays are emitted. The strong line at 0.511 MeV is due to the annihilation of electrons and their positively charged 'anti-particles', positrons. Radio-active nuclei, again produced by fast-proton collisions, are one source of the positrons. As with neutrons, the positrons and electrons must be slowed down before they combine to form two photons of energy 0.511 MeV. Other gamma-ray lines are due to the de-excitation of excited nuclear energy states. Just as atoms are raised to higher energy states by collisions with electrons, atomic nuclei can be excited by collisions with, e.g., protons. Thus, the relatively strong 4.4-MeV line is due to the de-excitation of an excited nuclear state in carbon.

The time profiles of gamma-ray emission during flares vary widely, but like hard X-rays there are generally one or more short-lived bursts near the onset, each lasting between 10 s. and 2 minutes. A series of bursts may last up to 20 minutes. These peaks are strongly correlated with the hard X-ray impulsive bursts – those at very high energies (more than 10 MeV) simultaneous with hard X-rays and those at smaller gamma-ray energies slightly delayed.

As already indicated, the release of a flare's energy results in the acceleration of electrons and protons as well as other particles. Some are directed down the legs of flare loops to give rise to electromagnetic radiation while others escape into space. The latter have been detected with instruments aboard spacecraft near the earth and on interplanetary missions. Their energy ranges are very wide: 0.3 MeV to several GeV for atomic nuclei and protons, and a few tens of keV to several MeV for electrons. Very energetic protons (more than about 500 MeV) directed at the earth can be detected at ground level.

A large flare may result in particle emission that, even at the distance of the earth, would result in a dosage that would be lethal for a human were it not for the shielding effect of the earth's atmosphere. There is consequently some concern by space agencies for the safety of astronauts during high levels of solar activity. Spacecraft observations indicate the presence of quasi-steady streams of particles from active regions as they rotate across the sun, as well as specific enhancements or 'events', many of which are definitely flare-associated, but with some having less obvious connections (e.g. because the particles were emitted by a flare behind the limb).

Energetic protons and electrons detected during specific events show an approximately power-law energy distribution, i.e. of the form E^{-n} (E=particle energy, n a number between 2 and about 5), so there are many more particles with lower energies. The instantaneous energy distribution of particles observed by a spacecraft instrument is not the same as the initial distribution when the particles started their journey from the sun, since the energy distribution is modified by the particle travel. The energetic particles (i.e. those travelling fastest) arrive first, followed by the less energetic ones, so the observed energy distribution (to use the analogy of hard X-ray spectra) is at first 'hard', then 'softens' with time.

Electron events are strongly associated with type III radio bursts, with a near-coincidence in the arrival of electrons and the occurrence of low-frequency radio emission detected by spacecraft near the earth. Some events show the presence of only electrons ('pure electron' events), while others have both protons and electrons ('mixed' events). Proton events have associations with very large flares, e.g. those with type II and IV radio bursts, while true 'proton flares' are those observed with ground-based instruments (see §7.5).

After protons, helium nuclei (or α-particles) are the commonest heavy particles. The ^{4}He form of the helium nucleus is normally, as expected, most abundant, but some particle events, not always flare-associated, have shown ^{3}He nuclei to be greatly enriched, the amount even equalling ^{4}He. Heavier ions such as those of iron also show large anomalies.

The direct detection of flare neutrons was made with the *SMM* Gamma Ray Spectrometer. Neutrons have a half-life of only $15\frac{1}{2}$ minutes – they decay to form a proton, electron and neutrino – so unless their time of flight from the flare to the earth is less than roughly this time, they cannot be directly observed. During a large flare in 1980, the *SMM* instrument detected a short-lived burst of gamma rays, following which was an extended enhancement in the counting rate, peaking about 8 minutes after the start of the impulsive event. This enhancement is due to neutrons rather than gamma-ray photons since the spectrum was quite different from other gamma-ray events. Neutrons were known in any case to be present in flares as they are responsible for the formation of the 2.2-MeV line, but this was the first direct detection of neutrons from a solar flare.

6.4 Coronal mass ejections

Rapid motions and brightness changes in the white-light corona have been noticed for a number of years. Spacecraft-borne white-light coronagraphs have allowed observations of these 'transients' to much greater distances from the sun than have ground-based ones, and also have the advantage of not being subject to the weather or atmospheric conditions. Sensitive electronic detectors are used in place of photographic film. These are easily overexposed, and careful provisions must be made to ensure that direct photospheric light never reaches them. Such provisions include having an occulting disc substantially larger than the sun's diameter. Observations with these coronagraphs have shown that a common form of coronal transient event is a large-scale ejection of mass, apparently lost by the sun to interplanetary space. Such coronal mass ejections (CMEs) were first seen in 1973 with a coronagraph on *Orbiting Solar Observatory – 7*. The subsequent *Skylab* mission (1973–74), carrying the High Altitude Observatory's coronagraph, recorded over a hundred transient phenomena, of which 77 were recognized as mass ejections, the remainder being 'rearrangements' of existing coronal structures. Much more data have been provided by the Naval Research Laboratory's SOLWIND coronagraph on the *P78-1* spacecraft, which operated from 1979 to 1985, and the High Altitude Observatory's Coronagraph/Polarimeter (C/P) on *SMM*, which operated in 1980 and between 1984 and 1989. The SOLWIND instrument viewed the corona over heliocentric distances between $2.5\,R_\odot$ and $10\,R_\odot$ with a resolution of 75 arc seconds, while the C/P instrument had higher resolution (10 arc seconds) but rather smaller distance range ($1.6–6\,R_\odot$). Data from two ground-based instruments at the High Altitude Observatory's Mauna Loa Station in Hawaii, the K-coronameter (observing over the range $1.2–2.2\,R_\odot$) and Hα prominence monitor, were used to complement the *SMM* data, giving views of the corona from just above the limb out to $6\,R_\odot$.

Most CMEs are associated with erupting prominences. As these occur throughout the solar activity cycle, CMEs have been evident both in the low-activity periods (e.g. during the *Skylab* mission and in 1984–85) and in the high-activity periods (1979–81 and 1986–89). There seems to be some increase in CME frequency when activity is high (about one or two CMEs per day), though various statistical studies do not completely agree on this point. CMEs occur over a wide range of latitudes when the sun is active, but are confined to low latitudes (as are coronal streamers generally) near sunspot minimum. The speeds of ejection have a wide range, with a maximum of more than 1000 km/s for the most energetic events, but down to about 10 km/s for events with no associated activity, generally seen near sunspot minimum. Some 10^{13} kg of mass are ejected, so that the kinetic energy ($\frac{1}{2}mv^2$)

may be 5×10^{24} J. To this should be added magnetic energy and enthalpy, giving a total that may exceed 10^{25} J. This makes them comparable to or more energetic than flares.

We will describe two well-observed CMEs which typify the pattern usually seen. The first is a flare-associated CME which occurred on 15 October 1986, and was observed with the K-coronameter and Hα prominence monitor as well as the coronagraph on *SMM* (Fig. 6.24). A sequence of images shows a bright rim around a dark cavity starting to move out from the sun at a speed of 650 km/s, greater than the solar escape velocity (618 km/s). Shortly after, the prominence monitor observed what seemed to be an Hα

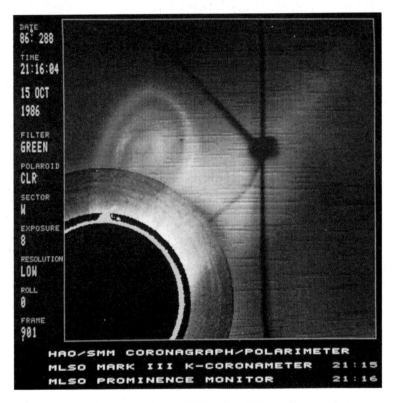

Fig. 6.24 Coronal mass ejection on 15 October 1986 observed with the Coronagraph/Polarimeter instrument on *Solar Maximum Mission* spacecraft (outermost field of view), the Mark III K-coronameter (intermediate field of view) and Hα prominence monitor (field of view including limb region). The latter two instruments are at the Mauna Loa Solar Observatory (part of High Altitude Observatory). The *SMM* and K-coronameter images show the outer bright rim, followed by the dark cavity and second loop structure, identified with an eruptive prominence, also seen with the Hα prominence monitor near the limb. The dark linear structures across the image are artefacts. (Courtesy of A. Hundhausen (1987) and investigators on the *Solar Maximum Mission* spacecraft)

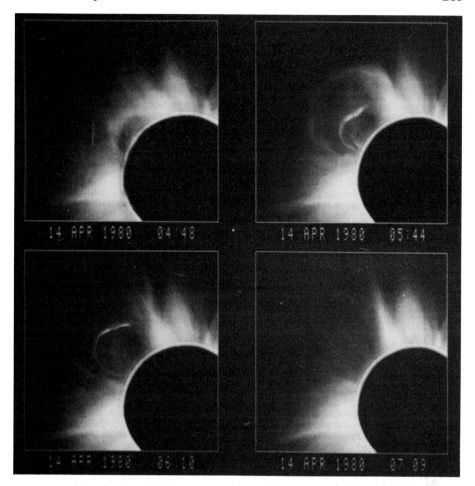

Fig. 6.25 Four stages in the development of the coronal mass ejection on 14 April
1980 (from upper left to lower right, the times are 04:48, 05:44, 06:10
and 07:09 UT). (Courtesy A. Hundhausen (1987) and investigators on the
Solar Maximum Mission spacecraft)

spray or erupting prominence. This moved out to appear as a bright, secondary loop in the K-coronameter and *SMM* images, connecting back to an
active region on the limb well to one side of the leading bright rim of the
CME. The speed of this loop-like feature was less than that of the bright rim.
At the time of the image in Fig. 6.24 (21:16 UT), the bright rim had moved
out to a heliocentric distance of about $3 R_\odot$.

The second CME occurred on 14 April 1980, and was associated with a
prominence eruption. Figure 6.25 is a sequence of four images from the
SMM instrument showing how first a bright outer loop with dark cavity
behind moves outwards, followed by a second loop which may be an erupting

prominence. The two loops are seen together in the second and third images, but by the time of the fourth image (2 hours 21 minutes after the first image), only the prominence loop is visible. The speed of the outer rim was about 270 km/s.

The pattern of an outer bright rim followed by a dark cavity and finally a more slowly moving erupting prominence is common to many CMEs. The bright rim seems to be material swept up by the advancing disturbance, deriving its mass from pre-existing coronal structure, while the erupting prominence may be the result of the CME, not as previously thought simply occurring with it. After the passage of a CME, radial coronal structures are left, the remnants of the bright rim of the CME where it was connected to the sun, forming a pair of 'legs'. These may persist for a day or two.

While the association of CMEs with erupting prominences is undeniable, what is less clear is that of CMEs with flares. Some studies have indicated that as many as 40% of CMEs are flare-associated, with flares having Hα ejecta (sprays or surges) being especially prone to produce CMEs. As with the 15 October 1986 CME described above, the ejection is already in progress when the flare occurs. The time corresponding to when the CME is estimated to have left the sun's surface is generally at least a few minutes before the flare.

The three-dimensional form of a CME is still disputed: are the leading bright rims bubbles or loops? Evidence that at least some are bubbles was provided by the observation of a CME by the SOLWIND instrument in 1979 that was apparently directed towards the earth. The images (Fig. 6.26) show the CME in the form of a bright halo expanding symmetrically around the occulting disc. It was associated with an Hα filament disappearance near sun centre.

There has been much speculation and modelling of CMEs since their discovery. In one picture, CMEs are due to ropes of magnetic field that rise by magnetic buoyancy, the gas being less dense on the inside of the rope than outside (see §6.5), while another view is that CMEs result from an untwisting of the field lines. The possibility that flare-associated CMEs are propelled by a pressure pulse due to the flare is now thought unlikely in view of the fact that the CME seems to precede the flare. In another approach, outward flow of coronal mass is considered to be the normal situation, while the confining of parts of the corona in loops is less usual. The corona would release mass everywhere in the form of a 'wind' were it not for its magnetic field, since solar gravity is insufficient to hold back the hot gas which tries to expand against it. The magnetic field when bipolar helps gravity by tending to confine the gas through a generally downward-directed tension. A CME occurs by a large-scale departure from equilibrium when the magnetic and gravitational forces no longer counter the coronal gas expansion, and lifts off with constant speed, either slow or fast. Accordingly, the CME is not the 'result' of a flare

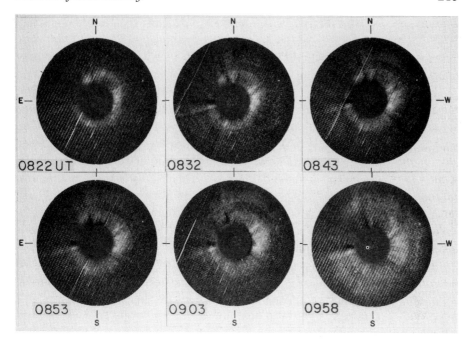

Fig. 6.26 Images obtained with the SOLWIND instrument on the *P78-1* spacecraft of a coronal mass ejection on 27 November 1979 directed at the earth. The CME has been enhanced by subtracting the emission of the background corona. The emission occurs all round the occulting disc of the instrument, but is obscured by a pylon supporting the disc for a portion in the upper left of each image. (Courtesy N. R. Sheeley; in Howard *et al.* (1982). Reprinted with permission from *The Astrophysical Journal*)

or erupting prominence, but all three may be due to the loss of equilibrium. This may go some way towards explaining the different start times of flare and CME and different speeds of coronal and prominence material.

6.5   Theories of solar activity

All the forms of activity on the sun – sunspots, active regions and the transient phenomena such as flares and coronal mass ejections – are strongly related to magnetic fields and indeed it is clear, from both observations and theory, that fields are responsible for their occurrence. Nevertheless, observations often lack the needed time and space resolution and give us no indication of the coronal magnetic field, while the theories have the weakness that they cannot be tested for the conditions met with in the solar atmosphere. With these thoughts in mind, we now outline the main ideas current in explaining how fields give rise to what is observed.

Firstly, we take up the concept described in §2.4 for the eleven-year cycle of solar activity. In this, magnetic field at the base of the sun's convection zone having direction parallel to the solar equator (the toroidal field) is constantly regenerated by the action of convection and differential rotation; convection produces a field component parallel to lines of longitude (the poloidal field), and differential rotation stretches the poloidal field lines to produce a toroidal field again, but with direction opposite to the original toroidal field. The time taken for a complete cycle of field regeneration process is identified with the solar magnetic cycle of 22 years, or two sunspot number cycles of 11 years.

The field lines of the toroidal field in the solar interior are not in the form of parallel lines because convection will cause them to be twisted round each other to form rope-like tubes called flux ropes. There is a balance of external pressure exerted on the flux rope and internal pressure. External pressure is simply the gas pressure, but the internal pressure is the sum of the gas and magnetic pressures, so that the internal *gas* pressure is less than the external pressure. If the temperatures outside and in the rope are equal, this must mean (since gas pressure is proportional to density times temperature) that the density inside the rope is less than outside, i.e. the rope is 'buoyant', and floats towards the photosphere. Near the photosphere, there is no longer a pressure balance as the external gas pressure decreases, and as a result of this individual flux rope 'strands' become detached from the main rope structure and bulge out above the photosphere to form the arch filaments seen at the birth of an active region. Eventually, the main flux rope breaks through the photosphere to form a sunspot pair with leader and follower components of opposite magnetic polarity. The leader and follower spots slowly separate, as observed, as the rope continues to rise.

There remain a number of detailed problems with this picture, and individual sunspot models have particular solutions to them. Not all theorists, for example, accept the need for the twisted, or rope-like, magnetic field pattern to form spots. The twist in the field lines prevents an instability (known as fluting) that untwisted field lines are subject to when forming a large structure like a sunspot. This instability causes the field lines to break apart into many smaller tubes; a large spot would never form because of fluting, only numerous small pores or much smaller spots. However, some theorists, while admitting that fluting does occur, maintain that there is a reclustering of individual tubes just below the photosphere; as the root of the field lines lies deep down, near the base of the convection zone, the individual tubes are constrained to lie close to each other and actually merge again at the photosphere to form a sunspot, rather like a group of balloons tethered to a single piece of string (Fig. 6.27).

The most obvious characteristic of a spot is that it is darker and therefore cooler than the surrounding photosphere. On the flux-rope picture, the cooling is due to decreased convection. In one model, a spot consists of not a

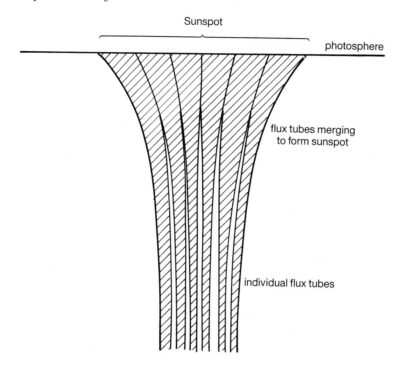

Fig. 6.27 Magnetic flux tubes, tied together at a large depth, forming a sunspot. (After Spruit (1981); © AURA, Inc.; courtesy National Solar Observatory, Tucson, Arizona)

single rope but many hundreds of smaller strands, each made up of twisted field lines, and themselves twisted to form the main flux rope. Convection is suppressed along a particular strand and only occurs between the strands. The spot umbra is, in this model, the place where the strands are squeezed together so that only the cool gas within each strand is visible; in the penumbra the strands are not so tightly packed but have lanes between them where convection can occur, these being observed as the light filaments. The radial outflow from a spot at the photospheric level – the Evershed effect – is explained by the continued convective motions along the light filaments of the penumbra, which are directed outwards rather than upwards. The decay of a spot is explained by the unwinding and breaking away of rope strands (see Fig. 6.28), convected outwards by the Evershed flow. The strands separate from the main flux rope at increasing depths until the rope eventually disintegrates.

The magnetic field emerging through the photosphere in the form of flux tubes or ropes passes into the chromosphere and corona, where, because of the greatly reduced external pressure, they are free to expand. As with the quiet-sun field at the chromospheric network, a nearly horizontal canopy of

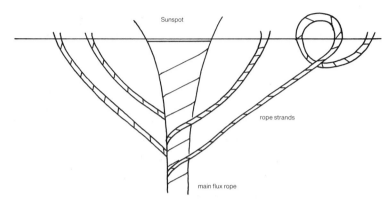

Fig. 6.28 The magnetic flux rope below a sunspot breaking apart as the spot decays, with separate strands of the rope carried away from the spot by the Evershed flow. (Piddington (1976). Reprinted by permission of Kluwer Academic Publishers)

field is formed (Fig. 4.2), though at a lower altitude – about 500 km. Above this height, the field is shaped like giant loops. This is evident from X-ray photographs that show coronal active regions as hot (about 2 000 000 K), X-ray-emitting plasma, confined to one or more magnetic loops, with the more complex regions having many such loops that are even hotter (up to 4 000 000 K). The plasma is confined to the magnetic loops for exactly the same reason as for the quiet coronal loops, viz. the electrons and protons gyrate in tight spirals along field lines (§5.5).

The heating of these coronal counterparts of active regions is in principle no different from the heating of the quiet corona: active regions are simply heated more. The magnetic field is associated with the heating: apart from sunspots, locations of intense field within active regions (and outside them) are hotter than their surroundings, i.e. have more energy deposited there. As with coronal heating, this argues strongly against non-magnetic heating mechanisms such as sound waves. A number of processes associated with magnetic fields have been considered for active-region heating, many of them similar to those for the general corona (see §5.7), but as with the corona there is still a lack of consensus about which is the most important. These include dissipation of electric currents in thin sheaths around active-region loops and Alfvén waves, propagated up from the convection zone in regions of high field, dissipating in coronal active-region loops after coupling to fast MHD waves. For the smaller, hotter active regions and the X-ray bright points, additional processes, such as small-scale magnetic field reconnections and the untwisting of magnetic flux ropes that have emerged from below the photosphere, have been considered.

Many attempts have been made to understand sunspots and the heating of active regions, but few topics in solar physics have engaged the minds of theorists so much as the phenomenon of the solar flare, the fascination being how the energy can be released so suddenly. It is important to draw a distinction between the initiating burst of energy, known as the primary energy release, and the diverse phenomena that are the result, which, interesting and spectacular though they may be, are secondary processes. An important difference between the two is that we can easily observe the secondary processes but the primary energy release may well take place in a region – presumed to be in the corona – too small to be seen with present instruments. There is another major difficulty in our understanding of the primary energy release. It can be shown that the only possible source of the energy in a flare is that stored in magnetic fields in the coronal part of an active region. But our knowledge of the strength and detailed configuration of such fields where the primary release occurs (as with all coronal fields) is at best rudimentary. Thus, the field patterns assumed by theorists as the starting point for the occurrence of flares are not directly observed at all, though there is often some justification for such assumptions.

The reconnection of magnetic fields is thought to be the cause of the primary energy release. In this, it is envisaged that a volume of coronal plasma with magnetic field lines in one direction is brought into close contact with another such volume with magnetic field lines in the opposite direction (Fig. 6.29). Such a field configuration is far from 'potential' and in fact can only exist if there is an associated current located along the boundary which has a direction perpendicular to (and out of) the page. The boundary region is for this reason spoken of as a current sheet (or, if the two sets of field lines are exactly parallel, a 'neutral' sheet). This current flows almost without resistance because the conductivity of coronal gas is extremely high. Nevertheless, there may locally be a small amount of resistance. If so, just as a current flowing in a solid such as copper wire encounters resistance leading to the dissipation of the current and to heating, so the current in the current sheet in Fig. 6.29 is dissipated. The magnetic field associated with this current correspondingly disappears or 'diffuses away', as indicated in the figure. Magnetic field reconnection, then, is accomplished through the small amount of resistivity (the inverse of conductivity) that coronal plasma has. The energy that the now-dissipated current had reappears as heat energy or goes into acceleration of particles.

Now, the resistivity of a solid is due to atoms arranged in a rigid lattice structure which impede the flow of electrons that constitute an electric current. For the thinly ionized coronal plasma, the resistivity is normally due to sparsely located ions or other electrons which act as much weaker obstacles to a current, largely through their electrostatic charge. This resistivity is called 'classical' resistivity. The resistivity may, however, be enhanced by the presence of waves or turbulence in the plasma since these also impede the electron flow. The resistivity is then 'anomalous'. The rate at which the current of Fig. 6.29 dissipates depends on the width of the region where field diffusion occurs (L metres, say), the plasma temperature and resistivity; for classical resistivity and a temperature of 10^6 K, it will occur in a time of the order of L^2 seconds. Thus, a primary energy release in a time of a few seconds, as indicated by, e.g., flare hard X-ray observations, requires a current sheet of only a few metres in thickness! If the resistivity is anomalous, it will be larger, but still probably only of the order of a kilometre. This is far smaller than the observational limits of present-day instrumentation and even of those being planned, so that there is little chance that the primary energy release if due to magnetic reconnection will be observed directly in the foreseeable future.

Reconnection mechanisms have therefore been studied entirely theoretically, but some predict consequences which may be observable. Two such

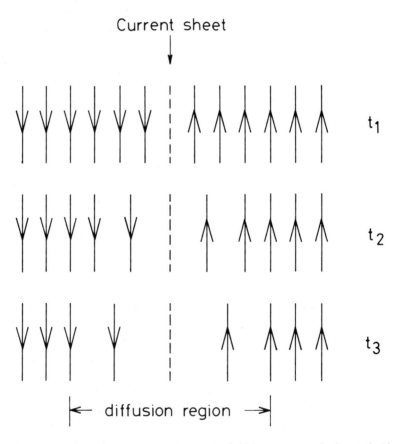

Fig. 6.29 Stages in the diffusion of magnetic field lines, oppositely directed either side of a boundary layer. (After E. R. Priest (1972). Reprinted by permission of Kluwer Academic Publishers)

mechanisms have received considerable study. In the model of H. Petschek, proposed in 1964, two volumes of plasma containing oppositely directed field are pushed together in the way indicated by Fig. 6.30. The two field systems, supposed slightly convex to each other, interact such that the central field lines form an X-shape at the centre of the diffusion region. Extending away from the diffusion region is a pair of slow-mode MHD shock waves, which remain stationary or 'standing' as the plasma flows through them from either side. Reconnection takes place in the diffusion region, with the now-heated plasma carrying the new field lines away along the length of the current sheet and between the pairs of standing waves. A steady state may occur in which the plasma flows in from each side at a rate balanced by the outward diffusing plasma.

In the tearing-mode instability (Fig. 6.31), two regions with oppositely directed magnetic field are in close proximity forming a current sheet, with plasma flowing along the field lines. The regions are not pushed together, as in the Petschek mechanism, but reconnection is more spontaneous. At the current sheet itself, any tendency for the plasma to move perpendicularly to the field lines is opposed by a restraining force that is large because of the small resistivity; but if by chance the field strength becomes zero at some point along the sheet, the restraining force is much reduced and an instability develops. Plasma is driven towards this point in the current sheet by non-uniformities in the field outside the sheet, and this driving force overcomes the restraining force within the sheet. An X-shaped neutral point develops, and the sheet 'tears'. It does so repeatedly along the length of the sheet, forming magnetic 'islands' that contain separate strands of current. This 'linear' phase of the tearing-mode instability does not seem to have the energy release appropriate for flares, but the 'non-linear' development of the instability, in which the magnetic islands coalesce, results in a much greater energy release, and shows greater promise of explaining flares.

Various suggestions have been offered for how in practice a field configuration might give rise to reconnection. Two of them are depicted in Fig. 6.32. In P. A. Sturrock's picture (Fig. 6.32a), reconnection takes place above a helmet streamer, consisting of mostly closed field lines but with some open in the way shown in Fig. 5.15. Photospheric motions at the foot of the configuration lead to a stressing of the field and the formation of a current sheet which is subject to the tearing-mode instability and so reconnection of field lines. Plasma is ejected in the form of a plasmoid, driven outwards by magnetic forces, and particles are accelerated in the closed-loop system left below, which travel to the footpoints and give rise to the hard X-ray emission.

In the model of J. Heyvaerts, E. R. Priest and D. M. Rust (Fig. 6.32b), emerging magnetic fields (or flux) near an existing active-region loop configuration lead to a current sheet if the emerging flux has opposite polarity to the existing system. Such flux systems can be identified with the small satellite

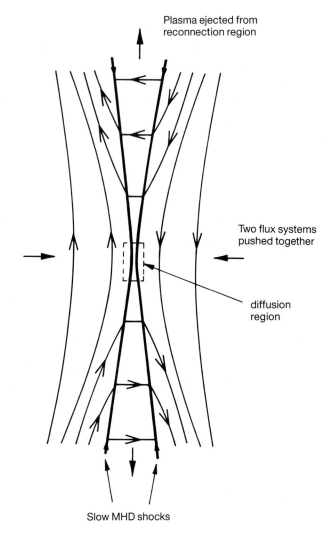

Plasma ejected from
reconnection region

Two flux systems
pushed together

diffusion
region

Slow MHD shocks

Fig. 6.30 Magnetic field configuration for the Petschek mechanism. Plasma motions
push two sets of field lines from the sides (arrows), and plasma is pushed
away from the diffusion region between the MHD shock fronts, taking
with them the now-reconnected field lines.

magnetic features and spots near large spots (§6.1). The current sheet formed
between the two flux systems in this model has a rapid-expansion phase due
to the onset of small instabilities which raise the plasma resistivity from a
classical value to an anomalous one. The flare's explosive phase is identified
with this expansion of the current sheet. Energetic particles travel down the
legs of the small loop to form the hard X-ray impulsive bursts.

Reconnection need not occur in a sheet geometry. It can take place in a

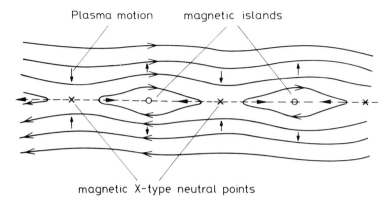

Fig. 6.31 The tearing mode. Magnetically neutral X and O points are formed at the boundary between regions of oppositely directed field, with plasma flow in the directions indicated by the small arrows. (After Furth *et al.* (1963))

loop having field lines that are 'sheared', i.e. have twists of increasing angle to the direction of the loop going towards the loop's core. As a result of a tearing-mode instability, the current sheets, which are in the form of cylinders, tear into ribbons. This is the basis of a model by D. S. Spicer in which a sheared, current-carrying loop is initially subject to an MHD instability (known as a 'kink' instability) which then leads to one involving resistivity. It is envisaged that the kink instability leads to a part of the loop expanding which may explain mass motions during flares, shock waves as

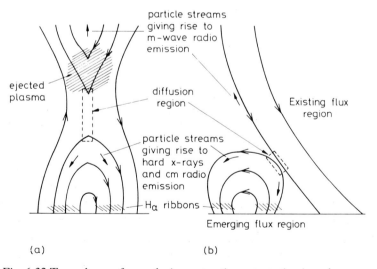

Fig. 6.32 Two schemes for producing magnetic reconnection in a sheet geometry in the solar corona: (a) Sturrock (1980); (b) Heyvaerts *et al.* (1977).

indicated by type II radio bursts and Moreton waves, and possibly ejection of a plasmoid.

Observations of homologous flares suggest that, under certain circumstances, flares can repeatedly occur having very similar characteristics. It would seem that a magnetic configuration has energy fed into it for some time (a few hours, maybe up to a day), then becomes unstable, releasing this energy as the flare, then is recharged to supply the energy of a second flare, and so forth. The occurrence of the flare in these circumstances does not alter the configuration appreciably, so that all the flares have the same appearance. This all suggests the existence of a flare 'trigger'. That is, energy is continuously fed into the configuration by some means that does not disturb it appreciably until a certain threshold is reached when the configuration loses its equilibrium and its energy content is released. The nature of the trigger is not known, but it might, for example, be some process that results in a rapid increase of plasma resistivity.

Whatever the correct theory of primary energy release is, it is widely accepted that upon release much of the energy goes into accelerating particles. An outstanding theoretical problem is explaining how particles are accelerated to such a vast range of energies: hard X-rays, for instance, suggest the presence of electrons up to a few hundred kilo-electron volts, while gamma rays, white-light emission and direct spacecraft observations indicate the presence of extremely energetic, relativistic particles. Doubt has been expressed whether this could be accomplished with a single process, and it has therefore been proposed that there are two acceleration steps: one to produce non-relativistic energies, and a second to produce relativistic energies. In the tearing mode, for instance, particles could be accelerated in a first stage by electric fields in the diffusion region to produce non-thermal beams of particles, with energies up to non-relativistic values. In a second stage of acceleration, a small fraction of these now-energized particles are further accelerated to relativistic energies. This might occur by the so-called Fermi process, in which particles repeatedly collide with moving magnetic-field structures, gaining energy at each collision at the expense of the field, or by shock waves.

The model shown in Fig. 6.18 and already described in § 6.3 shows how the particles might lead to the various forms of radiation. Thus, hard X-rays are emitted by Bremsstrahlung as the accelerated particles – thermalized or not – are stopped by the dense material at the loop footpoints which they meet as they travel down the loop legs. The microwave emission may arise as synchrotron radiation as the electrons move along field lines on their way down to the footpoints. Soft X-ray emission results from hot gas filling the loop, convected from the transition region or chromosphere on the impact of energetic particles at the loop footpoints. The Hα and other visible line emission on this picture is due to heating of the chromosphere by conduction of the hot soft X-ray-emitting plasma in the flare loop, while the occasional white-light emission, if at the temperature minimum, could be due to protons penetrating to this level. Protons are also required to give rise to the gamma-ray line emission apart from the 0.511-MeV line. Electrons that escape from the flare loop may give rise to the type III radio emission, shock waves generated by the flare to the type II burst and ejected plasmoids to the moving type IV bursts.

This scheme has not received unanimous support. It has been proposed (by G. M. Simnett and D. Heritschi) that protons, not electrons, accelerated by the primary energy release, give rise to the hard X-ray bursts and many of the other phenomena. Also, the reality of chromospheric 'evaporation', i.e. the convection of chromospheric or transition-region gas to give rise to the shift of the soft X-ray lines, has been disputed. It is pointed out that not all flares show the impulsive hard X-ray bursts (though this might be simply due to the bursts being below the instrumental threshold) and further the existence of a soft X-ray rise *before* the impulsive hard X-ray bursts is difficult to explain. The soft X-ray line shifts might, as an alternative, be due to small loops that rapidly balloon out as energetic particles dump their energy at their footpoints.

Our perception of the solar flare has drastically changed over the past few years as observations have improved in quality, but it would seem that we still lack some vital pieces of information, most notably we do not yet have coronal magnetic field measurements or X-ray images of the required quality at especially the impulsive stage. The models that have been worked on, particularly those involving magnetic field reconnection and particle acceleration, give a reasonable description of what is observed, but unfortunately what is observed is really the aftermath of the primary energy release processes. Reconnection is largely an untested area, at least for the sun, and may be in effect untestable, in view of the fact that current sheets seem to have dimensions far below the limit of present instruments. Possibly the way forward is the study of plasmas in laboratory fusion devices, since they have densities, temperatures and field strengths not too unlike those thought to exist in solar flares.

Chapter 7

The sun and the solar system

The sun is the dominant body of the solar system by virtue of its mass and energy output, while its atmosphere in the form of a steadily expanding solar wind extends throughout the solar system. In this chapter we review some of the aspects associated with the sun's gravitation (leaving solar energy to Chapter 9), and the solar wind and its interaction with the solar system. We then describe the particular case of the atmosphere of the earth, and how it is affected by the sun, together with considerations of terrestrial climate and weather, including some non-solar effects on the earth's atmosphere.

7.1 The sun's gravitation and the solar system

The sun contains 99.87% of the mass of the solar system and as a result gravitationally controls all bodies in the solar system. The nine major planets, the asteroids and many comets move round the sun in orbits which Kepler found obeyed certain laws of planetary motions. These state:

1. Each planet moves in an elliptical orbit with the sun at one focus;
2. The line joining each planet with the sun sweeps out equal areas in equal times;
3. The square of the period a planet takes to revolve about the sun is proportional to the cube of the planet's mean distance from the sun.

Figure 7.1 shows an idealized planetary orbit in the form of an *ellipse*. An ellipse is obtained when a plane surface intersects a circular cone at an angle less than the angle the cone's side makes with its base. The two foci of the elliptical orbit (F_1, F_2) are such that the sum of the two distances from them to

226

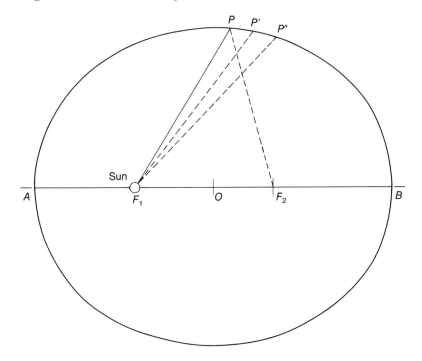

Fig. 7.1 Idealized planetary orbit illustrating Kepler's first two laws. The orbit is an ellipse, centre O, with the sun at focus F_1, while the other focus is F_2. The planet P moves such that the radius vector F_1P sweeps out equal areas in equal times (areas PF_1P' and $P'F_1P''$ are equal). A is the perihelion, B the aphelion of the orbit.

the planet at any point P on its orbit (F_1P+F_2P) is constant. The sun in Fig. 7.1 is considered to be at F_1, and the points A and B are points where the planet is respectively nearest (perihelion) and furthest (aphelion) from the sun. With O the centre of the ellipse, the eccentricity of the orbit is defined by the ratio OF_1/OA. The planet's mean distance from the sun is AO (or OB). The orbits of the nine major planets are all close to circles, with eccentricities greatest for Pluto (0.25) and Mercury (0.206). Several minor planets have highly eccentric orbits, most notably Icarus (0.827), whose orbit carries it nearer to the sun than Mercury at perihelion. Many comets ('periodic' comets) also have highly eccentric elliptical orbits, while some ('non-periodic' comets) have orbits that are 'open' in the sense that they appear to come into the solar system from outside and after perihelion leave the solar system entirely. The shapes of these orbits are parabolic or even hyperbolic – sections of a circular cone when the intersecting plane makes an angle equal to or more than the angle between the cone's side and base.

Kepler's empirical laws of planetary motion are consequences of Newton's universal law of gravitation, which states that two bodies having masses m_1

and m_2 and separated by a distance r are attracted to each other by a gravitational force equal to Gm_1m_2/r^2, where G is called the gravitational constant: this is the *inverse-square* law of gravitation. G can be accurately measured with torsion balance experiments in the laboratory. A planet's motion, then, is described in terms of two gravitationally attracting bodies, one the planet itself, the other the very much more massive sun. Elliptical orbits follow from Newton's law, and so, too, does the possibility of parabolic and hyperbolic orbits that non-periodic comets have.

The orbit of a planet can be found from repeated observations of its position in the sky. The orbit is completely specified when six 'elements', defining its shape, orientation and the time of the planet's perihelion passage, are known, allowing the planet's position at any time to be derived. Without any other information, the orbit's size can only be given in units of the earth's orbit. The absolute size of the orbit in, e.g., kilometres, and indeed the scale of the solar system, are only known if the size of the earth's orbit can be determined, i.e. the length of the astronomical unit (AU, the earth's mean distance from the sun) in kilometres. Up to the early 1960s measurements of this were made by observations of asteroids that passed close enough to the earth to show a 'parallactic displacement', i.e. having perceptibly different positions against the sky background when simultaneously seen from two different locations on the earth. This gave the distance of the asteroid in terms of the well-known radius of the earth (6371 km). Much more accurate determinations have since been made by reflecting radar signals off Venus; the time between transmitting and receiving the signal, which travels at the speed of light, gives twice the distance to Venus. The most recent determinations give the astronomical unit as 149 597 870 km.

The sun's mass can be found from Kepler's third law applied to the earth's orbit. The cube of the earth's mean distance, i.e. 1 AU, divided by the square of its period, i.e. one year of 365.2564 days (a 'sidereal' year, measured with respect to the stars) gives G times the product of the sun's and earth's masses. Neglecting the earth's mass, which is extremely small by comparison, gives the sun's mass as 1.989×10^{30} kg.

Although the planets have masses that are small compared with the sun, they do nevertheless measurably affect or 'perturb' each other in their orbital motion: the most massive planet Jupiter is most notable in this respect. A comet that happens to pass near a massive planet may have its orbit radically altered: Comet Lexell, a periodic comet till 1779, passed so close to Jupiter in that year that it was accelerated out of the solar system altogether. There are also interesting examples of continual interference between two planets orbiting the sun when the ratio of the orbital periods is a simple number, so that a 'resonance' is set up. As a result, particular configurations of the planets in their orbital motions repeatedly occur. In the asteroid belt, between Mars and Jupiter, there is a relative *absence* of asteroids at distances from the sun corresponding to resonances with Jupiter, the 'Kirkwood gaps'. These resonances have been investigated using computer techniques, and it is found that, were an asteroid to be located in a resonance, it would be subject to the gravitational pulls of both the sun and Jupiter in such a way that large and unpredictable changes in its orbit would occur, so explaining why no asteroids are observed at these positions.

For planets near the sun, especially Mercury and Icarus near its perihelion, there are small departures from Keplerian motion due to general relativistic effects associated with the sun's gravitational field: the sun's mass produces an appreciable 'curvature' in space–time (see also §1.8). The main effect (predicted by Einstein) is that the perihelion advances by a very small angle, so that the planet's orbit is not quite a closed ellipse. In the case of Mercury's orbit, the predicted advance is 43.0 arc seconds per century, much less than the advance of Mercury's perihelion produced by the gravitational pull of other planets (532 arc seconds per century) and the apparent shift that results from the precession of the earth's rotation axis (5600 arc seconds per century). However, this small amount can be measured by optical and radar means, and the amount is almost exactly (to an accuracy of 1%) what is predicted by general relativity.

The possibility has been raised that the sun is slightly oblate, i.e. its equatorial diameter is greater than its polar. R. H. Dicke and colleagues in the 1960s suggested that the difference was about 0.08 arc seconds, or 0.004% of the diameter. Solar oblateness is relevant as it also leads to a perihelion advance, the amount measured by Dicke and colleagues leading to an advance of 3.4 arc seconds per century for Mercury. Of the measured advance of 43.0 arc seconds per century, then, only 39.6 arc seconds can be attributed to general relativity, or 92% of what the theory predicted. However, more recent measurements indicate greatly reduced values for the oblateness, perhaps insignificantly different from zero, and it has been suggested that the Dicke result arose from an effect of solar activity, viz. the presence of an excess brightness at the equatorial limbs.

7.2 The solar wind

The outermost parts of the hot (K component) corona can be observed out to several solar radii from the sun (the F or dust corona to much more). We might ask where the corona ends and interplanetary space begins. Some evidence is provided by observations of comets passing by the sun. The most spectacular have tails stretching out for millions of kilometres away from the sun, consisting of dust and plasma 'boiled off' from a small solid nucleus. In 1953, Ludwig Biermann noticed that parts of a comet's tail would occasionally show accelerations along the tail's length, and he suggested that they were due to a continuous outflow of ions from the sun. According to this view, the corona exists even at the distance of such comets (typically a large fraction of the sun–earth distance, or 1 AU) and seems to be in a general state of expansion.

A little later, Sydney Chapman made a theoretical investigation of an entirely static corona, in particular the way in which its temperature and density would vary at large distances from the sun. This was still several years before it was fully realized that the corona was far from being a structureless layer of the solar atmosphere. The investigation assumed hydrostatic equilibrium, i.e. gas pressure balanced by gravity, and that the corona at large distances from the sun was heated by conduction from the million-degree, inner part of the corona. With the density of protons (or electrons) assumed to be $10^{14}/m^3$ for the inner corona, it was estimated that, extrapolated to a distance of 1 AU, the density was 10^8 to 10^9 protons/m^3 (agreeing with what Biermann had derived from comet tail accelerations) and the temperature 200 000 K. But what was most surprising was that even at much larger distances, indeed at the edge of the solar system, the slow decline of density with distance from the sun implied that the gas pressure was 10^{-4} Pa, ten million times more than what was thought to be the pressure of the gas between stars, the 'interstellar medium'.

E. N. Parker in 1958 proposed that the inconsistency of these pressures could be explained by abandoning the notion of hydrostatic equilibrium: i.e. supposing that the sun's gravitation is unable to hold back the pressure force of the extremely hot coronal gas, leaving the corona to expand freely. With similar assumptions to those used by Chapman, Parker was able to derive four families of solutions characterizing the outflow of the outer solar atmosphere. Figure 7.2 shows these solutions graphically, being plots of expansion speeds against radial distance from the sun. Three of these could be eliminated by the fact that they did not correctly describe the behaviour of the extreme inner or outer regions, leaving one possible solution (class 2 in the figure). This solution predicted an expansion with flow speeds that increased continuously with distance from the sun. On the assumption that coronal temperature is constant with radial distance, expansion speeds like those plotted in Fig. 7.3 result, each curve corresponding to a particular temperature. The curve for 2×10^6 K, for instance, indicates flow speeds of 700 km/s at 1 AU. This is considerably greater than the sound speed for this temperature (about 150 km/s), so the flow at 1 AU must be supersonic. Since the expansion speed increases from zero near the sun to this value, there must be a radial distance at which the expansion starts to become supersonic. Taking the expansion to begin at about 1.4 solar radii from the sun's centre (or 1.4 $R_{\odot}$), this level is at about 5.6 $R_{\odot}$. It is this supersonic flow that Parker called the 'solar wind'.

Confirmation of the solar wind's existence came a few years after Parker's analysis, with the NASA *Mariner 2* space probe to Venus in late 1962. Soviet

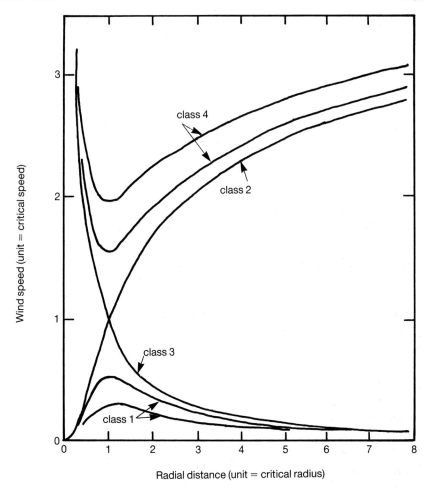

Fig. 7.2 Parker's solutions for the solar wind, describing wind speed with radial distance from the sun. The critical radius is approximately the heliocentric distance beyond which the wind flow exceeds the speed of sound and the critical speed is the wind speed at the critical radius. Only the class 2 solution correctly describes the wind's behaviour at very small and very large distances. (After Parker (1963))

spacecraft had earlier detected ionized gas in interplanetary space, but the instrumentation was not able to distinguish whether it was moving or stationary. The *Mariner 2* measurements over three months indicated a flow of plasma with widely varying speeds (about 300 to 700 km/s). Measured proton densities were far less (about $5 \times 10^6/m^3$) than what had been inferred by Biermann from comet tails.

We might expect that the magnetic field of the corona would be carried out by the solar wind. An 'interplanetary magnetic field' has indeed been

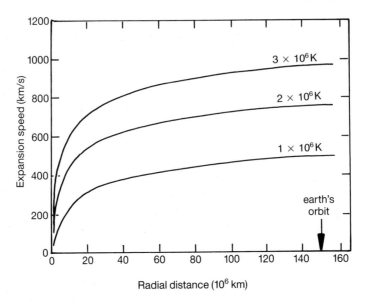

Fig. 7.3 Calculated expansion speed of the solar wind plotted against radial distance
for the three temperatures indicated. (After Parker (1963))

observed and measured by spacecraft magnetometers. Its strength at 1 AU is
about 5 nT. As with the corona, the field is frozen into the expanding plasma,
with field lines dragged out into interplanetary space. But, because the sun
rotates, the field lines form a spiral pattern (Fig. 7.4). For an observer on the
earth or other planet watching the rotating sun, the plasma expands radially,
while the field lines form spirals with the footpoints anchored to the sun itself.
(For an observer situated on the solar surface viewing interplanetary space,
however, the plasma does not flow radially but rather along the spirals formed
by the magnetic field.)

    The solar wind is heated by conduction from the non-expanding inner
corona and by other processes such as MHD wave dissipation that are
presumed to be the same as those heating the inner corona. Heat conduction
in a magnetized plasma occurs along field lines only, not across them (§5.5),
and so the heating of the solar-wind plasma by conduction of the hot inner
corona proceeds along the magnetic field spirals. Now, the heat conductivity
for electrons is much larger than for protons, so that electrons are heated
much more effectively than protons. In the relatively dense inner corona, and
even denser active region and flare loops, collisions between electrons and
protons occur often enough to ensure exact equality of the temperatures
describing the energy distribution of electrons and protons. However, the
solar-wind density at radial distances comparable to that of the earth is
extremely low, and so correspondingly is the frequency of collisions. This
allows the electron temperature to exceed the proton temperature. Rough

averages measured near the earth on occasions when the solar wind has a relatively low speed (about 300 km/s) are 150 000 K for the electron temperature, 40 000 K for the proton temperature.

Measurements of the solar wind are possible with ground-based radio receivers called interplanetary scintillation detectors. They observe the scintillation, or 'twinkling', of radio signals from very distant objects like galaxies and quasars. The twinkling of starlight that may be seen on any clear night is due to turbulence in the earth's atmosphere, but the scintillation of radio signals from distant objects arises from the passage of fast-moving plasma in interplanetary space, between the distant object and the observer. The detector at Cambridge University, designed by Antony Hewish, consists of over two thousand dipole antennas connected with wires and covering a large field (sheep are used to crop the grass!). It has recently been overhauled, and is being used for monitoring regions of enhanced density in the solar wind as they pass by the earth.

An extensive detector at the Kashima Space Research Centre in Japan has allowed M. Tokumaru and colleagues to see the acceleration region of the

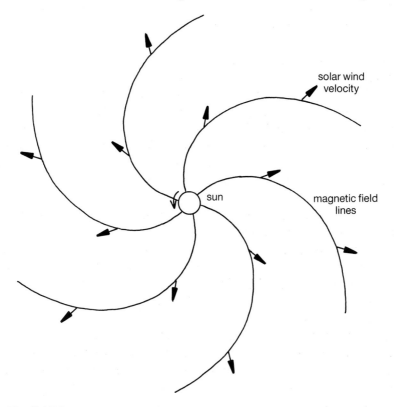

Fig. 7.4 Solar wind velocity (short arrows) and magnetic field lines as seen by a planet like the earth. Because the sun rotates, the field lines emerge as spirals.

solar wind (Fig. 7.5). By observing the scintillation of various radio sources when near the sun, they have been able to determine the solar-wind speed at very small heliocentric distances. The speed shows a sudden increase between 10 and $30\,R_\odot$, suggesting that this is where the solar wind is accelerated. Such distances are larger than in Parker's original model. Enhanced turbulence in this region is explained by Alfvén waves.

Measurements of the interplanetary field show that it has an orderly structure near solar cycle minimum; field lines alternate direction (towards or away from the sun) every few days, the reversals being quite sudden. The pattern is

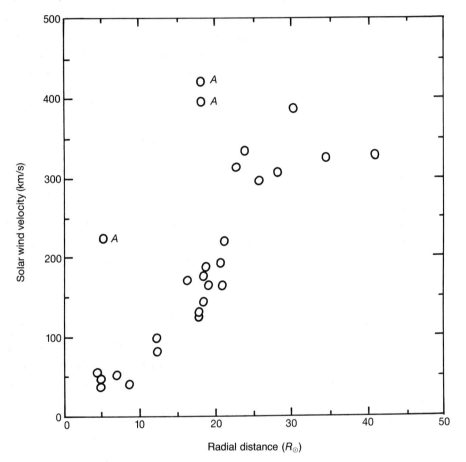

Fig. 7.5 Solar wind velocity at various radial distances from the sun's centre determined by interplanetary scintillation studies with the Kashima 34-m radio telescope in Japan. The points correspond to particular cosmic radio sources observed when in the vicinity of the sun. Most show a strongly increasing trend with radial distance, indicating that the region of the solar-wind acceleration occurs between 10 and $30\,R_\odot$. The three high points (marked A) are thought to be related to the passage of interplanetary disturbances. (Courtesy M. Tokumaru *et al.* (1991))

repeated every 27 days, the synodic rotation period of the sun. These single-polarity regions are known as 'magnetic sectors', and are related to the occurrence of 'high-speed streams', or parts of the solar wind with enhanced flow speeds. These were first observed with the *Mariner 2* spacecraft in 1962, and were also found to have a 27-day periodicity. When a high-speed stream passes by a spacecraft near the earth, the flow speed sharply increases, from about 300 to 700 km/s or more. The flow speed generally remains at this high value for a few days, then rather more slowly declines. This is illustrated by Fig. 7.6, showing flow speeds measured from the *Vela* and *Interplanetary Magnetic Platform* (*IMP*) spacecraft over three-month intervals between 1967 (near solar maximum) and 1974 (near solar minimum). The presence of high-speed streams becomes progressively more marked as solar minimum approaches. Measured particle densities increase at the onset of a high-speed stream at a spacecraft, but soon decrease to the level before the stream has completely passed. Proton temperatures, on the other hand, are enhanced over the entire passage of the stream. These facts can be explained by a stream emanating from the sun along a spiral path (as in Fig. 7.4), sweeping past the earth every solar rotation: as the stream's speed is enhanced relative to the remaining solar wind, the expanding material within the stream over-takes the more slowly moving solar wind, pushing against it and producing a zone of compressed, heated plasma at its leading edge (the 'interaction' region) and a zone of rarefied plasma following.

Figure 7.6 shows that, in 1974, the solar wind was dominated by two well-defined high-speed streams, with maximum flow speeds of 750 km/s. The interplanetary magnetic field similarly showed a simple pattern of two sectors. It was around this time that the *Skylab* mission operated, and data from its instruments enabled the M-regions to be identified with coronal holes. A giant high-speed stream seen in early 1973, with flow speeds of 700–750 km/s and persisting for several solar rotations, was found to be strongly associated with a large coronal hole observed in May 1973. The hole was a low-latitude extension of the south polar hole. The giant high-speed stream was found to emerge from a range of longitudes on the sun corresponding to this polar hole extension. The magnetic polarity of both the high-speed stream and the hole was negative, suggesting that the interplanetary field lines connected with those of the hole on the sun, and that plasma flowed out from the hole to become the giant high-speed stream.

Solar activity significantly alters the solar wind and produces interplanetary disturbances which consist of ejected material usually preceded by a shock front. They have been detected by spacecraft, both earth- and sun-orbiting. Thus, two German spacecraft, *Helios 1* and *Helios 2*, launched in 1976 into eccentric orbits round the sun with heliocentric distances of between 0.3 and 1.0 AU, observed moving clouds of gas with instruments designed to detect zodiacal-light emission at large distances from the sun. Particle detectors and magnetometers on various interplanetary space missions have also observed sudden enhancements of energetic protons or changes (in direction and intensity) of the interplanetary magnetic field that are most likely to be clouds of gas ejected from the sun. Figure 7.7 shows the sharp variations of density and velocity of protons and magnetic field recorded by the *ISEE-3* spacecraft (located just inside the earth's orbit) as a shock front passed across in a two-day period in August 1979. Interplanetary scintillation devices have also detected such disturbances (see Fig. 7.5: the three points with exceptionally high wind speeds indicate their presence). Images of disturbances are being made routinely with the Cambridge receiver as they travel outwards from the sun.

It continues to be a moot point how interplanetary disturbances are related to specific solar activity, e.g. flares, disappearing filaments or coronal mass ejections. The difficulty is that there is some uncertainty in the travel time to

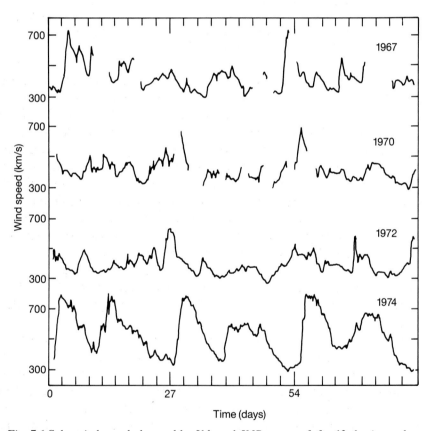

Fig. 7.6 Solar wind speed observed by *Vela* and *IMP* spacecraft for 63-day intervals in four separate years, showing the development of high-speed streams as sunspot minimum approaches. (Bame *et al.* (1976). Reprinted with permission from *The Astrophysical Journal*)

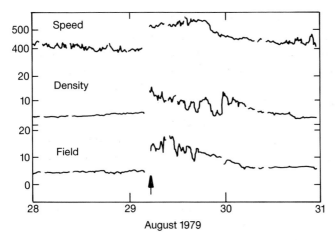

Fig. 7.7 Solar wind speed (km/s), particle density (in 10^{-6} particles/m^3) and magnetic field strength (nT) over a three-day period in August 1979, showing the sudden increase of each when an interplanetary shock passed the *ISEE-3* spacecraft which took the measurements. (After Gosling and McComas (1987); © American Geophysical Union)

either the earth or spacecraft where it is detected since the cloud's trajectory is not known precisely. The travel time is from one to a few days, so there is presumably some spiralling motion as with the normal solar wind or the high-speed streams. Though one can roughly estimate the time a cloud left the sun from the cloud's speed at the earth and assuming it to have been constant during its trajectory, there is usually so much activity occurring on the sun that it is impossible to pinpoint which particular event gave rise to the cloud. There is not much consensus on the topic at present, with Professor Hewish, for example, believing that coronal holes are the sources of the disturbances. But there is growing evidence that coronal mass ejections observed in the low corona can be identified as interplanetary clouds. As we saw in the last chapter (§6.4), CMEs are frequently associated with seemingly insignificant filament eruptions rather than the large flares that often follow; this may explain some of the past confusion in relating interplanetary clouds with flares.

Bearing the above in mind, we will try to envisage how a mass ejected by the sun propagates into interplanetary space. It travels out at a speed of around 500 km/s, and in doing so either moves against the more slowly expanding ambient solar wind, so compressing it, or else pushing it aside. The interplanetary magnetic field lines are likely to be compressed also, and probably 'drape' round the ejected mass. The mass may well be moving supersonically, in which case a shock will form ahead of it. The material ejected from the sun moves into a magnetized, almost fully ionized plasma, for which not only the local sound speed but also the Alfvén speed are important. Both these speeds decrease going out from the sun. In the low corona, the sound speed is about 150 km/s, and the Alfvén speed is 500–1000 km/s (the Alfvén speed depends on magnetic field strength and density). A mass ejection travelling at 150 km/s or more, then, will be supersonic, but its speed may be less or more than the Alfvén speed. If more, and the motion is along field lines, a fast MHD shock forms, shaped like a sonic shock, the outward-facing surface being convex; if less, a slow MHD shock with concave outward-facing surface forms. It is tempting to identify the bright, leading rim of the CME (which is always observed to be convex) as this shock front. Further into interplanetary space, the sound and Alfvén speeds decrease, so that a mass ejection which has sustained its original speed (and most seem to do so) is more likely still to have a fast MHD shock.

The enhancements of magnetic field and plasma density associated with interplanetary disturbances give rise to a scattering of cosmic rays that enter the solar system, produced perhaps by supernova explosions or other energetic events in our galaxy. There is, as a result, a decrease (known as a *Forbush* decrease) of galactic cosmic-ray intensity as a disturbance passes by the earth. The cosmic-ray intensity is for this reason *anti*-correlated with the solar eleven-year cycle.

The solar wind, as we have seen, can be thought of as the outer corona in a state of steady expansion. As the corona is not entirely composed of hydrogen but has trace amounts of other elements, notably helium, we should expect the solar wind to be composed not only of protons and electrons (i.e. fully ionized hydrogen) but small amounts of the nuclei of other elements. These have been detected using both analysers on interplanetary spacecraft and with pieces of aluminium foil left on the lunar surface during the *Apollo* moon landings, which, being directly exposed to the solar wind (the moon has no atmosphere), collected various ions whose impacts in the foil were measured on return to earth. From such measurements, the solar-wind composition does not seem to be too different from that in the low corona or photosphere, determined spectroscopically. Helium provides an exception, which is under-abundant in the solar wind (4% of hydrogen, compared with about 10% in the chromosphere or corona). It is usually detected in its fully ionized state (He^{2+}). However, there are surprisingly wide variations, particularly during activity-related disturbances, when the helium-to-hydrogen ratio is as much as 40%. There are occasions when the abundance of the ^3He form of helium (two protons, one neutron) is unaccountably high. Calculations suggest that helium will accumulate in the low corona, as it does not have the same tendency to escape into the solar wind as hydrogen, so that this may explain why the abundance of normal helium in the low corona is greater than that in the solar wind. There are also trace amounts of heavier ions such as oxygen and iron, but in a highly ionized state, reflecting the temperature of the inner corona from where they originated.

Because of the solar wind, the sun loses a small amount of mass and energy. The first can be estimated from the observed number of protons passing per second per square metre at 1 AU ('particle flux') and the second from the observed amount of energy (kinetic and potential) per second per square metre at 1 AU ('energy flux'). Both particle and energy fluxes are enhanced by the incidence of the high-speed streams and interplanetary clouds, though the particle flux by surprisingly little: perhaps between 1% and 20%. Rough averages are, for the particle flux, 3.6×10^{12} protons/m^2/s and, for the energy flux, 0.003 W/m^2. The total mass the sun loses in a year corresponding to this particle flux is 5×10^{16} kg, or only about 2.5×10^{-14} of the solar mass itself. The energy flux is also very small, only about one-millionth of the solar irradiance.

Until recently, most of our knowledge of the solar wind came from space-craft at distances of about 1 AU or less from the sun. The 1980s have seen measurements extended to more than 40 AU, beyond the distance of Neptune, notably by the NASA missions *Pioneers 10* and *11* and *Voyagers 1* and *2*. At these large distances, the spirals formed by the interplanetary magnetic field have wound round the sun several times, so that the field direction is much more nearly tangential. The high-speed streams which nearer the sun produce interaction regions as they collide with slower solar-wind flow tend to disappear further out leaving behind pressure waves. These have been observed out to 30 AU. The density here is extremely low – around 1000 particles per cubic metre.

Finally, we may ask where the region dominated by the solar wind – the so-called *heliosphere* – ends and interstellar space begins. The solar-wind press-ure decreases going outwards so there must be a distance where it can no longer balance that in interstellar space. At this distance, predicted to be around 100 AU (far beyond the known major planets), there is a 'terminal' shock front beyond which the solar wind becomes subsonic, as it is at extremely small heliocentric distances. Since the sun is moving in interstellar space, possibly an 'interstellar wind' blows against the heliosphere so that the terminal shock is not spherical but highly distorted. The *Pioneer* and *Voyager* spacecraft have not directly observed this transition, though there is tantaliz-ing evidence of its existence in the form of extremely low-frequency (3 kHz) radio signals seen from the *Voyagers*, and evidence also that the shock could be as close as 40 AU from the sun.

7.3   Interaction of the solar wind with the planets and comets

The solar wind, carrying with it the interplanetary magnetic field, flows out from the sun into the solar system and so encounters planets and other bodies orbiting the sun. How does the wind interact with them, particularly those planets like the earth that have magnetic fields of their own?

Measurements from spacecraft and the ground have given us a very detailed picture in the case of the earth, one which can be broadly applied to the other magnetized planets that have been explored by spacecraft in recent years. What happens is that, as the solar wind moves through the lines of the earth's field, electric currents are induced that alter the field's configuration, compressing it on the sunward side and causing the wind flow to be diverted, avoiding a region surrounding the earth. The result is a cavity, known as a *magnetosphere*, having a boundary (the *magnetopause*) that is nearly spherical on the sunward side but drawn out into a long cylinder, the *magnetotail*, on the night side (see Fig. 7.8). Expressed in units of the earth's radius ($1 R_E = 6371$ km), the distance of the magnetopause from the earth along the sunward direction is $10 R_E$ and the magnetotail's diameter is $50 R_E$. Outside the magnetopause, the solar wind continues to flow, but within it the terrestrial magnetic field dominates. The magnetosphere is filled with plasma having a large range of densities and temperatures; the charged particles constituting the plasma, unlike the particles making up the other regions of the earth's atmosphere, are bound by electromagnetic, i.e. non-gravitational, forces.

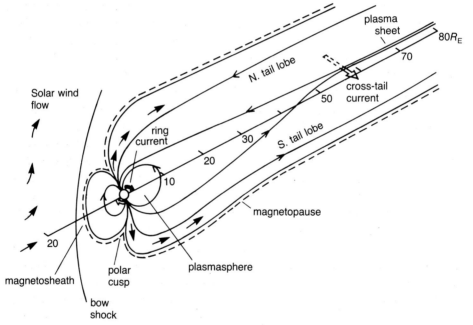

Fig. 7.8 Section through the earth's magnetosphere through N and S magnetic
   poles. Terrestrial magnetic field lines are pushed by the solar-wind flow to
   form the magnetosphere. Upstream of the sun-side magnetopause is the
   bow shock. Solar wind flow and plasma convection flows within the
   magnetosphere are indicated by the short arrows. A ring current flows
   round the earth and a cross-tail current flows across the magnetotail
   (currents shown by open arrows). The scale of the figure is shown by the
   diagonal line (units of earth radius, $1 R_E = 6371$ km).

Near the earth's surface, the terrestrial magnetic field is dipole-like, i.e. like that produced by a bar magnet near the earth's core, and is tilted by about $20°$ to the earth's rotation axis. On the sunward side of the magnetosphere, the terrestrial field lines are closed, as are those on the night side out to between 8 and $15\,R_E$ from the earth. The density of the plasma filling this, the *inner magnetosphere*, decreases sharply above a level corresponding to the magnetic field lines that are 4–5 R_E distant from the earth along the sun–earth line. Inside this level, the plasmapause, is a region called the *plasmasphere* where the gas has a density of 10^7–10^8 particles/m^3 and originates from the ionosphere, a region of the earth's atmosphere that is partly ionized by the sun's radiation (§7.5). Outside the plasmapause the density is only 10^6 particles/m^3. Much of this plasma, too, is ionospheric, but particles are present with much higher energies that come from the solar wind, particularly via the polar regions, or *cusps*; the reason is that some field lines at the cusps connect with regions very near the magnetopause, allowing the solar-wind particles to leak into the magnetosphere.

In the magnetotail beyond about 8–$15\,R_E$, the field lines are open in their normal configuration and parallel to the earth–sun direction. The magnetotail has a northern *lobe* with field lines pointing towards earth, a southern lobe with field lines pointing away from earth, and an intermediate region called the *plasma sheet* where oppositely directed field lines lie close to each other forming a cross-tail current, in the form of a sheet. The plasma sheet eventually converges to a neutral line some $100\,R_E$ from the earth. It is the site of various plasma processes giving rise to geomagnetic and auroral phenomena: we will deal with these in §7.4. Because of the tilt of the earth's magnetic axis to its rotation axis and the tilt of the rotation axis relative to the orbital plane, the plasma sheet is generally north or south of the ecliptic plane by up to $4\,R_E$.

The magnetopause is actually a layer with a thickness of several hundred kilometres rather than a sharp boundary to the magnetosphere. There is slight motion in it according to the state of the solar wind, and in particular there is a rippling motion along the magnetotail boundary.

Ahead, or 'upstream', of the magnetopause by about $3\,R_E$ on the sunward side is a standing shock front, known as the *bow shock*, whose presence is due to the fact that, relative to the solar-wind flow, the earth and its magneto-sphere are moving faster than either the sound or Alfvén speed. Its properties, like the magnetopause, depend on the solar wind, in particular the interplanetary magnetic field carried with it: when the direction of this field is almost parallel to the shock, the shock front is thin and well defined, but when nearly perpendicular it is rather wide and ill-defined. Some solar-wind particles incident on the shock are scattered back into a sunward direction. Between the bow shock and the nose of the magnetopause, a region known as the *magnetosheath*, the solar-wind flow continues at reduced speed, and is somewhat compressed and irregular.

There is a convection of plasma in the magnetotail as a combined result of a strong electric field set up across the magnetotail and the magnetic field there. The convected plasma moves towards the earth, then round it, leaving the magnetosphere on its sunward side. Some of the electrons and ions in this convected plasma become trapped in the field of the inner magnetosphere, following helical paths along the field lines and mirroring near the terrestrial magnetic poles. The motion between one mirror point and the other typically takes a fraction of a second. Owing to the decrease of field strength going away from the earth, there is a drift motion, with electrons drifting eastwards, ions westwards (Fig. 7.9). This separation of charge sets up a current – the *ring current* – flowing round the earth. The direction of an electric current is conventionally in the opposite direction to the flow of electrons, so the ring current flows westwards.

Our knowledge of the interaction of the solar wind with most of the major planets of the solar system has been investigated by several spacecraft, and they have revealed a great variety of properties. Of the so-called terrestrial planets – Mercury, Venus and Mars – Mercury has a measurable magnetic field, producing a tiny magnetosphere, and Mars may have a weak field also. Venus appears to have none, so that the solar-wind flow encounters its atmosphere directly. Bow shocks have been detected upstream of both Venus and Mars.

The giant planets – Jupiter, Saturn, Uranus and Neptune – have been explored by the *Pioneer 10* and *11* and *Voyager 1* and *2* spacecraft over the past 15 years or so. All the giant planets have large magnetic fields with extensive magnetospheres. Of these, the most interesting is that of Jupiter. Its field is

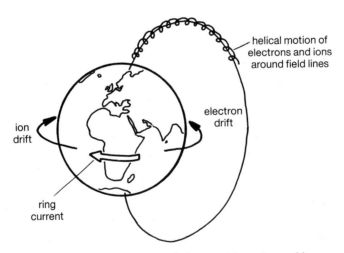

Fig. 7.9 Helical motion of ions and electrons along the earth's magnetic field lines. The electrons drift eastwards, the ions westwards, to produce the westward-flowing ring current (open arrow).

roughly dipolar, like the earth's, the dipole axis being inclined by $10°$ to the rotation axis. The field is much larger than the earth's, the field strength at the poles being $800\,\mu T$ (compared with $60\,\mu T$ for the earth). Measured in terms of what is called the dipole moment – equal to half the field strength at a pole times the planet's radius cubed – it is $20\,000$ times that of the earth, and is the largest in the solar system apart from the sun. The magnetosphere extends out to 100 Jupiter radii (R_J) on the sunward side, and the magnetotail is 300–$400\,R_J$ in diameter. A very large ring current causes the magnetosphere to be more flattened than the earth's. Jupiter rotates once every 10 hours, and whirls its giant magnetosphere around with it. Indeed, the energy supplied to the magnetosphere derives mostly from Jupiter's rotational energy rather than the solar wind, as with the earth's magnetosphere. Pictures from the *Voyager 1* encounter showed that the satellite Io, the innermost of the four largest satellites, with distance $6\,R_J$ from Jupiter, has several active volcanoes. The plumes from these volcanoes, containing mainly sulphur compounds, feed a hot 'plasma torus' encircling Jupiter; emission from this torus was detected spectroscopically from the earth before the *Voyager* mission. This is apparently the main source of plasma in Jupiter's magnetosphere.

Saturn has a weaker field (dipole moment 500 times the earth's) with only a $1°$ inclination to the rotation axis. There is no equivalent to Jupiter's Io torus. The magnetosphere is rather smaller than Jupiter's, less flattened and apparently variable in extent. Uranus' field is weaker still (dipole moment 43 times the earth's), but highly inclined (by $60°$) to the rotation axis and off-set from the planet's centre. The planet's rotation axis is itself highly inclined to the perpendicular to the orbital plane, and at the time of the *Voyager 2* encounter in 1986 was directed towards the sun. This resulted in the magnetotail being rotated about its axis once every 17 hours, Uranus' rotation period. Neptune's field is weaker than Uranus' and is only very roughly dipolar (dipole moment 26 times the earth's), with a $47°$ inclination to the rotation axis. At times, the magnetic poles are almost sunward, the plasma 'sheet' in the magnetotail then being cylindrical rather than sheet-like.

The interaction of the solar wind with comets gives rise to several different phenomena, one of which is the acceleration of features in comet tails which Biermann noted (§7.2). The central part of a comet is a solid but loosely bound nucleus, perhaps a few kilometres across, consisting of frozen water, dust and some complex molecules. When near the sun, the nucleus is heated and releases the complex molecules and dust which form a cloud, or 'coma', of up to $100\,000\,km$ diameter. The dust is forced away from the coma by the sun's radiation pressure, giving rise to the so-called dust tail. The complex molecules released are broken up by solar radiation and ionized; these ions interact with the solar-wind flow to form the 'ion' (or plasma) tail which may be present as well as the dust tail. The interplanetary magnetic field moving out with the solar wind interacts with the comet as indicated in Fig. 7.10: field lines drape round the ionized gas of the coma and are dragged back by the solar wind. Cometary ions then move along the draped field lines, forming the ion tail.

Changes in the solar wind, such as the onset of high-speed streams, are sometimes reflected in the shape of the ion tail which may show bends or kinks. A striking phenomenon observed with some comets is the disconnection of the ion tail and formation of the new one. This occurred with Comet Halley when it was near perihelion in 1986. Figure 7.11 shows the comet on two successive nights with the disconnected tail moving along the length of the smoother dust tail, stretching several degrees across the sky. These disconnection events are explained by a reversal of the interplanetary magnetic field as the comet crosses a magnetic sector. Figure 7.12 shows the

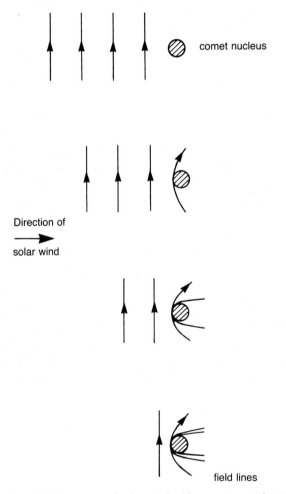

Fig. 7.10 Interaction of solar wind with cometary nucleus, with interplanetary magnetic field draping round nucleus (shown as the small hatched circle).

Fig. 7.11 Comet Halley on two occasions (9 and 10 March 1986), showing a disconnection event, with the separated tail moving away from the nucleus. (UK Schmidt Telescope Unit; © 1986 Royal Observatory, Edinburgh)

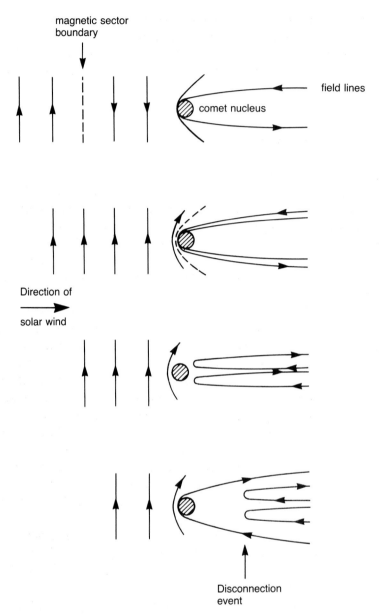

Fig. 7.12 Disconnection event in comet's ion tail explained by the interaction of two oppositely directed sets of interplanetary magnetic field lines, separated by a sector boundary (dashed line). The direction of the solar wind is indicated, and the comet's nucleus is the small hatched circle.

interplanetary magnetic field containing a sector boundary passing the comet's coma. Field lines drape round the nucleus as in Fig. 7.10, but when those with reversed direction encounter the nucleus, reconnection of field lines occurs, resulting in a separation of one ion tail and the formation of a new one.

7.4 Geomagnetic disturbances and the aurora

The magnetic field of the earth near its surface is made up of internal and external components. The internal or main field is the dipolar field that probably originates with electric currents associated with fluid motions in the earth's core, and varies only very slowly. There are local variations of this field over the earth's surface that are due to mineral deposits. The external field is due to electric currents flowing in the earth's ionosphere and magnetosphere, and may show rapid variations in time, amounting to as much as 10% of the internal field, but usually much less.

Measurements of the earth's magnetic field have been made since the time of Gauss in the 1830s (§1.9). There is now a world-wide network of almost 200 magnetic observatories, with three major ones in Britain (Hartland, N. Devon; Eskdalemuir, Dumfriesshire; and Lerwick in the Shetland Islands). The field strength and direction are measured by three magnetic elements, horizontal H and vertical Z components of field strength and declination D (i.e. the direction of compass north with respect to true north). The mean values of each for Hartland in 1987, for instance, were $H=19\,\mu T$, $Z=44\,\mu T$ and $D=6.6°$ west. Variations are continuously recorded, and the data summarized in the form of character indices, rather more detailed versions of the C index mentioned in §1.9. Thus, observatories record a K index, representing the amount of variation in a three-hour period in the H, D and Z elements on a scale of 0 (smallest variations) to 9 (largest variations), and a 'planetary' index, K_P, is derived from an average of the K indices from selected observatories distributed over the earth. Another index, the A_P index, describes the daily magnetic activity from the actual variations of field strength recorded at various observatories.

The variations of field at a particular point are both regular and irregular. The regular variations include a daily one which results from atmospheric tides due to solar heating and (to a smaller extent) the gravitational pull of the moon. Each causes atmospheric winds that move a conducting layer in the lower ionosphere (at an altitude of 110 km) across the steady terrestrial field, a motion that induces electric currents. The fields associated with these currents give rise to variations in the total field as measured by an observer on the earth's surface rotating beneath the current systems. Those due to the sun are known as the 'solar quiet' (or S_q) variations and those due to the moon the 'lunar quiet' (L_q) variations. On a magnetically undisturbed day in Britain, for example, the daily variations in declination consist of a slow oscillation with a range of about 12 arc minutes that is most easterly early in the morning, most westerly just after midday; at night there is hardly any variation. Figure 7.13 shows the changes for the three magnetic elements at Hartland on a magnetically quiet day.

Other quasi-regular variations occur over 27-day periods, associated with solar active regions or high-speed solar-wind streams; over six-monthly intervals, with maxima near the spring and autumnal equinoxes, whose origin is uncertain; and an eleven-year period associated with the solar activity cycle.

The irregular variations are associated with solar activity. Small changes in the geomagnetic field, consisting of a sudden initial phase and more gradual return to the pre-flare field, are known as crochets. They occur in local daylight hours only, and the field changes during them are generally an enhancement of the daily (S_q) variations at the time of their occurrence. Crochets are ascribed to increased ionization in the lower ionosphere produced by soft X-ray emission (wavelengths less than about 1 nm) during solar

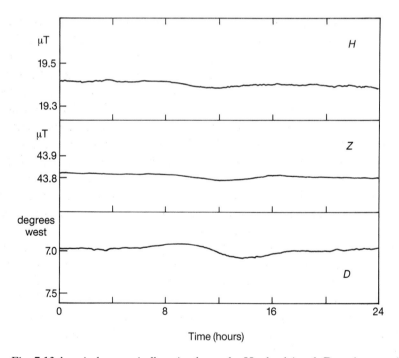

Fig. 7.13 A typical magnetically quiet day at the Hartland (north Devon) magnetic observatory. The plots show the variations in the magnetic elements H and Z (horizontal and vertical magnetic field strength in μT) and D (the magnetic declination, or direction of compass north with respect to true north) on 21 March 1985. (Courtesy British Geological Survey)

flares: this increased ionization causes an intensification of the currents giving rise to the S_q variations. A typical change in either the H or Z elements during a crochet observed at Hartland is a few tens of nanoteslas, or less than about 0.1% of the total field strength.

Much larger than the crochets are the *magnetic storms*. They are associated with solar-wind disturbances, such as mass ejections and high-speed wind streams, in which there are enhanced plasma densities and speeds. When a solar-wind disturbance arrives at the earth's magnetosphere, there is often a sharp increase in field strength (by about 10 nT), known as a sudden commencement, with the increase maintained for around 30 minutes. This field change is attributed to a compression of the geomagnetic field by the enhanced solar wind, such that the distance of the magnetopause from the earth in the sunward direction is decreased from the usual value of $10 R_E$ to about $6 R_E$. A shock wave precedes the wind disturbance. A magnetic storm is world-wide, i.e. unlike a crochet it does not occur only on the sunlit side of the earth, although the sudden commencement is not simultaneous at all observatories; rather, there is a spread of a few minutes in its occurrence, attributable to the finite time the wind disturbance takes to encompass the earth.

After this initial increase in the geomagnetic field, there is a prolonged period when the field is extremely disturbed, with variations of several 100 nT (or about 0.2% of the total for a British observatory). The field finally returns to its pre-storm levels after a day or so. Quite often, the field strength shows a *decrease* after the sudden commencement. This is illustrated by the changes in the field strength components H and Z, shown in Fig. 7.14, during an enormous storm beginning 13 March 1989, associated with a large solar flare with X-ray rating X4 on 10 March. The field decrease after the sudden commencement is due to an enhancement in the ring current encircling the earth. The number of trapped particles giving rise to the ring current is greatly increased during a magnetic storm because of an injection of solar-wind particles as well as the upward acceleration by electric fields of particles (e.g. oxygen ions) from the earth's ionosphere.

The great storm in March 1989 was the second largest since 1932 as measured by the A_P index. Large magnetic storms give rise to induced currents in electric power lines, which can lead to voltage and frequency variations and the tripping of protective relays in power systems. The March 1989 storm caused voltage reductions on distribution lines in Sweden and a complete failure of the Hydro-Quebec power system in Montreal which was not restored for several hours.

Unlike magnetic storms, *substorms* are magnetic disturbances occurring in a limited region of the earth, near the geomagnetic poles. They last for about half an hour, with a burst-like initial phase, and may occur within the main phase of storms or in otherwise undisturbed periods. They are strongly

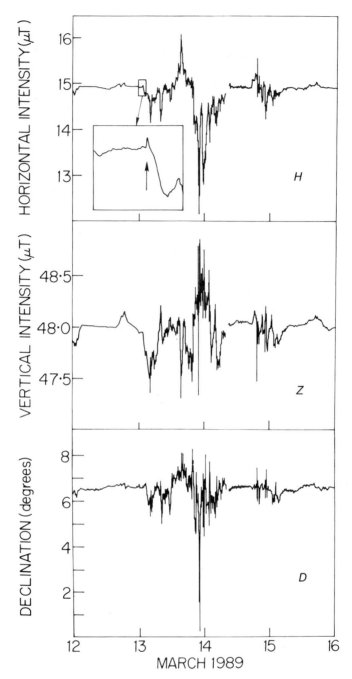

Fig. 7.14 Variations in *H*, *Z* and *D* during the great magnetic storm of 13/14
March 1989 recorded at Lerwick Observatory. The sudden
commencement is indicated in the *H* variations. There was a decrease of
total field strength immediately after this, followed by large, extremely
rapid changes. (Courtesy British Geological Survey)

correlated with the direction of the interplanetary magnetic field carried by the solar wind. When the component of this field parallel to the earth's magnetic axis is directed northwards, substorms are infrequent and weak; but on a reversal, i.e. at the boundary of a magnetic sector, when the field's component parallel to the earth's magnetic axis is southwards, there are many strong substorms. The reason for this is thought to be reconnection of field lines on the sunward side of the earth's magnetosphere. Figure 7.15, illustrating this, shows the interplanetary magnetic field, considered newly reversed to a south–north direction, incident on the magnetosphere. It permeates within the magnetopause to allow reconnection with the geomagnetic field. As the solar wind carries the interplanetary field past the earth, the points of recon-

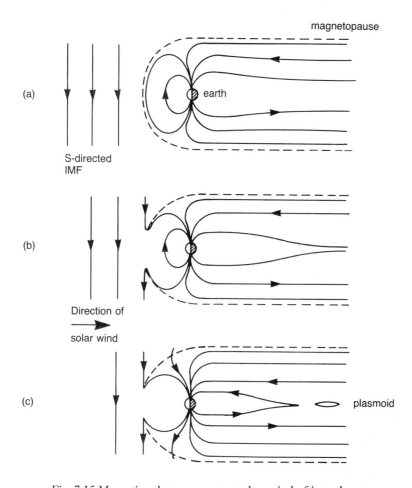

Fig. 7.15 Magnetic substorms occur on the arrival of interplanetary magnetic field (IMF) from the sun which is southward-directed. This is explained by the erosion of terrestrial field on the sunlit side of the earth and addition of field to magnetotail lobes, with the creation of plasmoids in the magnetotail.

nection pass from the sunlit side to the magnetotail. Thus, magnetic flux is 'peeled off' or eroded on the sunlit side and transferred to the lobes of the magnetotail. There is no corresponding situation for interplanetary field with component parallel to the earth's magnetic axis that is directed northwards.

A substorm is thought to be initiated by processes in the plasma sheet. A thinning of the sheet occurs, allowing reconnection that results in plasma convected both towards the earth on the earthward side of the reconnection site and away from the earth in the form of plasmoids (see Fig. 7.15c). This accords with spacecraft observations of fast plasma flows during substorms.

The most obvious effect of a substorm is the occurrence of auroral displays near the north and south magnetic poles (Fig. 7.16). As was discovered in the last century (§1.9), aurorae are most frequent not at the magnetic poles themselves but in almost circular zones some 4000 km in diameter centred on the magnetic poles, known as the auroral ovals. The visible emission seen in an aurora occurs 90–130 km above the earth's surface, and is due to the excitation of atoms in the atmosphere by energetic electrons. The characteristic red and green colours of strong auroral displays arise from spectral lines of neutral oxygen atoms in the earth's atmosphere, with wavelengths of 630.0 nm (red) and 557.7 nm (green).

Fig. 7.16 A discrete aurora photographed just before dawn at the Andoya Rocket Range in northern Norway. (Courtesy T. Edwards, Rutherford Appleton Laboratory)

Techniques in the observation of aurorae have advanced greatly in recent times. Descriptions and photographs of aurorae cannot do justice to the extraordinarily rapid changes in auroral forms, which have been captured using all-sky survey cameras at locations within the auroral zone. The entirety of an auroral display cannot be seen from a single station, however, and use has therefore been made of instruments that image the earth from spacecraft. Spectacular results have been obtained from one of these, an imaging photometer made by the University of Iowa, on the NASA satellite *Dynamic Explorer 1* (*DE1*). This satellite was launched in 1981 with a companion satellite *DE2*, carrying several instruments to study particles and fields in the magnetosphere; the orbits of both spacecraft are polar, that of *DE1* being highly elliptical (largest distance from earth $3.6\,R_E$) while *DE2* has a much lower orbit. The imaging photometer on *DE1* has ultraviolet and visible spectral ranges. A series of images of the earth in ultraviolet light is shown in Fig. 7.17, taken as *DE1* was approaching the most distant point in its orbit. The daylight side of the earth appears light because of strong ultraviolet lines emitted by neutral oxygen in the atmosphere, while the glow beyond the earth is due to the hydrogen Lyman-α line, emitted by the sun and scattered by diffuse hydrogen atoms in the earth's exosphere (see §7.5). The bright circle is the auroral oval near the earth's north magnetic pole. A global view of auroral activity, then, is possible by such imaging from a spacecraft.

Images of the auroral oval during substorms from *DE1* have been particularly revealing. The auroral oval consists of a poleward (inner) and equatorward (outer) part, as illustrated in Fig. 7.18, a view of the southern auroral oval in ultraviolet wavelengths. The poleward edge is formed by so-called discrete aurorae, which mark the footpoints of magnetic field lines forming the boundary layer of the plasma sheet in the magnetotail (Fig. 7.8). The equatorward edge is formed by diffuse aurorae, marking footpoints of field lines that are much closer to the earth, associated with the ring current. A substorm is observed to start with a brightening at a particular location in the discrete auroral edge that rapidly expands and intensifies. The aurora moves polewards at speeds of a few hundred metres per second. The poleward expansion occurs as the site of reconnection in the magnetotail recedes from the earth causing a thickening of the plasma sheet, with field lines forming its boundaries connecting to higher latitudes at the earth's surface. Often, a westward 'surge' of auroral activity occurs. Soon after the discrete auroral activity, the diffuse edge brightens and expands eastwards. During a substorm's recovery stage, the auroral oval's intensity and latitudinal width diminish gradually.

The development of an auroral display is now known to be strongly related to a complex system of electric currents connecting the magnetosphere with the ionosphere. Figure 7.19 shows this system schematically. One current system (inner field-aligned currents) enters the ionosphere at the dawn side of the poleward boundary of the auroral oval, and leaves it on the dusk side. It flows along field lines that directly connect to the solar wind. A second, weaker system (outer field-aligned currents) leaves the ionosphere at the outer edge of the auroral oval on the dawn side, enters it on the dusk side, again travels along field lines and is due to plasma convection towards the earth from the magnetotail. Between these current systems are the eastward and westward auroral electrojets, current filaments that flow parallel to the earth's surface along the night side of the auroral oval. It is the intensification

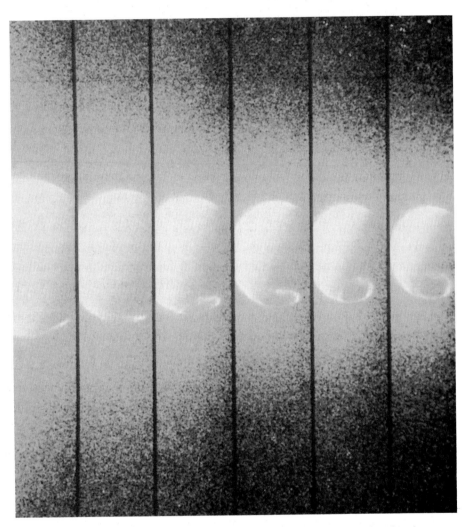

Fig. 7.17 Sequence of six *Dynamic Explorer* images showing the auroral oval at the north pole. The diffuse emission around the earth is due to solar Lyman-α emission scattered by neutral hydrogen atoms in the earth's exosphere, forming a 'geocorona'. (Frank and Craven (1988); © American Geophysical Union)

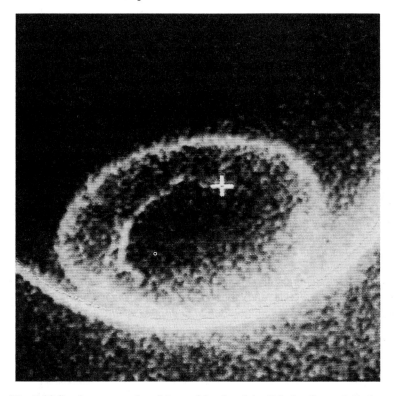

Fig. 7.18 Southern auroral oval imaged in ultraviolet light by *Dynamic Explorer*, showing the discrete (poleward edge) and diffuse (equatorward edge) aurorae. (Frank and Craven (1988); © American Geophysical Union)

of the auroral electrojets during a substorm that is strongly associated with the westward surge of discrete aurorae and the eastward expansion of diffuse aurorae.

When the interplanetary magnetic field is directed northwards, substorm activity as has been mentioned is almost absent, but *DE1* images at such times have shown the auroral oval to take on a remarkable appearance: see Fig. 7.20. A bright arc occurs across the polar cap from the day to the night part of the oval, so that a configuration like the Greek letter theta (θ) is formed. The transpolar arc possibly arises from distortion of the magnetotail plasma arising from the northward-directed interplanetary magnetic field.

7.5 The sun and the earth's atmosphere: the ionosphere

An obvious example of the sun's control of the solar system as far as human life is concerned is the influence on the earth's atmosphere and in particular on climate and weather. The amount of solar radiant energy per second received at the top of the earth's atmosphere at the point where the sun is overhead (the 'sub-solar point') is $1368\,W/m^2$, the solar irradiance (§§1.10, 9.1). Averaged over the entire surface area of the earth, night as well as sunlit hemispheres, this is reduced to one-quarter, or $342\,W/m^2$. Of this amount, some 24% ($82\,W/m^2$) is absorbed by gases making up the atmosphere (mostly water vapour but also carbon dioxide), dust and clouds. About 30% of incoming radiation is reflected by clouds or particles in the atmosphere, or by the earth's surface, particularly where covered by snow and ice: this amount is known as the earth's albedo. The remainder, around 45% or $153\,W/m^2$, is absorbed by the sunlit side of the earth's surface which is warmed as a result. The measured mean surface temperature is around $15\,°C$. At such a temperature the earth radiates like a black body, with this radiation peaking in the infrared part of the spectrum. The earth's average surface temperature is not too different between day and night hemispheres, so this radiation is emitted by the entire earth's surface. The amount emitted is far larger – $384\,W/m^2$ – than the $153\,W/m^2$ directly absorbed from the sun.

This apparent contradiction of energy conservation is largely explained by

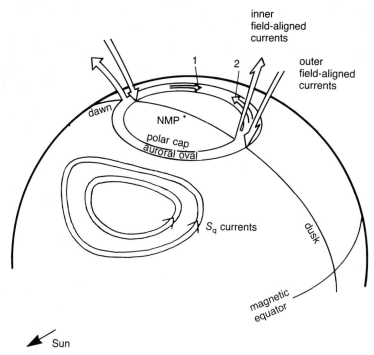

Fig. 7.19 The current system connecting the earth's magnetosphere and ionosphere. The field-aligned currents connect to the auroral oval on the dawn and dusk side of the earth, and the auroral electrojets (current filaments) flow along the night side of the auroral oval (1=W electrojet, 2=E electrojet). NMP=North magnetic pole. The S_q currents that flow on the day side give rise to the daily variations of magnetic elements.

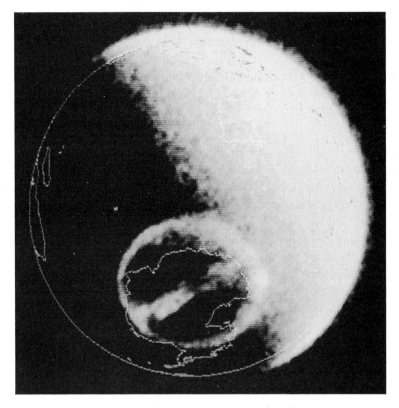

Fig. 7.20 *Dynamic Explorer* image of a theta aurora over the southern pole (land areas have been added). (Frank and Craven (1988); © American Geophysical Union)

the fact that most of the infrared radiation is absorbed by the atmosphere, in particular by the so-called *greenhouse* gases, of which carbon dioxide (CO_2) and water vapour (H_2O) are the most important. They have a similar property to that of glass in a greenhouse: they transmit visible radiation from the sun but absorb infrared radiation emitted by the earth. Being good absorbers, by Kirchhoff's law, they are also good radiators, and much of the infrared radiation that they absorbed from the earth is returned to it, amounting to $330\,W/m^2$, while $181\,W/m^2$ escape to space. Energy is also lost by the earth's surface to the atmosphere by the evaporation of surface water and by the warming of air next to the ground by conduction: this amounts to $99\,W/m^2$.

There is, then, quite a complicated interchange of the energy received from the sun between the earth's surface and its atmosphere. We can clarify this by writing the energy balance out in full. Firstly, for the earth's surface, we have:

Energy gains ($153\,W/m^2$ incident from sun$+330\,W/m^2$ received from atmosphere)=energy losses ($384\,W/m^2$ infrared radiation+ $99\,W/m^2$ water evaporation and warming of air by conduction);

and, secondly, for the earth's atmosphere:

> Energy gains (82 W/m^2 incident from sun+330 W/m^2 received from earth's surface+99 W/m^2 received from the earth by evaporation and conduction)=energy losses (181 W/m^2 infrared radiation lost to space+330 W/m^2 infrared radiation emitted towards earth and absorbed by it).

Incidentally, if the atmosphere did not absorb the earth's infrared radiation, the earth's surface would be much colder – about −45 °C – and could not support most life-forms.

The above represents the average situation for the earth, and shows that there is, after all, a balance of received solar radiation and energy lost by the earth and its atmosphere: if this were disturbed (by, e.g., a change in the solar radiant energy), the earth would heat or cool to produce a new balance. However, there are wide variations from this simplified picture, both from place to place over the earth's surface and in time. Thus, the amount of solar energy received at a particular location depends on various factors, most especially its latitude. At the equator, where the sun is never far from over-head at local midday, the annual solar energy received is $2\frac{1}{2}$ times that at the poles, where during winter the sun is below the horizon and during summer has a low altitude and where there is much reflecting ice and snow cover. By contrast the infrared radiation the earth emits is much more uniform over its surface. As a consequence regions near the equator (35°S to 40°N, in fact) receive more energy from the sun than is emitted by the earth, and are therefore warmer, while remaining regions receive less and are therefore colder. Also, the average cloud cover affects the receipt of solar radiation as clouds are partly reflecting. The time effects are principally seasonal, i.e. arise because of the $23\frac{1}{2}$° tilt of the earth's axis. Any location receives more solar radiation during the summer because of the higher altitude of the midday sun then and the longer period of daylight. Seasonal variations of temperature depend on geography, a continental location having greater extremes of temperature than a maritime one since water has a greater capacity for storing the solar energy. This accounts for the northern hemisphere having warmer summers and colder winters on average than the southern where there is a greater proportion of sea area. This more than offsets the counteracting effect whereby 7% more solar radiation is received in early January because the earth is then closer to the sun (distance 147 100 000 km) than in early July (distance 152 100 000 km), which tends to make summers cooler and winters warmer in the northern hemisphere.

The extremes of cold near the poles and heat in the tropics, arising from the unequal warming of the earth's surface, are tempered by a transfer of heat towards the poles, primarily by a general atmospheric circulation but with ocean current systems also contributing. The atmospheric circulation is complex. At low latitudes it is primarily vertical, with convection or 'Hadley' cells formed by hot air rising at the equator, and sinking as it cools at around latitudes 30°N and S. Beyond this zone, the circulation is mainly horizontal, in the form of travelling anticyclones and cyclones ('highs' and 'lows').

The temperature of the earth's atmosphere varies vertically over a wide range: see Fig. 7.21 for the atmosphere's temperature profile with altitude (this depends on latitude: the figure is the profile for a latitude of 10°N at the time of the spring equinox, where the mean surface temperature is 17°C). The lowest layers are warmed by the absorption of terrestrial infrared radiation, principally by carbon dioxide and water vapour, but for progressively higher altitudes this warming is reduced and the temperature decreases. It reaches a minimum of about $-75\,°C$ at 17 km, beyond which it rises again. The temperature minimum is known as the tropopause, and the part of the atmosphere below it the troposphere. For higher latitudes the tropopause occurs at lower altitudes – 8 km at the poles. The troposphere contains three-quarters of the mass of the earth's atmosphere, and almost all the water vapour. It is the region where all weather phenomena and most cloud formations (those composed of water droplets such as cumulus and the higher cirriform clouds composed of ice crystals) occur. Molecular nitrogen, N_2 (78% by volume), and molecular oxygen, O_2 (21%), are the principal constituents of the troposphere; water vapour may account for up to 2%. There are also small amounts of argon (0.9%) and carbon dioxide (0.03%) and traces of other gases.

The temperature rise which starts at the troposphere continues up to an altitude of about 50 km, where it reaches a maximum of 0 °C (again for a latitude of 10°N). At this altitude, the pressure has fallen to a thousandth of the surface pressure. This region of the atmosphere is known as the stratosphere, and the surface of maximum temperature the stratopause. The warming is due to the absorption of solar ultraviolet radiation by molecules of ozone, made up of three oxygen atoms (O_3). Stratospheric ozone is formed in two stages. Firstly, an oxygen O_2 molecule is split by solar ultraviolet light (wavelengths less than 243 nm) into two oxygen atoms:

$$O_2 + UV\ photon \rightarrow O + O$$

This is followed by one of the oxygen atoms combining with an oxygen molecule,

$$O + O_2 \rightarrow O_3.$$

Ozone can be transformed back to oxygen by the action of various compounds, and can also be transported to other regions of the atmosphere. Its

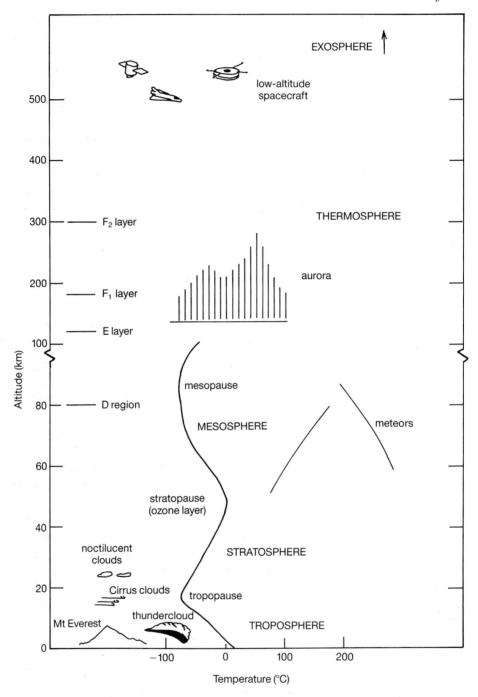

Fig. 7.21 Section through the earth's atmosphere, showing the variation of
temperature for altitudes up to 100 km (latitude 10°N at the northern
spring equinox, from Houghton (1986)) and the altitudes of ionospheric
layers, aurora and other features.

concentration is thus determined by a balance between creation and destruction processes. At high altitudes, the solar ultraviolet radiation is abundant but the atmosphere is too rarefied for many atom–molecule collisions, while lower down where the density of atoms is much greater, so allowing more frequent collisions, all the ultraviolet radiation has been absorbed. The creation of ozone from the splitting of oxygen molecules occurs most especially over equatorial regions where the ultraviolet radiation penetrates to lowest altitudes. There is therefore a zone of maximum ozone concentration, which occurs in the lower stratosphere (15–35 km altitude), though most ozone is created in the upper stratosphere. Stratospheric winds carry ozone over the entire earth to produce the 'ozone layer'. Though ozone's maximum concentration is minute (a few parts per million), it is primarily responsible for absorbing the solar ultraviolet radiation that is harmful for life-forms. It totally absorbs radiation at 240–290 nm and partly absorbs that at 290–320 nm.

Above the stratopause, the temperature again decreases with height, reaching a minimum of about −79 °C at an altitude of 85 km (latitude 10°N). This region is called the mesosphere and the surface of minimum temperature the mesopause. Above the mesopause, there is an extensive but very tenuous region of the atmosphere: that up to about 500–750 km altitude is the thermosphere, and that beyond is the exosphere. There is a second rise of temperature over these altitudes owing to heating by solar X-rays and extreme ultraviolet radiation (wavelengths less than about 150 nm, including the intense Lyman-α line of hydrogen). As the intensity of this radiation strongly depends on solar activity, the thermospheric and exospheric temperatures vary with the eleven-year activity cycle. There is also a 24-hourly heating and cooling cycle as this outer part of the atmosphere passes into and out of the earth's shadow. The temperature variation is considerable: at around 250 km and beyond (where the temperature is almost constant with height) the range is roughly 500 K to 2000 K.

This variation of temperature means that the outer atmosphere in a sense expands with greater solar activity; or to be more precise, the atmosphere's pressure scale height (to use the concept introduced for the solar atmosphere: §3.4), being proportional to temperature, increases. This is illustrated by the decreasing altitude of certain low-orbit satellites. The spacecraft *Solar Maximum Mission*, for instance, was launched in early 1980 into an orbit with altitude of 574 km, i.e. in the thermosphere. This was at a time of high solar activity, and the orbit quickly decayed to about 500 km in 1983, but much more slowly for the next four years when there was low solar activity (Fig. 7.22). Thereafter, when there was a strong increase of activity, the orbit decayed more rapidly again until the spacecraft was burned up in the lower atmosphere in December 1989. The final decay was extremely rapid as the amount of drag a spacecraft experiences depends on the atmospheric density and the square of the orbital speed which progressively increases (by Kepler's second law) for decreasing altitude.

Below the level of the mesosphere the atmosphere's chemical constituents are mixed by turbulence, but above it they begin to separate. The separation occurs according to scale height, which for an individual gas is inversely proportional to its atomic or molecular weight. Thus, the lightest gas, hydrogen, is predominant in the exosphere. A small amount escapes from the top of the exosphere, to be replaced by hydrogen atoms produced lower in the atmosphere by the break-up of water and methane (CH_4) molecules. Figure 7.17 clearly shows the hydrogen of the earth's atmosphere, visible by its scattering of solar Lyman-α radiation to form a 'geocorona'. Helium and atomic oxygen also occur in the thermosphere and exosphere, but very little of these gases escapes.

The solar X-ray and extreme ultraviolet radiation is sufficiently energetic to produce ionization in the thermosphere between altitudes of about 50 and 300 km. The free electrons and ions thus formed are distributed with altitude in various layers making up the *ionosphere*. These layers have been studied by the reflection of transmitted radio signals, either from ground-based *ionosondes* for the lower layers or from space-based *top-side sounders* for the upper layers. It is found that, for the sunlit hemisphere, there is a peak in the electron concentration at about 250–300 km altitude, but there are other less marked peaks or changes of slope lower down (see Fig. 7.21). The main peak

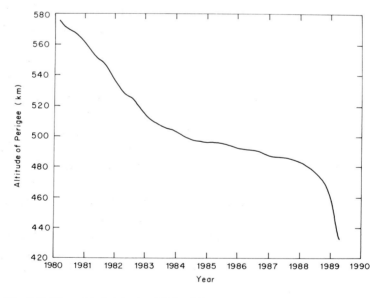

Fig. 7.22 The orbital altitude of *Solar Maximum Mission* during its lifetime (1980–89). The perigee altitude is plotted, i.e. that of the least distance in the satellite's orbit to the earth's surface (though the orbit was close to circular). The variation of the altitude is related to solar activity because of the activity-dependent extent of the earth's thermosphere. The high levels of solar activity in 1980–81 and 1988–89 led to a rapid decrease in the spacecraft's altitude.

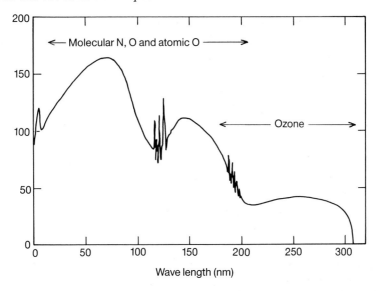

Fig. 7.23 Penetration of solar ultraviolet and soft X-ray emission in the earth's atmosphere as a function of wavelength, with principal absorbing gases indicated. (After Hertzberg (1965))

at 250–300 km altitude is known as the F_2 (or Appleton) layer, and smaller peaks or slope changes at 170 km the F_1 layer and 120 km the E (or Kennelly–Heaviside) layer. The D region is between 50 and 90 km altitude. On the night-time hemisphere, the F_1 layer is often absent.

The solar radiation producing the ionization has a penetration depth that depends on the wavelength. This is shown in Fig. 7.23. Extreme ultraviolet radiation with wavelengths between 14 and 80 nm reach an altitude between the F_1 and E layers, but soft X-rays with wavelengths less than 14 nm and extreme ultraviolet radiation with wavelengths between 80 and 102 nm reach rather deeper, to the E layer. The ionosphere is above the region of turbulent mixing, i.e. where individual chemical constituents separate according to their scale heights. At the level of the F layers, molecular nitrogen (N_2) and atomic oxygen (O) are the main constituents, and electrons are primarily produced by their ionization. At the E layer, molecular oxygen (O_2) is as important as O, so electrons are also produced by the ionization of O_2. In the D region, electrons are produced by the ionization of oxygen and nitrogen by soft X-rays of wavelengths less than 1 nm, which are strongly enhanced during periods of solar activity, especially flares. The strong radiation due to the hydrogen Lyman-α line also penetrates to the D region, and results in the ionization of the nitric oxide (NO) molecule. In addition, cosmic rays from outside the solar system may produce significant ionization in the D region.

The distribution of free electrons in the ionosphere is determined by a balance between creation and loss processes in the atmosphere, like that of ozone. The above creation processes, i.e. ionizations, are balanced by two principal loss processes, recombinations (with either atoms or molecules) or a drift of the electrons out of the region where they were created.

The great practical significance of the ionosphere is that it allows long-distance communication by high-frequency (greater than $20\,\text{MHz}$) radio waves. A location R can receive radio waves from a transmitter T beyond its horizon by virtue of the reflection such waves undergo in the ionosphere, normally at the E layer. However, when a strong X-ray flare occurs giving rise to enhanced ionization in the D region, these signals often fade out as they are absorbed before they can undergo reflection. Communication is recovered almost as soon as the flare is over, since the enhanced D-region ionization soon disappears. Such 'short-wave fade-outs' (SWFs) are but one example of sudden ionospheric disturbances (SIDs) connected with solar activity. Further examples include sudden enhancements of atmospherics (SEAs) due to numerous, distant lightning flashes which emit very-low-frequency (10–$20\,\text{kHz}$) radiation. The enhanced D-region ionization, while absorbing high-frequency radiation, gives improved reflectivity for such low-frequency radiation. A sudden phase anomaly (SPA) is an SID that occurs with very-low-frequency (around $20\,\text{kHz}$) radio waves from a near-by transmitter. This may be observed at a receiving station which detects low-frequency waves from a nearby transmitter both by a direct route and via reflection from the E layer. There will be a phase difference between the two signals which remains constant until a flare occurs, giving rise to D-region ionization and thus to a sudden alteration of the phase difference, or phase 'anomaly'.

Records of these SIDs thus give information about flare soft X-ray emission, a useful source before spacecraft were available to observe it directly (the *GOES* satellites now monitor this radiation continuously). An SID that gives an indication of flare soft X-ray and ultraviolet radiation is the sudden frequency deviation (SFD), recorded when high-frequency radio waves reflected off the E and F_1 layers of the ionosphere undergo a sudden frequency increase, owing to enhanced ionization in these layers. Solar-flare emission in the 1–$103\,\text{nm}$ wavelength range gives rise to this enhanced ionization, so that the time development of an SFD indicates the amount of emission in this range.

Not only electromagnetic but also particle radiation from solar flares has ionospheric effects. A 'polar cap absorption' (PCA) is a reduction of radio emission from sources in our galaxy or beyond it observed in polar regions of the earth. It is due to extra ionization in the D region arising from extremely intense 'proton' flares on the sun; if the emitted protons have relatively low energies, they reach only the area of the earth's polar 'cap', i.e. where terrestrial magnetic field lines are open to distant regions of the magnetosphere. The continuous monitoring of this cosmic radio emission by instruments called riometers gives a 'proxy' measurement of low-energy solar protons.

7.6   The earth's weather and climate

We have seen that solar activity has an appreciable effect on the earth's upper atmosphere – the heating of the thermosphere and the ionization of the ionosphere. A perennial question of solar–terrestrial relations is: does solar activity affect the lower atmospheric layers, in particular climate and the weather? The well-defined eleven-year solar cycle has led many to search for similar periodicities in weather patterns in attempts to look for connections. One approach taken some years ago was to examine tree-rings of very old wood, looking for changes in the ring widths connected with solar activity. Norman Lockyer in the early years of this century had suggested that the annually formed rings reflected local weather conditions and so might be a key to solar activity in the remote past, if this were related to weather. The science of dendrochronology was born out of the attempts made along these lines, and is strongly associated with the name of A. E. Douglass in Arizona. His investigations seemed to show several instances of eleven-year patterns, but equally there were many examples showing no periodicity. Interesting cyclical patterns continue to come to light, though sceptical observers have concluded that they are all chance occurrences.

The discovery of similar patterns in the sediments of sandstone rock excavated in South Australia by G. E. Williams in 1981 seemed to indicate a solar influence, but more recent investigations lead to quite different conclusions. The sediments were thought at first to have been formed annually, being laid down by spring melt-water into an ancient lake, and a cyclical variation in the thicknesses was identified with the eleven-year solar cycle. The apparent connection with solar activity was explained by the possibility that, when the sediments were laid down around 700 million years ago, the earth's lower atmosphere was much more sensitive to solar ultraviolet radiation owing to reduced amounts of ozone. More recently, it has been proposed that the sediment cycles are not solar in origin but are due to a monthly cycle of spring and neap tides.

Recent evidence has been put forward that suggests perhaps more definitely than hitherto a link between weather patterns and the eleven-year cycle. This involves a wind system in the tropical stratosphere, at 15 km altitude, whose direction alternates every 26 months or so between eastwards and westwards: for this reason it is called the quasi-biennial oscillation (QBO). It may be due to the interaction of gravity waves propagating up from the troposphere. Although occurring at low latitudes, it has an effect on the stratosphere at the poles. Measurements by Karin Labitzke of the Free University of Berlin between 1952 and 1988 indicate that the QBO is related to solar activity. When the QBO blows from the west, the north-pole stratosphere is relatively warm when solar activity is high, but is cold when solar activity is low. On the other hand, when the QBO is from the east, the north-pole stratosphere is cold when activity is high, warm when activity is low (see Fig. 7.24). At the level of the troposphere, where weather systems occur, the relation tends to be that pressure is high over the North American continent and low in the oceans either side of it for westerly QBO, and vice versa for easterly QBO. The flow of air, and thus ground temperature, is controlled by these pressure systems, so there is some correlation of temperatures with solar activity, again separating the years when the QBO is from either the west or east. A more recent finding (by B. A. Tinsley of the US National Science Foundation) is that the latitude of North Atlantic storms is correlated with

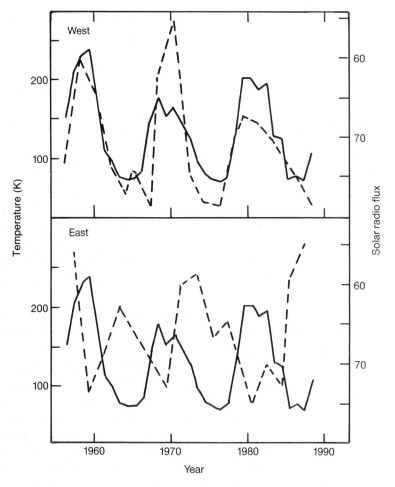

Fig. 7.24 North polar stratospheric temperatures (dashed lines) compared with solar activity as indicated by the level of 10.7-cm radio flux (solid lines). A correlation is indicated for years when the quasi-biennial oscillation blows from the west (top graph) but an anti-correlation when from the east (bottom). (After Van Loon and Labitzke (1988))

solar activity for westerly QBO, being nearer the equator at solar maxima than at minima.

There is geological evidence for extensive changes in the earth's climate, with periods of major glaciation or 'Ice Ages' lasting a few million years every 250 million years since about 1250 million years ago. The reasons for these Ice Ages is not altogether clear, but some outside influence – the solar system passing through some particularly dusty region of interstellar space has been suggested – combined with the drift of continents over the earth's surface seems to be responsible. Even over the past 1 000 000 years the earth's climate is known to have altered greatly, with several episodes of wide-spread glaciation, each lasting for about 10 000 years and coming to an abrupt end. The latest peaked 17 000 years ago and ended 12 000 years ago. During each period of glaciation, ice sheets in both hemispheres extended down to latitude 60° (further in some continental areas). Studies of oxygen isotope ratios in sea-bed sediments have given an accurate record of these periods for the last few hundred thousand years. Major glaciations have occurred roughly every 100 000 years, but several other periodicities are discernible. Their occurrence may be linked, as was suggested many years ago and refined in a theory due to the Yugoslav Milutin Milankovitch in the 1930s, to three astronomical cycles that affect the amount of sunlight received at any one location: a period of 23 000 years for the precession of the equinoxes (earth's rotation axis making a complete circle around the sky); one of 41 000 years for variations in the tilt angle of the earth's axis (from $21\frac{1}{2}°$ to $24\frac{1}{2}°$); and one of 100 000 years for variations in the eccentricity of the earth's orbit. These factors together cause the amount of summer sunshine in high northern latitudes to vary over a range of 20%, which is sufficient to alter the polar ice cover appreciably. It is apparently the changes in the northern hemisphere glaciation that most influence the global climate.

The occurrence of even more recent glaciations seems to be attested to by historical records. Attempts to relate them to solar activity have been made by dendrochronology in a new guise, in which the radio-active carbon content in tree-rings is studied. This is more certainly related to solar activity. Most naturally occurring carbon is in the form of ^{12}C (nucleus containing six protons and six neutrons), but there is a radio-active form ^{14}C, or 'radio-carbon' (nucleus containing six protons, eight neutrons), which is produced when cosmic rays from outside the solar system interact with atmospheric nitrogen. The radio-carbon combines with oxygen to form carbon dioxide ($^{14}CO_2$) which trees eventually absorb during photosynthesis; the radio-carbon is then preserved by cells of trees as cellulose in newly formed wood. A tree sample can thus be analysed ring by ring to find the amount of radio-carbon compared with the more common ^{12}C form. The ratio $^{14}C/^{12}C$ for a particular year can be estimated, taking account of the fact that radio-carbon has a half-life of 5700 years, and so the cosmic-ray flux can be indicated. As

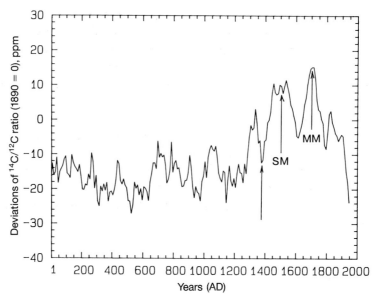

Fig. 7.25 Radio-carbon record from tree-ring samples for the last 2000 years. The vertical axis represents deviations of the $^{14}C/^{12}C$ ratio (in parts per million, ppm) from its level in 1890, with increasing amounts of ^{14}C downwards: in this way, increasing solar activity is indicated by any upward deviation. The arrows mark possible prolonged minima or maxima, notably the Maunder minimum (MM, 1645–1715) and Spörer minimum (SM, 1450–1540). (Courtesy M. Stuiver and B. Becker)

the latter inversely depends on solar activity (§7.2), one might expect the $^{14}C/^{12}C$ ratio to reflect solar conditions. But because radio-carbon takes an appreciable time – about 20 years – to diffuse to the lower part of the earth's atmosphere where it can be absorbed by trees, any eleven-year modulation in the radio-carbon is smeared out. Longer-term variations can be recognized, however, and one of the most interesting is an increased amount of radio-carbon (i.e. indicating decreased solar activity) at the time of the Maunder minimum, around 1645–1715 (see §1.2). Figure 7.25 shows a 2000-year-long radio-carbon record from tree-ring data. The radio-carbon record would seem to confirm the apparent sparseness of sunspots and auroral displays at that time. Other major deviations (first pointed out by H. DeVries in 1958) include a radio-carbon maximum (and thus a solar minimum) in 1450–1540 (the 'Spörer' minimum) and a radio-carbon minimum (solar maximum) in the twelfth century. Unfortunately, pre-telescopic records of naked-eye sunspots from the Orient (§1.2), the most complete in existence, are rather too infrequent to indicate adequately the level of solar activity at these times.

From the present point of view, it is interesting that the two prolonged solar minima indicated by the radio-carbon data corresponded with a lengthy cold spell, from 1550–1700, known as the 'Little Ice Age'. At this time Arctic ice sheets attained their greatest extent since the last major glaciation period. Further, the possible increase of solar activity in the twelfth century occurred at a time (AD 1000–1250) when northern latitudes at least seem to have been warmer than at present, as shown, e.g., by the colonization of Greenland by the Vikings. How could solar activity lead to such large-scale climatic changes? Recent spacecraft measurements of solar energy output show variations over a range of 0.15%, greatest at sunspot maximum (§9.1). Perhaps larger variations occurred during the periods when the radio-carbon record shows deviations. A solar output decrease of 0.4–1.4% could account for the occurrence of the Little Ice Age. All this could have relevance to future generations if a recent claim for a 200-year periodicity in the radio-carbon record is confirmed, since this points to a Little Ice Age some time in the twenty-first century.

7.7   Human activities and the earth's atmosphere

The sun overwhelmingly controls processes in the earth's atmosphere, but it has been evident in recent years that human activities are affecting the atmosphere to a considerable extent, leading to unwelcome and even dangerous consequences. As these relate to the processes described in §7.5, we will outline them here.

Much publicity has been given to the discovery of a depletion in the ozone layer above the Antarctic – the 'Ozone Hole'. Ozone screens out solar ultraviolet radiation, as we saw in §7.5, which is harmful to humans (increasing the incidence of skin cancers), to certain aquatic life-forms and to crops. Ozone is formed in the stratosphere mainly near the equator and is carried to higher latitudes by stratospheric winds. In the northern hemisphere, the circulation is assisted by atmospheric motions due to large mountain ranges like the Rockies and Himalayas, so equatorial air circulates right to the north pole itself. In the southern hemisphere, the circulation only reaches to about latitude 60° for much of the year. The stratosphere over Antarctica is, as a consequence, relatively isolated, and during the southern winter an extremely cold air mass forms, spinning round to form a 'polar vortex'. Up until a few years ago, the ozone concentration in the Antarctic stratosphere remained almost constant throughout the southern winter months, but showed a rapid rise in October, the southern spring, when the polar vortex starts to break down, allowing ozone-rich air from lower latitudes to invade.

Measurements of the Antarctic ozone concentration have been made by the British Antarctic Survey from their base at Halley Bay (75°S) since 1957. In 1982, J. C. Farman and colleagues found some evidence of an ozone depletion which by October 1984 had become much more pronounced. Their findings were published in May 1985. It was found that there was a sharp decrease in ozone concentration in October lasting for a month or so, followed by a rise. Whereas before the ozone concentration had, during the winter, maintained a steady value of 300 'Dobson' units, rising to 400 during the spring, since 1984 it has fallen to less than 200 Dobson units before the spring rise, which now occurs roughly a month later. (One Dobson unit corresponds to a thickness in units of 0.01 mm that the ozone layer would have at sea level.) Figure 7.26 shows the Halley Bay measured ozone amounts in October since 1957. The large depletion has been confirmed by images from space, made by an instrument called the Total Ozone Mapping Spectrometer (TOMS), on the *Nimbus 7* spacecraft (Fig. 7.27).

A clue to the ozone depletion was provided by an experiment carried out in the 1970s showing the harmful effects of chlorofluorocarbons, or CFCs (also known as freons). These are artificial compounds widely used as aerosol propellants, cleansers for electronic components and refrigerator coolants. In

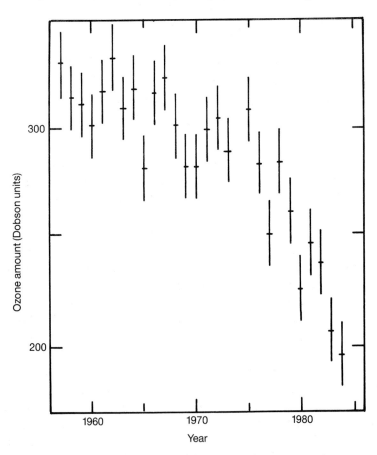

Fig. 7.26 Variation of October ozone amounts (Dobson units) from 1957 to 1984 measured at Halley Bay (76°S) by the British Antarctic Survey. (After Farman *et al.* (1985))

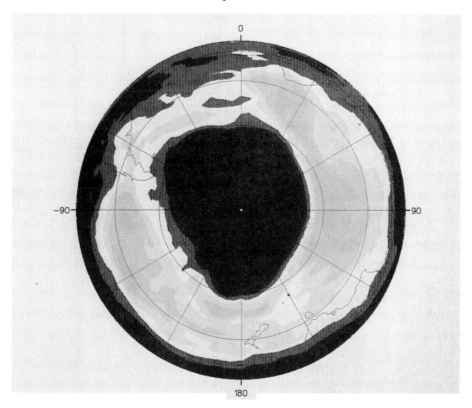

Fig. 7.27 The ozone hole over the Antarctic on 1 October 1987 as indicated by ozone amounts in Dobson units measured by the TOMS instrument on the *Nimbus-7* spacecraft. (Courtesy RAL Geophysics Data Facility)

the troposphere, CFCs are inert and as a result extremely long-lived, varieties such as F-11 ($CFCl_3$) and F-12 (CF_2Cl_2) persisting for nearly a century. They can find their way up to the stratosphere, above the ozone layer, where they are broken up by solar ultraviolet light to form chlorine. The released chlorine in time then sinks down to the ozone layer and reacts with ozone to produce chlorine monoxide (ClO) and an oxygen molecule (O_2). The chlorine monoxide is highly reactive and readily combines with an oxygen atom (O) to form an oxygen molecule (O_2) and free chlorine once more, which is now able to react with another ozone molecule and so repeat the cycle. Thus, an ozone molecule is destroyed for each cycle.

Whether this process is the chief one leading to ozone destruction in the stratosphere was for some time disputed, but evidence from *in situ* measurements taken in 1987 points strongly to this being so. High-flying aircraft crossing over the Antarctic showed that the amount of chlorine oxide present was enough to account for the ozone depletion. Also, balloon-borne instruments have provided measurements of ozone amounts over a range of altitudes; they show that, while the Antarctic ozone layer is well developed at about 16 km altitude in August, it is almost totally removed at this altitude in October.

There are many unknowns in this very recent discovery – one being the rôle of possible natural reactions leading to ozone depletion. But inescapably it has been shown that CFCs are major contributors to this profound environmental change. International co-operation will be needed to curb their production. A step in this direction was made by a protocol signed by 40 countries in Montreal in September 1987, under the terms of which CFC production will be halved by 1999. There is, however, considerable concern that such reductions are much too modest and more stringent controls will be required to halt the ozone depletion, let alone reverse the trend.

Another major concern in recent years has been the global warming expected by an increase in the atmospheric concentrations of greenhouse gases as a result of human activities. The average global temperature over the past century has increased by 0.5–0.7 °C, with large increases in the past decade. Up to 1989, the six warmest years globally since the beginning of this century were all in the 1980s. The warming is also indicated by rising sea levels, caused by melting polar ice, and retreating glaciers. It has been accompanied by increasing concentrations of carbon dioxide, methane and other greenhouse gases, which trap terrestrial infrared radiation and so maintain a relatively high temperature near the earth's surface (§7.5). The concentration of CO_2 has increased by about a fifth (from 0.029% to 0.035%) since the early years of industrialism in the mid-nineteenth century, that of methane (CH_4), much less abundant but a far more effective greenhouse gas, has doubled over that time.

There are four major reservoirs containing terrestrial carbon dioxide – the oceans, plant-life (or biosphere), fossil-fuel deposits and the atmosphere. There are interchanges between the oceans and biosphere on the one hand and the oceans and atmosphere on the other by processes such as plant respiration (conversion of atmospheric oxygen to carbon dioxide) and evaporation or condensation at ocean surfaces. In the normal way these processes balance each other over a sufficiently long period. However, the burning of fossil fuel (coal, natural gas and oil) and forests by humans over the past century has led to a marked imbalance, causing increased levels of atmospheric CO_2. At present, these processes add up to 10 gigatonnes (1 gigatonne$=10^{12}$ kg) per year of CO_2 to the atmosphere, the present total CO_2 content of which is 700 gigatonnes.

The global warming resulting from an increase of carbon dioxide is larger than what one might expect. It appears that global warming increases plant respiration and organic decay and thus accelerates the CO_2 production: the warming itself induces more warming. Furthermore, the production of methane – from swamps and wet regions – is then increased, so adding to the warming. Many details, such as the increased cloud cover that may result, are as yet poorly known, but it is already clear that the trend is alarming enough. The concentration of CO_2 could be effectively doubled by the year 2030, and the probable consequences would include the increase of the average global temperature by 1.5–4.5°C and a raising of sea levels by 0.2–1.4 metres, with major effects on coastal and estuary areas.

Counteracting the greenhouse effect is likely to be much more complex and difficult than counteracting the stratospheric ozone depletion. One cannot suddenly cool the earth: the only means available are reductions in fossil-fuel burning and reversing the trend of de-forestation. Most fossil-fuel burning occurs in industrialized countries, but that in developing countries is likely to increase. Conservation measures and reduced dependence on fossil fuels (e.g. by shifting to solar power systems) are becoming steadily more urgent, as are measures to stop the huge rate of de-forestation which is now occurring. Conferences have been held periodically to encourage countries to take effective measures of this sort but until there is international agreement, even on the level of the Montreal Protocol, the future prospects are very uncertain.

Chapter 8

The sun and other stars

The sun is very special to the earth and all life-forms on it, but in a more general context it is just one star among countless millions of others. In this chapter, we examine how the sun compares with these other stars in its properties, the nature of its birth and how eventually it will die. Finally, we shall look at examples of stars having solar-like chromospheres, coronae and types of activity such as spots and flares.

8.1   The sun as a star

Some 100 000 million stars, including the sun and nearly all the stars visible to the unaided eye on a clear night, make up our own galaxy. As the sun and solar system are entirely within this huge system of stars, it is not immediately evident what form the galaxy has, but astronomers have deduced that it is similar to certain nearby 'spiral' galaxies, an example being the Andromeda Galaxy, also named M31 (Fig. 8.1). There is, as with M31, a central condensation, or nucleus, of relatively densely packed stars that are predominantly red in colour, and spiral arms radiating from and wrapping round the central nucleus, also composed of stars, which are bluish and rather less densely spaced. In cross section, the galaxy is a flattened disc, with the nucleus forming a bulge. A sprinkling of stars distributed above and below the plane form a galactic 'halo'. Tenuous gas and dust pervade the space between the stars of the spiral arms, but none is apparent within the central nucleus. The diameter of the galaxy is estimated to be about 10^{18} km. The sun is situated in a spiral arm, about two-thirds of the way out from the nucleus to the edge. Looking out from our location, we see the galaxy as a faint band of light passing right round the sky – the 'Milky Way' – which has a concentration in the constellation of Sagittarius corresponding to the direction of the nucleus,

274

Fig. 8.1 The Andromeda Galaxy, M31, which is thought to be very similar to our galaxy, the Milky Way.

while the stars we see in directions away from the Milky Way are relatively close ones lying above and below us in the disc structure.

Not all galaxies are like our own or M31. Some are elliptical concentrations of stars without spiral arms, others have irregular forms. A few have an unusual degree of activity in the form of X-ray and radio emission, while some seem to be in the act of colliding with each other. Clustering of galaxies occurs, and there are even recognizable superclusters of clusters. Our galaxy belongs to a small Local Group that comprises the Andromeda Galaxy, and the Local Group appears to be at the edge of a large cluster of galaxies known as the Virgo Cluster.

All galaxies, whatever their type, have a common and very remarkable feature: they are receding from each other at speeds that are proportional to their distances. Extremely distant galaxies and more particularly the peculiar quasi-stellar objects (or quasars) are receding from us at sizeable fractions of the speed of light. This fact, discovered for galaxies by Edwin Hubble in the 1920s, has given rise to the notion that the entire universe is expanding, with our galaxy sharing in this motion. In remote times, then, the universe must have been more condensed than at present and, assuming that the observed recessional speeds of galaxies have always been maintained, we can deduce that, some 16 000 million years ago, the universe was reduced to a point. A vast explosion – the 'Big Bang' – set the universe expanding. Cosmologists calculate that, within a tiny fraction (about 10^{-30}) of a second, all the matter and energy in the universe was created. In the first three minutes or so, the matter, in the form of hydrogen nuclei, was sufficiently hot that some 10% was converted to helium and small amounts of some other light elements by nuclear fusion processes. The rapid expansion and cooling of the universe prevented heavier elements or any more helium from being formed. Any density fluctuations below a certain size were smeared out by the random motion of particles making up the matter, but those above this size survived and became the material out of which galaxies and clusters of galaxies formed. For the first few hundred thousand years of the universe's life, matter and radiation were completely coupled to each other, with photons continually interacting with atoms and other elementary particles. Eventually, as the universe expanded further, the interaction between matter and radiation decreased and the radiation began to propagate freely. Today this radiation still exists, and is observable as a uniform background of microwave and infrared emission.

We can make a comparison of the sun with other stars in several different ways. Firstly, we should summarize the most important quantities or properties that describe the sun. These have been discussed in Chapters 2–7, along with the various ways of determining them, or in some cases inferring them from models. A list of these properties is given in Table 8.1 (references to appropriate sections where they are discussed are included). Most are very well determined because the sun is so close to the earth and to the various space probes from which measurements have been made. We will now see how some of the corresponding data can be derived for stars.

8.2   Properties of stars

An immediate problem in trying to compare the sun with other stars is the fact that stars are enormously remote. Even the nearest star – Proxima Centauri, a very faint red star – is 4×10^{13} km from the earth, or 271 000 times the distance to the sun. Most of the bright stars in the night sky are much more distant still. Consequently, until very recently it was impossible to see stars as

Table 8.1. *Observed solar properties*

Property	Magnitude
Mass (from Kepler's laws: §7.1)	1.989×10^{30} kg
Radius (of photosphere) (from direct measurement: §3.3)	695 970 km
Luminosity (from solar irradiance: §3.4)	3.85×10^{23} kW
Effective temperature (from luminosity and Stefan–Boltzmann law: §3.4)	5778 K
Chemical composition of photosphere (from analysis of Fraunhofer lines: §3.7) proportion of elements	
(by number)	91% H, 9% He, 0.1% other
(by mass)	71% H, 27% He, 2% other
Age (from age of oldest meteorites: §2.1)	4.6×10^9 yr
Rotation period (latitude-dependent: from sunspots and Doppler-compensator measurements: §3.2) (equatorial rotation speed is 2 km/s)	25–35 days
Magnetic field (from Zeeman measurements: §3.8)	0.4 T in sunspots, 0.1 T in filigree and other small-scale concentrations; both small-scale bipolar groups and large-scale weak unipolar areas exist
Atmospheric (average quiet-sun) temperature structure (from Fraunhofer lines and ultraviolet and X-ray emission lines: §§4.5, 5.4)	
photosphere	6400 K
temperature minimum	4400 K
chromosphere	6000–20 000 K
transition region	20 000–2×10^6 K
coronal large-scale structures and holes	1×10^6–2×10^6 K
Solar wind (from spacecraft measurements: §7.2)	expansion of outer corona with speeds of 300–750 km/s
mass loss due to solar wind	5×10^{16} kg/yr (2.5×10^{-14} of solar mass/yr)
Solar activity (Chap. 6)	Sunspots, active regions and flares occurring with a frequency cycle of 11 years (magnetic cycle of 22 years)
Solar interior (theoretical description: §2.1) standard model (describing temperature and density with distance from sun's centre) predicts	

Table 8.1 (*cont.*)

Property	Magnitude
central temperature	15.6×10^6 K
central density	$148\,000$ kg/m^3
energy source	proton–proton reactions synthesizing helium
	Radiative transfer for inner two-thirds, convection for outer third of sun (§2.1)
	Magnetic fields at surface thought to be regenerated by dynamo in lower convection zone (§2.4)
Solar interior (observed properties)	neutrinos from nuclear reactions less than predicted from standard model (§2.2)
	Global oscillations broadly confirm standard model
	Inner rotation rate appears to decrease with depth in the sun (§2.3)

anything more than pinpricks of light; a star the size of the sun would subtend only 0.007 arc secs at the distance of Proxima Centauri, quite beyond the theoretical optical resolving power of, say, the 200-inch (5-m) reflecting telescope at Palomar Observatory (about 0.025 arc secs), let alone the much larger limit imposed by atmospheric turbulence. Nevertheless, astronomers have succeeded in making measurements of the most important character-istics of many stars and, often using what is known about the sun, have developed an understanding of stars in general. Determining some of these characteristics for a star requires the knowledge of its distance.

Distances to nearby stars can be measured with good accuracy by the method of trigonometric parallax, i.e. from their displacements against the background of more distant stars due to the earth's orbital motion round the sun. The diameter of the earth's orbit, 2 astronomical units (2 AU) or almost $300\,000$ km, is used as the 'baseline' for this measurement (Fig. 8.2). Seen from the earth, a star that lies in a direction perpendicular to the plane of the earth's orbit (at the 'ecliptic pole'), moves in a small circle that has an angular radius, or parallax angle Π, given by $\sin \Pi = 1$ AU/d, where d is the distance to the star. Actually, d is always so large (and Π so small) that we can approximate this to $\Pi = 1$ AU/d, with Π measured in radians. Stars in other directions apparently move in ellipses with semi-major axes equal to Π. Measurements of such a star's position six months apart, with the earth at opposite ends of its orbit, then give the parallax angle and thus distance. It can be seen that a star

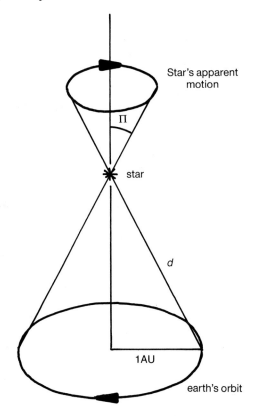

Fig. 8.2 As the earth moves in its orbit, a nearby star in the vicinity of the ecliptic
pole appears to move in the sky in a circle. The angular radius of this
circle is the parallax angle Π, which, in radians, is given by 1 AU divided by
d, the distance to the star.

with $\Pi = 1$ arc sec has a distance of d equal to 206 265 AU. This is used as an
astronomical distance unit and is called 1 parsec (pc). For Proxima Centauri,
Π is 0.76 arc seconds, so that its distance is 1.3 pc. A light-year is also
frequently used as a distance unit, the distance to Proxima Centauri being 4.3
light-years. These astronomical units are discussed in the preface. With
conventional telescopes, parallaxes as small as 0.05 arc sec (i.e. distances as
large as 20 pc) can be measured with reasonable accuracy. There are several
thousand stars closer than this, but only a few hundred are bright enough for
accurate measurements to be possible. The European Space Agency (ESA)
spacecraft *Hipparcos*, launched in 1989, should in time enable trigonometric
parallaxes to be measured to an accuracy of 0.002 arc seconds for as many as
100 000 stars. The trigonometric parallax method of finding stellar distances
is the only absolute one, giving distances in terms of the known length of the
astronomical unit. There are other methods for finding distances, and they

can be applied to more remote stars, but they are all indirect and depend on trigonometric parallax distances to nearby stars.

From several points of view, the most important properties of a star are its mass, radius (if spherical), luminosity, effective temperature, chemical composition and age, the first six of the observed properties listed for the sun in Table 8.1. Knowledge of the distance to a star is essential for determination of its mass, radius and luminosity.

Reasonable estimates for stellar *masses* can be made for certain types of systems called binaries in which two stars are gravitationally bound and orbit each other. The method is in principle the same as used to find the solar mass from planetary motions (§7.1). Though it might be thought otherwise, the chances of a star's belonging to a binary or even multiple system are very high: 85% of nearby stars are components of such systems. The case of the sun as a single star seems, then, to be rather unusual.

Binary stars in which the two components are clearly visible with a telescope are called 'visual' binaries. The two components perform elliptical orbits about a common centre of gravity, though measurements are generally made of the fainter, or secondary, star relative to the brighter or primary: the secondary then appears to orbit the primary in an ellipse with the primary at one focus. The apparent orbit (i.e. projected against the background sky) of the secondary can be plotted if the orbital period is reasonably short (say, a few years), and from this the true orbit (i.e. that in space) derived. The dimensions in kilometres of the true orbit can be determined if the distance to the binary can be found by parallax measurements. The sum of the masses of the two stars can be derived from Kepler's third law (§7.1).

For some binaries, known as spectroscopic binaries, the stars are so close that they cannot be resolved telescopically, but their orbital motion may nevertheless be revealed spectroscopically. A spectroscopic binary has a spectrum with two sets of Fraunhofer lines: the star with velocity directed towards the observer has its set of Fraunhofer lines Doppler-shifted towards the violet, while the other star with velocity directed away has its Fraunhofer lines Doppler-shifted towards the red. Continued observations of these Doppler shifts as they vary with the stars' orbital motion give limited information about the masses, viz. the sum of the two masses multiplied by a quantity that depends on the orbital angle of inclination. Individual star masses can be derived for 'eclipsing' binaries, i.e. spectroscopic binaries for which the observer is conveniently in or very near the orbital plane as revealed by periodic eclipses of one star by the other.

The masses of several dozen stars which are members of these different types of binary systems have been measured, the values ranging from 0.1 to 50 times the sun's mass.

Stellar *radii* are generally as difficult to measure as stellar masses, the problem being the extremely large distances involved. Estimates of a star's radius in kilometres can be made for eclipsing binaries consisting of stars of unequal size, with the smaller star passing across the disc of the larger to produce a temporary decrease in the combined brightness of the two stars; the duration of this decrease can be combined with other information to give the dimensions of the two stars in kilometres. Alternatively, the radius of a nearby star may sometimes be obtained by measuring its angular radius or diameter. As mentioned, stars have angular dimensions far smaller than what can ideally be observed with a conventional telescope, the usual problem being not the limitation of the telescope but seeing. Interferometer techniques have been successfully applied to the measurement of stars with relatively large angular diameters, as have measurements of the interference pattern produced when a star is occulted by the sharp edge of the moon. For stars having measurable angular diameters and parallaxes, it is a simple matter to derive linear diameters (or radii) in kilometres. Figure 8.3 shows how the sun compares in size with a few stars measured by interferometric techniques.

The *luminosity* of a star may be found from the total amount of its radiated energy incident at the earth and its distance if known. For many stars, much of this energy lies in the visible range of wavelengths, so the visible brightness of a star approximates the total radiated energy. By a long-standing tradition in astronomy, the apparent brightnesses of stars are expressed as magnitudes, taking over and making more exact a system used by the ancient Greek astronomers. In this the apparently brightest stars in the sky are of the first magnitude, the less bright of second magnitude and so on to stars of the fifth magnitude which are just visible to the naked eye (note magnitudes increase for decreasing brightness). Now, the eye has a roughly logarithmic sensitivity to brightness. Thus, a star of magnitude one is brighter than that of magnitude two by an amount which is equal to that by which a magnitude-two star is brighter than a magnitude-three and so on. Astronomers use the convention that a difference of five magnitudes corresponds to a brightness ratio of exactly 100: i.e. a star of magnitude m is 2.512 (the fifth root of 100) times as bright as a star with a magnitude $m+1$. Exceptionally bright stars on this scale have magnitudes less than 1.0, even (for a few stars) negative magnitudes: thus, Sirius, the apparently brightest star in the night sky, has a magnitude of -1.5. Also, the magnitude scale has been extended to stars below naked-eye visibility, having magnitudes 6, 7, 8, etc. The sun's apparent brightness can, incidentally, be placed on the stellar magnitude scale: it is -26.7.

Generally, photoelectric devices are used by astronomers to measure stellar magnitudes. A system of stellar magnitudes and colours, introduced by H. L. Johnson and W. W. Morgan in 1953, is now widely used. This expresses a star's brightness by three magnitudes, labelled U, B and V as measured through filters that have a peak sensitivity in the near-ultraviolet (360 nm), blue (420 nm) and near the midpoint of the visible spectral range (550 nm).

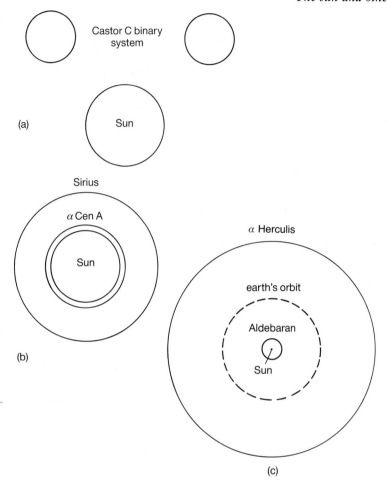

Fig. 8.3 Dimensions of some stars relative to the sun. (a) The Castor C binary
system (distance between the two stars is correct relative to their
diameters). (b) The sun compared with two bright stars. (c) The sun and
earth's orbit compared with a red giant (Aldebaran) and supergiant (α
Herculis).

The filters cover a broad band of wavelengths, between about 80 and 90 nm.
The magnitudes of the stars already quoted are V magnitudes.

A star's luminosity may be expressed using this system. A star's 'absolute
magnitude' is defined as its apparent V magnitude at a distance of 10 pc, and
is denoted by M_V. It measures a star's intrinsic visual brightness. Thus, Sirius,
which is at a distance of 2.7 pc, has an absolute magnitude of +1.4; i.e. at a
distance of 10 pc, it would be of the first magnitude though much less
impressively bright than it is in our skies. The sun's absolute magnitude is
only +4.8, so it is intrinsically much fainter than Sirius and indeed many of
the bright stars in the night sky, but on the other hand it is brighter than vast

numbers of very nearby stars. Since the V magnitude (and therefore the absolute magnitude) only measures radiation in a small part of the visible spectrum, a correction – the 'bolometric' correction – must be applied to get the absolute magnitude in the entire spectral range, from which the luminosity can be derived.

A star's magnitudes on the UBV system are usually quoted in catalogues by the three quantities V and the colours $B-V$ (the star's 'blueness') and $U-B$ (its near-ultraviolet emission). The bright star Vega (α Lyrae), which to the naked eye is a bright blue-white star, is used as a standard, having $V=0$, $B-V=0$, $U-B=0$. The $B-V$ and $U-B$ colours of Sirius are similar. A red star like Aldebaran (α Tauri) has $V=0.9$, $B-V=+1.5$, $U-B=+1.9$. The star's redness is indicated by the positive values of the $B-V$ and $U-B$ colours; the B and U magnitudes are larger, i.e. fainter, than the V magnitude, and so the star's light is relatively deficient in the blue. A yellow star like the sun has $B-V=+0.65$, $U-B=+0.1$, values intermediate between the red Aldebaran and blue-white Vega. The $B-V$ and $U-B$ colours of a star are independent of its distance (unless there is dust between us and the star which tends to redden the star's light). The wavelength distribution of a star's light generally resembles that from a black body, just as with the sun (§3.4). Stars cooler than the sun are reddish, stars hotter than the sun bluish. Thus, a star's colour is related to its *effective temperature*, and the $B-V$ and $U-B$ indices provide convenient and easily measurable indicators of effective temperature.

When stellar spectra were first examined in detail, an attempt was made to classify them, and the resulting 'spectral sequence', still used by astronomers, has a connection with both stellar temperatures and luminosities. The scheme of spectral types, denoted by letters, runs thus:

O–B–A–F–G–K–M

in order of decreasing temperature. The corresponding stellar colours progress from blue-white to red. Additional spectral types (W, R, N and S) indicate spectra with special features. Spectral types are subdivided (e.g. A0, A1, A2, etc.) to indicate fine spectral details. Table 8.2 shows the chief characteristics of each spectral type. The sun, with a spectrum marked by relatively strong H and K lines and rather less strong Balmer lines, has a spectral type G2. Vega and Sirius are both A stars, while the cool star Aldebaran is a K star.

Stars of the same spectral type, i.e. of the same temperature, show certain spectral differences that are related to luminosity. Thus, for B stars, the Balmer lines are much narrower for very luminous stars than for those with smaller luminosity. There are other luminosity indicators for cooler stars.

A graph of absolute magnitude against spectral type for a selection of stars shows a striking pattern: see Fig. 8.4, which is for nearby stars. In this graph, called a Hertzsprung–Russell (H–R) diagram after the two astronomers who

Table 8.2. *Spectral classification of stars*

Spectral type	Characteristics
O	Very hot stars (*c.* 30 000 K) with ionized helium (He II) lines
B	Hot stars (*c.* 10 000–30 000 K) with neutral helium (He I) and some hydrogen (Balmer) lines
A	Intermediate-temperature stars (7500–10 000 K) with Balmer lines very strong and ionized calcium (Ca II) H and K lines present
F	Intermediate-temperature stars (6100–7400 K) with weaker Balmer lines and stronger H and K lines and some metallic lines appearing
G	Coolish stars (5000–6000 K) with strong H and K lines and lines of metals such as Fe; Balmer lines weaker
K	Cool stars (3500–4900 K) with strong metallic lines and molecular (e.g. CH and CN) bands
M	Very cool stars (2400–3500 K) with strong molecular bands, especially due to TiO

first realized its significance, most stars fall along a curve going from top left to bottom right known as the *main sequence*. Sometimes the $B-V$ colour is used instead of spectral type to plot the graph, it then being a 'colour–magnitude' diagram. The H–R or colour–magnitude diagrams show that, for main-sequence stars, luminosity has a one-to-one correspondence with colour and therefore with temperature. The sun, with its G2 spectral type, falls near the middle of the main sequence in this graph, while Vega and Sirius fall on the main sequence in the top left of the graph. Aldebaran does not lie on the main sequence, but is one of the sprinkling of stars at the top-right part of the graph; it is called a 'red giant', having a much greater luminosity than stars with the same colour or temperature on the main sequence. Some stars show much larger luminosities still, and are called supergiants. Main-sequence stars are sometimes referred to as dwarfs because of their relatively smaller luminosities. Another feature of Fig. 8.4 is a small group of faint, hot stars below the main sequence, known as 'white dwarfs'.

The most luminous and hottest stars at the top end of the main sequence have the greatest masses and radii also. The $B-V$ colour of a main-sequence star thus determines luminosity, mass, size and effective temperature. This 'mass–luminosity relation' applies not only to main-sequence stars but also the red giants and supergiants.

There are, even to the naked eye, recognizable clusters of stars, well-known examples of which include the Pleiades and Hyades in the constella-

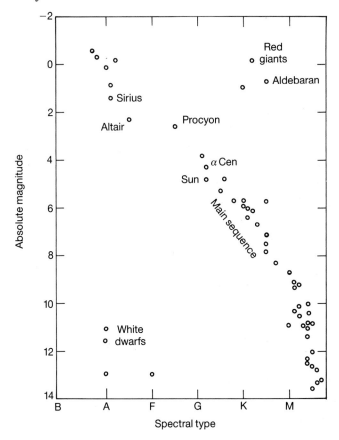

Fig. 8.4 Hertzsprung–Russell (H–R) diagram for nearby stars and several well-
known bright stars and the sun. Most stars fall on the main sequence
(curved line running diagonally from top left to bottom right), but there are
stars in the top right of the figure (red giants, such as Aldebaran) and
bottom left (white dwarfs).

tion Taurus. These are 'open' or 'galactic' star clusters, loose conglomera-
tions of a few tens to a few hundreds of stars. Each star in a cluster can be
regarded as being at approximately the same distance from the sun, so that a
colour–magnitude diagram for a cluster can be formed simply by plotting
$B - V$ against *apparent* magnitude V. Figure 8.5 shows this colour–magnitude
diagram for several open clusters, including the Pleiades and Hyades. A part
of the main sequence can be identified, but many of the stars form a branch
that departs from the main sequence. The colour–magnitude diagrams of
different clusters can be made to coincide by sliding them vertically with
respect to one another until the main-sequence sections form a single curve,
as in Fig. 8.4 for nearby stars: this has already been done in the case of Fig.
8.5. The amount of sliding in fact gives a measure of the relative distances of

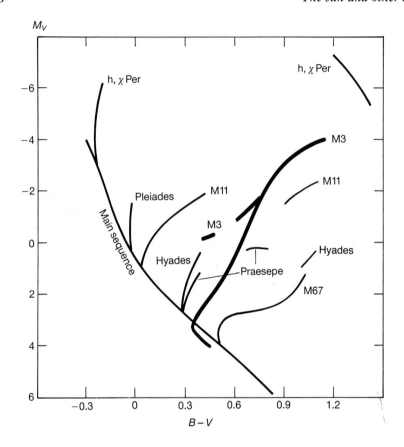

Fig. 8.5 Colour–magnitude diagram for several open clusters and the globular
cluster M3. For the open clusters, there is a main-sequence section with
an adjoining section of stars just evolving off the main sequence, and a
quite separate red giant section. For M3, there is only a short main
sequence, with extensive red giant and horizontal branches. The names of
the clusters are identified by each curve.

the clusters, since sliding in effect means adjusting the apparent magnitudes
of each cluster to give a common main sequence.

Certain other star clusters, called globular clusters, have very different
characteristics. They contain many more stars – some 10 000 to 10 million –
so many that individual members are only distinguishable in their outer parts.
Their colour–magnitude diagrams are quite different in character from those
of open clusters. Figure 8.5 includes the example of a globular cluster known
as M3 in the constellation Hercules. Only a small part of the main sequence is
present for it and other such clusters; most stars fall above the main sequence,
and are arranged in a segment going upwards towards the red giant stars ('red
giant branch'), and a horizontal one in the upper part of the diagram

('horizontal branch'). As with open clusters, the relative distances to globular clusters can be determined by matching the main sequences through a vertical sliding of their colour–magnitude diagrams.

In §8.3 we shall be referring to these characteristics of the H–R diagram when discussing the evolution of the sun and other stars.

The fifth of the important properties of stars listed earlier is the *chemical composition*. Element abundances are determined from visible-region spectra using the same procedure as that for the sun, viz. the curve-of-growth method (§3.7). Unlike the sun's photosphere, stars of spectral types O and B are hot enough that helium lines are excited, and from the strengths of their Fraunhofer lines the helium abundance can be found. The abundances of other elements can generally be determined to only an accuracy of a factor of two (i.e. the actual value could be in the range 50–200% of the measured value), but this is generally adequate for comparing abundances of different stars. Analyses show that most nearby stars have abundances like those of the sun (see Table 3.3): hydrogen is by far the most abundant (91% of all atoms by number), followed by helium (9%), with all other elements (carbon, nitrogen and oxygen in particular) accounting for only about 0.1%.

The abundances of heavy elements in the sun are very low, but some stars – those of globular clusters, the galactic nucleus and halo, and elliptical galaxies – have even lower heavy-element abundances. For such stars, elements like carbon, oxygen or iron are relatively under-abundant by factors of between three and about a hundred. They are known as Population II stars. Stars having heavy-element abundances like the sun (these include stars of the galaxy's spiral arms and those in open clusters) are known as Population I stars. These two broad classifications have been subdivided so that five classes, ranging from 'extreme Population I' to 'extreme Population II', are now recognized. It turns out that these stellar populations or rather their subdivisions represent a sequence of stellar *ages*. Extreme Population I stars are identified as being astronomically very young, while no Population I star is older than about 5×10^9 years; the sun (age 4.5×10^9 years) ranks as an 'Old Population I' star. From studies of their motions, it appears that Population I stars are generally performing nearly circular orbits round the centre of the galaxy. Population II stars, on the other hand, have ages of around 6×10^9 years, and from their motions appear to be travelling round the galactic centre in orbits that are highly inclined or elliptical; a few that are near the sun are apparently passing through the galaxy's plane and are moving at relatively high velocity with respect to the sun.

The picture that broadly explains these facts is that the galaxy originally had a nearly spherical structure. The first generation of stars formed from collapsing clouds of gas whose chemical composition was the same as that of the gas made in the Big Bang explosion, i.e. mostly hydrogen, some helium but no heavy elements. The galaxy later collapsed into its present disc-like structure with spiral arms, leaving behind the first-generation, or Population II, stars whose orbits were in general highly inclined to the galactic plane. Some of these stars lost mass or even underwent supernova-like explosions, returning their material to the interstellar gas, now enriched with heavy elements synthesized by nuclear reactions in their cores. The second-generation, Population I stars, including the sun, then formed out of this material, in regions close to the plane of the galaxy, and such stars continue to be formed at the present time. A star's chemical composition, then, in particular the abundances of the heavy elements, gives an indication of its age. We will deal with more definite estimates in the next section.

To conclude, let us briefly summarize how the sun compares with its stellar neighbours. The sun is like the vast majority of nearby stars in belonging to the main sequence, in which luminosity, mass and temperature are directly related; it is not a red giant, supergiant or white dwarf. It is an old Population I star which formed some 4.5×10^9 years ago from collapsing interstellar gas enriched with heavy elements, synthesized by nuclear fusion processes at the cores of first-generation, Population II stars. It is located in the galactic plane in or near a spiral arm, and is moving in a nearly circular orbit about the galactic nucleus. Its temperature and spectral type (G2) are intermediate between hot blue stars and cool red dwarfs. The sun's mass is not far from being 'average' in a logarithmic sense, but it is unusual in being single and not a member of a binary or multiple system and appears to be brighter than the vast majority of nearby stars.

8.3   Evolution of the sun and other stars

We now consider how stars including the sun come into being, how they maintain themselves in their lifetimes and what will ultimately happen to them. Although the changes a star undergoes in its lifetime are normally far too slow to be directly observed, astronomers have formed a picture of what happens by building theoretical models which use known laws of physics, and by identifying those stars undergoing stages which the models predict will occur. We will describe their findings with reference to the sun.

Now, between the stars of particularly the spiral arms of our galaxy there is extremely tenuous gas and dust, and it seems likely that this is the material out of which some future generation of stars will form – indeed, in certain regions, stars are apparently being born. Astronomers believe that the probable chain of events that lead to the formation of stars from this gas is as follows.

The interstellar gas, mostly of hydrogen with a density of from 10^5 to 10^7 particles per cubic metre, is confined to the plane of the galaxy by its own pressure. It is calculated that, if this pressure is relatively small, the gas will be highly rarefied and hot – up to about 10 000 K – since it is heated by various strong X-ray-emitting objects in the galaxy while there is no efficient cooling mechanism. A slight increase in pressure may lead to a sudden increase in the cooling of the gas caused by the recombination of various ions. This leads to a spontaneous breaking up into clouds of cold, relatively dense gas. Further increase of pressure causes the cold clouds to cool still further to extremely low temperatures – less than 20 K – the cooling mechanism now being the infrared radiation emitted by hydrogen molecules. A cloud of cold gas so formed now starts to collapse under its own gravity if its internal pressure is insufficient to withstand the gravitational force. Most likely, a very large cloud breaks up into many subclouds, each of which collapses to a 'proto-star'.

The gravitational energy released by the contraction heats the proto-star to a temperature of a few thousand degrees Kelvin. The proto-star contracts a hundredfold in only a few years: from 10 000 to 100 solar radii across. During this short period, the proto-star remains transparent to its own radiation, which increases sharply to give a 'flare'. After this, the proto-star becomes opaque and so fades. Convection now starts to transport energy from the centre of the proto-star, and when the convection reaches the surface, a second 'flare' occurs, within perhaps a few years after the first. Further contraction to the size of a star is much more gradual, occurring over the next few million years, with energy generated by gravitational collapse and transported to the proto-star's surface by convection. A fully fledged star is formed when the interior is hot enough to produce nuclear reactions which now become the energy-generating mechanism.

Observational evidence for the above is provided by large-scale associations of stars that seem to have been formed in the (astronomically) recent past. Some ('O-associations') are of hot, bright, very young stars. Others ('T-associations') are of much fainter objects called T Tauri stars, which show irregular brightness fluctuations including short-term flaring and ejections of material, and appear to be proto-stars in their final contraction phase before nuclear reactions begin (§8.5). Stellar associations probably arose from the break-up of very large clouds into subclouds that subsequently turned into proto-stars, as already described. The newly born stars of an association may eventually form an open cluster, though most rapidly disperse into the rest of the galaxy. Figure 8.6 shows one of the most extensive stellar associations, in the 'Belt' and 'Sword' regions of the constellation Orion. Stars in these regions have expansion velocities that indicate the association is only a few million years old. The mass of the association is some 100 000 solar masses.

There is also possible observational evidence for the flare-like brightenings predicted to occur as a proto-star undergoes its initial, rapid collapse. The young object FU Orionis, about 25 times the size of the sun, underwent such a brightening in 1936, and has been shining at about this level ever since. Like T Tauri stars, it shows irregular brightness variations, with small flares that are possibly due to infalls of blobs of gas as the object contracts.

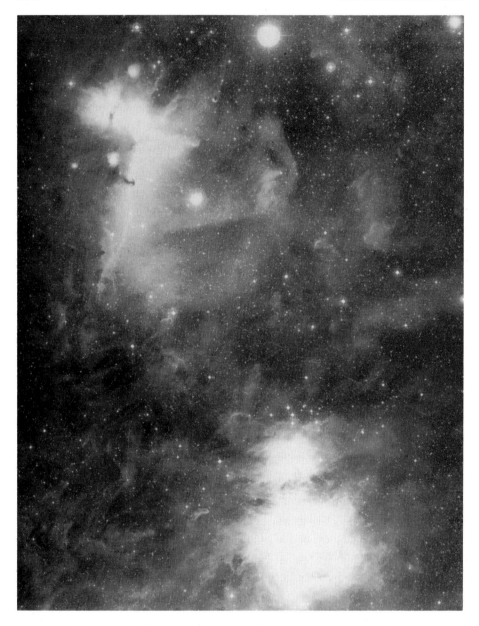

Fig. 8.6 Extensive stellar associations in a familiar part of the sky: the Belt and
Sword of Orion, including the Orion Nebula (bottom of figure) and the
Horse's Head Nebula (small dark notch against bright background, top
left). (Photography by D. F. Malin, UK Schmidt Telescope; © 1987 Royal
Observatory, Edinburgh)

A proto-star's luminosity and effective temperature can be calculated as it contracts, so that the track it takes across the colour–magnitude diagram can be plotted. Figure 8.7 shows the tracks of model proto-stars with various masses, including one of the solar mass, as calculated by C. Hayashi. (In this figure, the colour or horizontal axis has been converted to effective temperature, on a logarithmic scale.) All the proto-stars start their contraction phase well above the main sequence, i.e. are very luminous owing to their large radii, especially the more massive proto-stars. Eventually they move towards the main sequence, the massive proto-stars to the top end and the less massive towards the lower end; those of extremely small mass (less than 0.1 times the solar mass) never become stars as such because their central temperatures are not high enough for nuclear reactions to occur. A star's mass thus determines the position of a star on the main sequence. It also determines the time spent in the contraction phase: a proto-star of four times the solar mass reaches the main sequence in only about a million years, one with mass equal to the sun's takes 10 million years, and one with a fraction of the solar mass several times longer. Nuclear reactions start to occur a little before the proto-star arrives at the main sequence.

These are the probable circumstances of the birth of nearby stars as well as the sun. The sun has now reached a state of middle age, so to speak, like most of the stars we see in the sky and in particular those in the sun's neighbour-hood. On the H–R diagram, such stars have remained on the main sequence very near the point where they joined it from the proto-star stage. The sun's radiated energy nearly all derives from the proton–proton (pp) nuclear reactions at its core. Stars more massive than the sun are expending energy at a much more rapid pace, and for them a different set of nuclear reactions occurs.

Models for stars more massive than the sun can be calculated in a similar way to the standard solar model (§2.1). Physical laws are used in the calculations to describe (*a*) hydrostatic equilibrium (balance of gravity and pressure of gas and radiation in the star); (*b*) energy generated at the star's core; and (*c*) energy transport from the core to the star's surface. The temperature at the core of the star is large enough for a set of reactions to occur involving carbon and nitrogen nuclei; these reactions are more complex than the pp chain but essentially lead to the same net reaction: the fusion of four protons to form a helium nucleus. This 'CN cycle' starts with the capture of a proton by a ^{12}C nucleus to give an unstable nitrogen nucleus ^{13}N, which decays to give the carbon nucleus ^{13}C. A second proton is captured to give ^{14}N, and a third to give the unstable nucleus ^{15}O. The ^{15}O nucleus decays to ^{15}N, which finally captures a fourth proton to give a nucleus of the common form of carbon, ^{12}C, plus a helium (^{4}He) nucleus. Gamma rays, neutrinos and a positron are also products of some of these reactions. Most of the energy generated is available to heat the star's core, but a small fraction is carried away by the

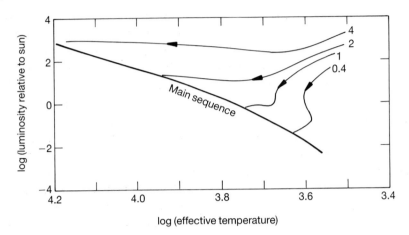

Fig. 8.7 Evolutionary tracks for proto-stars of various masses (indicated by each
curve, in units of a solar mass). The vertical (luminosity) and horizontal
(effective temperature) scales are both logarithmic. (After Kaplan (1982))

highly unreactive neutrinos. As with the pp chain (§2.1), we can write the
details of the CN cycle symbolically, as follows:

$$^{12}C + {}^1H \rightarrow {}^{13}N + \gamma$$
$$^{13}N \qquad \rightarrow {}^{13}C + e^+ + \nu$$
$$^{13}C + {}^1H \rightarrow {}^{14}N + \gamma$$
$$^{14}N + {}^1H \rightarrow {}^{15}O + \gamma$$
$$^{15}O \qquad \rightarrow {}^{15}N + e^+ + \nu$$
$$^{15}N + {}^1H \rightarrow {}^{12}C + {}^4He$$

The symbol γ indicates a gamma-ray photon, e^+ a positron and ν a neutrino.
The net result (obtained by summing the left and right sides) is the fusion of
four protons to form ^4He helium:

$$4{}^1H \rightarrow {}^4He + 2e^+ + 3\gamma + 2\nu$$

Some 4.0×10^{-12} J of energy are liberated and are available to the star,
slightly less than for the set of pp reactions (4.2×10^{-12} J) because of a larger
energy loss by neutrinos. The carbon and nitrogen nuclei are destroyed but
are then recreated and so are available for the sequence of reactions to occur
over again (hence the name CN *cycle*); they act as 'catalysts'.

The CN cycle must occur at higher temperatures than for the pp chain as
the protons being captured by carbon and nitrogen nuclei need to be much
faster in order to overcome the much larger electrostatic repulsion force. It is
the dominant mechanism for temperatures of more than 20×10^6 K. Such
temperatures occur at the cores of stars that are more luminous and therefore
(by the mass–luminosity relation) more massive than the sun.

A star's mass also determines the way energy is transferred from the core to the surface of a main-sequence star. As we saw in §2.1, energy generated by the pp chain at the core of the sun is transported by radiation for the inner part of its interior, since the gas has an opacity low enough to permit this, but where the temperature has cooled to 10^6 K, there is a large increase in opacity, resulting in convective energy transport. For stars lighter than the sun and with a cooler core temperature, the boundary between radiative and convective zones occurs progressively closer to the core until, for stars with a third of the solar mass, convection is the sole energy transport mechanism. For stars heavier than the sun, the energy-generating mechanism is the CN cycle. This mechanism is a highly effective one, so much so that radiation is insufficient to transport the great amount of energy; such stars, then, have convective cores. The greater the mass of the star, the greater is the extent of this core. On the other hand, radiation is an effective transport mechanism from the outside of this core to the star's surface, and so, unlike the sun, there is no convective envelope.

There is a very large difference in the rate at which hydrogen is 'burnt' (i.e. converted to helium) for relatively low-mass stars like the sun and high-mass stars. It is estimated that it will be another 5×10^9 years before the sun depletes the hydrogen in its core at the rate at which it is being consumed. A main-sequence star with three times the solar mass, on the other hand, converting hydrogen to helium by the CN cycle, consumes hydrogen at a rate vastly higher, depleting its hydrogen fuel in 2×10^7 years; its middle age is a tiny fraction of the sun's. Any massive star now observed to be on the main sequence must therefore have been formed relatively recently.

The evolution of a star after the hydrogen fuel at its core has been depleted depends on its mass and chemical composition. Model calculations are able to follow this evolution up to a certain point, but uncertainties steadily grow as the final evolutionary stages are reached. We will first describe the proposed evolution of the sun.

After the depletion of hydrogen, the sun's core will be almost entirely composed of helium. The temperature is too low to permit further reactions, so the helium remains inert. But in time the core begins to collapse under its own gravity, causing it to heat up. The region just outside the core, still largely hydrogen as it has not undergone nuclear reactions, is brought in closer to the sun's centre and heated to a temperature at which the nuclear reactions can occur to convert the hydrogen to helium. Thus, the sun now consists of a helium core, a hydrogen-burning shell and a hydrogen envelope. The helium produced by the hydrogen-burning shell is added to that in the core, so that the shell occurs progressively further out in the sun's interior. The core, meanwhile, grows hotter and denser, and the energy it produces interacts with the materials of the envelope, causing it to expand greatly, and in doing so to become cooler and less dense. Meanwhile, the convection zone is brought to a deeper level, down almost to the level of the hydrogen-burning shell. On the H–R diagram (Fig. 8.8), the sun moves off the main sequence to the right (because the envelope is cool) and upwards (because of the effect of the deeper convection zone), to become a *red giant*.

The sun will spend a few hundred million years as a red giant. Its envelope will have expanded to more than a hundred times its present size, larger than the diameter of Mercury's orbit, its luminosity will be a thousand times the present value, and its surface temperature reduced from 6000 K to 3000 K. All life-forms on earth as we know them will be destroyed by the extreme conditions, but if by some chance intelligent beings remained after some

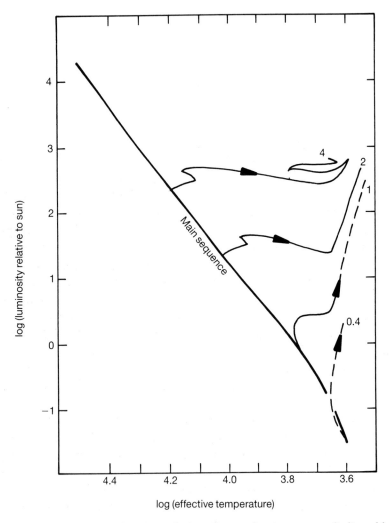

Fig. 8.8 H–R diagram showing evolution of stars of various masses (indicated by each curve, in units of a solar mass) off the main sequence up to the red-giant phase, and (for the 4 solar mass curve) beyond, to the helium-flash phase.

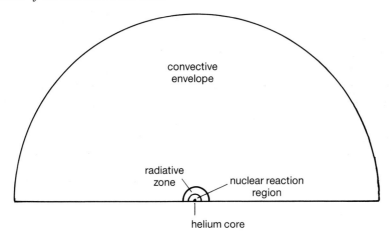

Fig. 8.9 Internal structure of a red giant.

fantastic evolutionary adaptations they would witness the huge red globe that our sun would have been turned into.

Figure 8.9 shows the expected structure of the sun in its red-giant phase. The tiny helium core, now at a temperature of 40×10^6 K, has contracted to a size of only a few thousand kilometres across, or 0.1% of the sun's total size, with a density of 3×10^8 kg/m^3. The material inside it is a fully ionized plasma consisting of helium nuclei and electrons, but it is at such a high density – the particles are squeezed more tightly than the atoms in a solid or liquid – that it no longer behaves like a conventional gas but more like a molten metal, with high heat conductivity so that it has the same temperature throughout. The material is said to be degenerate. Once in a degenerate state, the core cannot contract any further, even if pressure is increased. Surrounding it are successively the hydrogen-burning shell with a temperature of 25×10^6 K, a radiative zone and, finally, a distended envelope which contains 70% of the total mass and constitutes nearly all the volume. There is a continuous loss of mass at the surface as a powerful 'wind' expels the outer part of the envelope into space.

At the end of the sun's red-giant phase, the helium core becomes hot enough for the helium to react or 'ignite'. At first the reactions are principally two helium nuclei fusing to form a nucleus of ^{8}Be, a highly unstable isotope of beryllium which decays (in 10^{-14} s) into two helium nuclei again. But at temperatures approaching 10^8 K, reactions involving the fusion of three helium nuclei (the 'triple-alpha' process) are possible, in which the unstable ^{8}Be nucleus, formed by the fusion of two helium nuclei, absorbs a third helium nucleus before it has time to decay, forming a ^{12}C nucleus and liberating 1.2×10^{-12} J of energy. Such an energy release in a normal gas would result in expansion of the core. As the gas is degenerate, however, there is no expansion, only a strong heating, which leads to a greater and greater rate of triple-alpha processes which release more and more energy. This unstable situation ends with a sudden, very strong outburst of energy and a core temperature now so high that the pressure is inadequate to maintain the core's degenerate state: there is a sudden reversion to a normal, fully ionized gas, which unlike the degenerate state it was in is able to undergo a huge expansion. This 'helium flash' marks the end-point of the sun's red-giant phase.

Most of the energy released in the helium flash goes into the expansion of the core, but calculations suggest that a star like the sun will pass through the phase without being broken up or destroyed. The envelope apparently readjusts to the altered conditions in the core, though there is some mixing of the core contents. Some stars are observed that are thought to have undergone the helium flash without destruction.

After the helium flash, the sun will have two zones of nuclear burning – the core, where helium is being converted to carbon and oxygen, and a surrounding hydrogen-burning shell. In the H–R diagram, the sun moves to a new position, along the so-called horizontal branch. Its luminosity falls to about a tenth of what it was before the helium flash, while the surface temperature is not too different from before.

After some 150 million years, when even the helium fuel in the core has been exhausted, the sun evolves in much the same way as at the beginning of the red-giant phase: helium burning occurs in a shell surrounding the now inert carbon–oxygen core, with the hydrogen burning still occurring in a shell further out. The helium burning in the shell, which produces energy at a furious rate, is highly unstable and subject to periodic 'flash' phases that result in large energy outbursts and an expansion. Repeated shell flashes are calculated to occur as the sun expands and cools, then contracts and heats up again. Successive occurrences could result in convective mixing of the largely hydrogen envelope with the products – carbon and oxygen – of the helium burning 'dredged up' from the interior to appear at the surface. On the H–R diagram, this evolutionary stage is marked by a movement upwards from the horizontal branch to the so-called asymptotic branch.

The sun may then start to pulsate like the many 'long period variable stars' that are observed. There are two possible reasons why these stars pulsate. The first is that, if such a star were to contract, it would tend to increase the rate of nuclear reactions and so increase the energy output, causing the star to expand again; the expansion might 'overshoot', a contraction would then ensue, and the cycle is thus repeated. The second reason has to do with near-surface zones in the star's envelope where hydrogen and helium are both on the point of being ionized or recombining and which through their opacity properties are able to retain energy during a contraction phase, causing them to heat up and push the overlying material out and give rise to an expansion phase and so a repeated cycle of pulsations.

The final stage of the sun's evolution will be marked by a cessation of nuclear burning because of the continued large mass loss. The outer envelope in time will be entirely shed, and the hot, dense, degenerate core will be exposed. Most probably the objects known as planetary nebulae can be identified with this stage. Through a telescope these nebulae appear as small discs like planets, hence their name. Figure 8.10 shows the Helix Nebula (NGC 7293) in Aquarius. The ring or helix-shaped cloud of a planetary nebula is an expanding shell of tenuous gas. Its atoms or ions are excited by ultraviolet radiation that a faint, very hot central star emits (this star can be seen in the figure). The central star is a white dwarf, a low-mass star with extremely large density, some $10^9 \, \text{kg/m}^3$, and tiny size (10 000 km diameter), comparable to the earth's. A characteristic spectrum of highly broadened Fraunhofer lines on a continuum arises because of the presence of a thin atmosphere. The density of white dwarfs implies that their material is degenerate, and unlike ordinary stars, all nuclear reactions have ceased, and so they are in a state of inexorable cooling. Indeed, only recently formed white dwarfs are actually white – those formed long ago have cooled to become faint red objects. Eventually they will lose their energy altogether and not emit any light. On the H–R diagram of Fig. 8.4, white dwarfs are located well below the main sequence.

Such, then, seems to be the fate of the sun. The evolution off the main sequence for other stars may be rather different, depending mostly on their mass but for some stages their chemical composition as well. Stars that are less massive than the sun but still massive enough to burn hydrogen during their main-sequence stages generally evolve into red giants with extended cool envelopes and small helium cores. If the mass is less than about half of the sun's, there is insufficient material to form a hydrogen-burning shell, so no new helium is added to that in the core, as with the evolution of the sun. The temperature never rises to a point where helium burning can occur, and so the star finishes its life as a white dwarf. Stars more massive than the sun, on the other hand, end their lives more spectacularly. They evolve to the red-giant phase by moving in a horizontal direction across the H–R diagram (see Fig. 8.8). Their helium cores, unlike lower-mass stars, are not degenerate, and so the helium-flash instability does not occur for them. Helium burning results in a cooling of the star without much reduction in luminosity and the formation of a carbon core. Figure 8.8 shows how a star of four times the solar mass evolves during helium burning across the H–R diagram. The most massive stars may have core temperatures high enough to initiate carbon burning. Further and more exotic nuclear processes may then occur at an ever-increasing rate, the end-point of which is the synthesis of iron (^{56}Fe) nuclei. Reactions building still heavier nuclei no longer emit energy. Owing to energy loss by neutrinos, there is a sudden contraction of the core which raises the temperature high enough to break up the iron nuclei. The energy

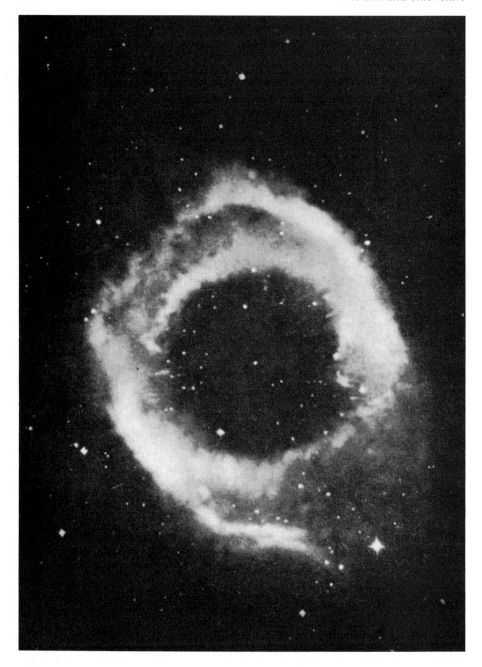

Fig. 8.10 The Helix Nebula (NGC 7293) in Aquarius, a planetary nebula, probably the eventual fate of the sun.

deficit caused by this process leads to a sudden implosion of the rest of the star, bringing in fresh material for nuclear processing. The result is a supernova, a catastrophic explosion of material into space, with the formation of an extremely small, dense remnant. This object is far smaller (a few kilometres) and denser (10^{13} kg/m^3) than even a white dwarf – at such densities, electrons and protons combine to form neutrons which compose much of the structure of this 'neutron star'. Several supernovae have been seen in historical times. That in AD 1054 in the constellation Taurus, recorded by oriental astronomers, is now marked by the famous Crab Nebula. The structure seen in this nebula is the material ejected by the explosion. At its centre is the neutron star observed as a pulsing radio source.

If the remnant left behind after a supernova explosion is too massive even for a neutron star to remain stable, there is a possibility that an object will be formed that is so dense that even radiation cannot escape from its surface – a *black hole*. Such objects have been identified with certain strong X-ray sources whose emission shows rapid fluctuations which are believed to be due to the accretion of matter into a black hole, the matter being heated to X-ray temperatures just before disappearing into a barrier beyond which it cannot be seen.

Model calculations could be used to follow the evolution of several stars, with a variety of initial masses but all assumed to be born at the same time, and their subsequent tracks across the H–R diagram followed with time. The results should correspond to the observed H–R diagrams of star clusters, like the composite one shown in Fig. 8.5. This is indeed the case. For the galactic clusters, there is a range of 'turning-off' points, below which the cluster members are still on the main sequence but above which stars have started to evolve into red giants. The different locations of the turning-off points show the different ages of the clusters. The comparatively young Pleiades cluster still has most of its stars on the main sequence though some of the more massive have evolved off to become red giants; comparison with model calculations gives the age of the Pleiades as 8×10^7 years. The cluster M67 is much older – 4×10^9 years – with just a few cool stars forming the main sequence, the rest having evolved off. Stars fully evolved to the red-giant phase form separate groups by themselves in the top-right corner of the diagram. All the clusters shown in Fig. 8.5 apart from M3 have stars belonging to Population I. The globular cluster M3 has a markedly different H–R diagram. As it is relatively very old – more than age 10^{10} years – most of the stars have evolved into red giants. There is a horizontal branch, consisting of stars that have undergone the helium-flash stage. Models predict that the position of a star after the helium flash depends on its chemical composition – a Population I star like the sun will occupy a position near to their former red-giant positions on the H–R diagram, while a Population II star with no heavy elements ends up well to the blue; thus, the Population II stars of M3 form a well-defined horizontal branch as observed.

The foregoing has described the evolution of stars as single objects, but as we saw in §8.2 a high proportion of stars are in binary or multiple systems. Sometimes the stars in such systems are so close that their structures interact, so affecting their evolution. Thus, if one star of a binary is more massive than the other, it will arrive at the red-giant phase first. Its outer envelope may swell out to such an extent that some mass is lost to the other star. This mass loss may result in the first star becoming not a red giant but what is called a subgiant, lying between the main sequence and red giants on the H–R diagram. We shall discuss some of these binaries in the next section.

8.4   Stellar chromospheres, coronae and winds

A remarkable feature of the sun's atmosphere is the rise of temperature with height through the chromosphere up to the corona. If we were to observe the sun from a nearby star, the evidence for this atmospheric structure would be much more subtle. The sun would appear as a pinprick of light, and we would therefore have to rely on its spectrum to gain any information about it. In visible light, the tiny calcium H and K line emission peaks (see §4.2, Fig. 4.4) would be virtually the only indication of the chromosphere. Ultraviolet and soft X-ray emission from the chromosphere and corona would, with present technology, only just be detectable. It would appear, then, that solar-like chromospheres, transition regions and coronae are unlikely to be observed for more than a few stars. In fact, they are observed for many main-sequence stars with effective temperatures comparable to or cooler than the sun's, and for stars with special characteristics.

Strong emission in the calcium H and K lines, a greatly exaggerated form of the weak emission (H2 and K2) peaks in the solar lines, is increasingly common for stars progressively cooler than spectral type F, becoming almost universal for red dwarfs of type M. As with the sun, there are two emission peaks either side of a self-reversed core (H3 or K3). For M dwarfs the emission peaks are much more intense than the neighbouring continuum, and other lines, notably Hα, may also be in emission, especially for stars cooler than M4. The presence of such emission lines strongly suggests the existence of solar-like chromospheres.

Those M dwarf stars with Hα in emission are called 'dMe' stars – 'd' for dwarf, 'e' for Hα line in emission – and those without are 'dM' stars (even if the calcium lines are in emission). Figure 8.11 is a spectrum of one such star, AT Microscopii (actually a close visual binary, consisting of two dM4e stars); the Balmer lines from Hβ to Hη and the calcium lines are all in emission and

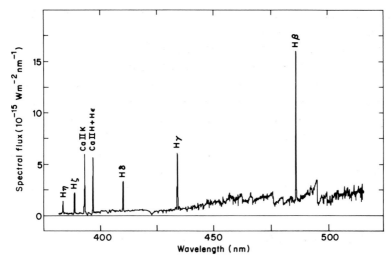

Fig. 8.11 The visible spectrum of the dMe star AT Microscopii in 1980. The cool
     star's photosphere emits the continuum which has absorption bands of the
     molecule titanium oxide; its chromosphere is responsible for the strong
     emission Ca II (H and K) and Balmer (Hβ to Hη) lines. (Bromage *et al.*
     (1982))

are chromospheric in origin, while a weak continuum that is crossed by
titanium oxide molecular bands is emitted by the cool photosphere.

Whether an M dwarf has dM or dMe characteristics depends on the nature
of its chromosphere, i.e. how steeply the temperature rises above the photosphere. A sharp temperature rise is equivalent to the chromosphere's being
brought deeper into the star's atmosphere. Model calculations of the
presumed atmospheres of M dwarfs have been calculated, making the
temperature rise a variable quantity and seeing what happens to Hα and the
calcium H and K lines. Now, the H and K lines in the solar atmosphere are
collisionally controlled (§4.2), i.e. electron collisional processes predominate
in their source functions, but the Hα line is photoelectrically controlled, i.e.
radiation from the photosphere governs processes in its source function. For
M dwarfs, the photosphere is much cooler (about 3000 K) than the sun's
(6400 K), so the effect of its radiation on the source function of the Hα line is
much reduced. Depending on the chromospheric structure of an M dwarf,
the Hα line can be collisionally controlled, like the calcium lines of an M
dwarf. As the temperature rise in the chromosphere becomes steeper, the
model atmosphere calculations show that the Hα line turns first from a weak
absorption line to a strong absorption line, as in the spectra of dM stars, and
then to a strong emission line with self-reversed core, as is observed in the
spectra of dMe stars.

Ultraviolet emission lines that in the solar spectrum are emitted by the chromosphere and transition region (see Table 4.1, §4.4) are prominent in the spectra of cool dwarf stars. Many of these stars have been surveyed with the *International Ultraviolet Explorer* (*IUE*) satellite which has on board a telescope and spectrometer to detect ultraviolet spectra in the range 110–300 nm. Chromospheric lines like Mg II (280 nm) and Si II (181 nm) are generally present for dwarf and giant stars cooler than spectral type F. For dM stars, the lines have intensities per unit area of the star's surface comparable to the quiet sun, but for dMe stars they are comparable to solar active regions. Lines characteristic of the solar transition region, like the C IV doublet (155 nm), are also prominent for such stars. Again, for dM stars, the intensities per unit area are comparable to the quiet sun, but for dMe stars they are much larger than even solar active regions.

The coronae of cool dwarf stars have been detected by soft X-ray instruments on board the NASA *Einstein*, the ESA *Exosat*, and most recently the German–US–UK *Rosat* spacecraft. The data indicate that possibly all dwarf stars between types F and M have coronae. The X-ray luminosities vary widely, from about 10^{20} W to 10^{27} W; the quiet solar corona, for comparison, is at the bottom end of this range. For dMe stars particularly, a far greater fraction of their total energy is in the form of soft X-rays than is the case for the sun.

Further evidence for coronae in dMe stars is provided by radio observations, now possible with the fine resolution and great sensitivity of especially the Very Large Array interferometer. At a wavelength of 6 cm, large brightness temperatures and circular polarization are observed, indicating a nonthermal (e.g. synchrotron) emission mechanism, as in the solar corona.

There are some important clues about stellar chromospheric and coronal activity. Firstly, there seems to be a rather sudden 'turn-on' of activity at spectral type F0 – very few stars hotter than this appear to have solar-type chromospheres or coronae, but many cooler stars do. This is almost precisely where outer convection zones appear in main-sequence stars. Secondly, stellar chromospheric and coronal activity increases the greater the observed rotational speeds. Convection and differential rotation are requisites for the dynamo thought to maintain solar magnetic fields. It is quite likely, then, that the combination of an outer convection zone and rapid rotation (which may imply rapid *differential* rotation as well) leads to magnetic fields in stellar atmospheres generally, and if fields are present, the outer atmosphere may be heated by magnetic processes. The coronae could therefore be heated by the processes suggested for the solar corona, viz. steady or impulsive dissipation of currents or the dissipation of MHD waves (§5.7), and may well be in the form of hot plasma in magnetic flux tubes and outflowing plasma in coronal holes. The chromospheres of dMe stars may also be heated by the dissipation of MHD waves, though sound waves transformed into shocks may be adequate for the heating of the chromospheres of dM stars. Heating of the dMe star chromospheres by the X-ray coronal emission may also be significant.

Chromospheric and coronal activity as well as rotation seem to decline with a star's age. This can be seen by comparing cool dwarfs in the nearby clusters for which ages can be assigned, in particular the Pleiades (age 8×10^7 years) and Hyades (age 4×10^8 years). This can be explained by the fact that stars are gradually 'spun down' in time by angular momentum loss due to solar-type winds, and so with the reduced rotation rate the dynamo generating magnetic fields decreases in efficiency.

Cool dwarf stars often occur in binary systems, and this fact affects their atmospheric activity. Frequently, for example, two K or M dwarf stars orbit each other, with one or both being a rapid rotator and chromospherically active: these are the so-called BY Draconis stars. The stars may be sufficiently well separated that there is no exchange of mass from one to the other, though one component may suffer tidal effects from the presence of the other or the two stars may be tidally 'locked', i.e. have equal rotation and orbital periods. With RS CVn (named after the prototype star known as RS Canum Venaticorum), an F or G main-sequence star orbits a K 'subgiant' that is dimmer but more chromospherically active. The orbital periods vary from 3 days to about 60 days. It seems that the K star evolved off the main sequence at some stage, but as it began to develop the extended envelope of the red-giant phase, material was transferred to the companion star.

RS CVn stars have visible spectra showing Hα and the H and K lines in emission, and the ultraviolet spectrum shows intense chromospheric and transition-region lines. They are powerful X-ray emitters, indicating the presence of coronae with much higher temperatures (several tens of million degrees Kelvin) than the solar corona. There are indications (from the observations of RS CVn binaries in which the stars periodically eclipse each other) that there are extremely large hot structures in the coronae, perhaps magnetic flux tubes, that may connect the two stars. Figure 8.12a is a sketch of the RS CVn binary AR Lacertae deduced from observations of its eclipses.

More remarkable are the contact binary stars, typified by the star W UMa (prototype star W Ursae Majoris). They are composed of two dwarf stars with spectral types between A and K that are in contact as they orbit each other, in periods of less than a day. Mass is transferred from the more massive star to its companion. Ultraviolet and X-ray observations suggest that both stars share a common chromosphere and corona in the way indicated by Fig. 8.12b.

Another group of stars with solar-like characteristics is the T Tauri stars, mentioned in §8.3. They are extremely youthful objects (ages of a few million years) with masses comparable to the sun's, still contracting in the proto-star stage and moving towards the main sequence. Their visible and ultraviolet spectra have emission lines like those of the solar chromosphere and transi-

(a) RS CVn system

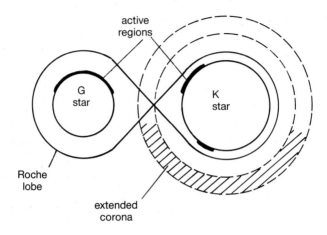

(b) W UMa system

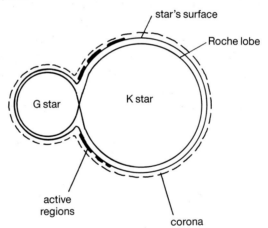

Fig. 8.12 (a) Model for the RS CVn binary system AR Lacertae; (b) Model for the W UMa (contact binary) VW Cephei. (After Walter *et al.* (1983) and Xuefu and Chengzhong (1983). Reprinted by permission of Kluwer Academic Publishers)

tion region. Some T Tauri stars also emit X-rays, though weakly compared with the ultraviolet line intensities. A strong wind-like mass outflow, indicated by the profiles of emission lines, is often present. Objects very similar to T Tauri stars, perhaps forming a subgroup, show outflows of cold, molecular gas, with the proto-star embedded in a cold, dust envelope that is revealed by its infrared emission.

Opinions about the nature of the T Tauri and related stars are changing very rapidly at present, so nothing more than a broad picture can be given. It seems possible that the visible, ultraviolet and X-ray emission arises, not from a chromosphere – transition region – corona structure as in the sun, but from a region between the proto-star and a slowly rotating disc-like structure made up of gas and dust and left over after the contraction. Material from this 'accretion' disc slowly falls on to the proto-star, forming a hot 'boundary layer', which gives rise to the observed emission (see Fig. 8.13). In time, the disc material diminishes, and with it the X-ray emission, giving what has been called a 'naked' T Tauri star; this appears to be a stage immediately before the contraction on to the main sequence. The outflows of cold molecular gas seen for some T Tauri stars seem to be due to a stellar wind that is driven along the poleward directions by the accretion process, and which pushes against and sweeps up the cold gas out of which the star formed.

It can be concluded from the above that cool dwarf stars and certain types of binary star systems have atmospheric structures similar to the sun's. What can be said about stars in other parts of the H–R diagram? Strong X-ray emission is seen for O and B main-sequence stars (but not for A stars): it is thought to arise from strong, hot winds in these stars rather than solar-type coronae. These winds are not driven thermally (i.e. by virtue of the high temperature of the gas), as in the solar wind, but by the pressure exerted by the strong ultraviolet continuous radiation from the hot photospheres of these stars. The winds are detected by characteristic shapes of ultraviolet emission lines, which also show evidence of periodic discrete ejections not unlike the solar coronal mass ejections. X-ray emission is absent for stars that have

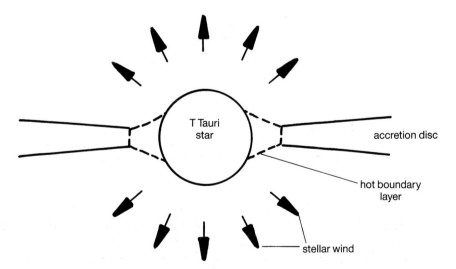

Fig. 8.13 Model for a T Tauri star, in which ultraviolet and X-ray emission comes from a hot boundary layer between the star and an accretion disc of cold material.

evolved off the main sequence and are cooler than type K (for giants) or type G (for supergiants). The acceleration due to gravity is much lower in the greatly distended atmospheres of these stars than at the sun's surface, and so any hot coronal gas much more easily overcomes gravity to form a solar wind-type flow, though with lower speed than the solar wind (50 km/s instead of 400 km/s). This wind is too tenuous or too cool to produce observable X-ray emission.

The schematic colour–magnitude diagram in Fig. 8.14 (originally drawn by J. L. Linsky) shows the locations of stars mentioned in this section having solar-like coronae or winds. There is a sharp division at around spectral type F0 ($B-V=0.2$) separating main-sequence and giant stars showing no solar-type coronae or chromospheres (hotter than F0) from stars (including the sun) that do. A dividing line at around spectral type K0 ($B-V=1.0$) separates giants, supergiants and RS CVn stars that have solar-type coronae and chromospheres (hotter than K0) from giants and supergiants that instead show extended chromospheres but no coronae. Hot main-sequence stars of type O and B (like ζ Puppis) do not have solar-type characteristics but rather a hot, X-ray-emitting wind which is radiation-driven.

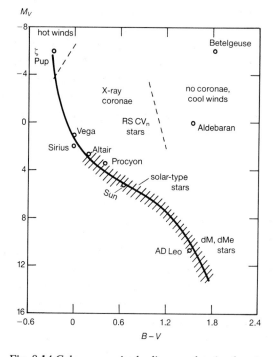

Fig. 8.14 Colour–magnitude diagram showing locations of types of stars mentioned in the text. Stars with solar-type chromospheres and coronae include all main-sequence stars between spectral types F and M, and giants, supergiants and RS CVn stars between types F and K. (After Linsky (1985))

8.5 Stellar activity

Stars with solar-like chromospheres and coronae frequently seem to show evidence for spots, active regions, prominences and flares, i.e. phenomena like those of the active sun. Since we cannot resolve individual features on these stars, such activity can only be recognized from spectra or variations in visible, ultraviolet or X-ray emission.

Long-term variations, including periodic behaviour akin to the eleven-year solar cycle, have been identified in cool main-sequence stars and binary systems like the RS CVn stars. Monitoring of the calcium H and K lines in their spectra as indicators of the presence of solar-type plages or active regions has been carried on for many years to see whether the emission peaks in these lines vary as in the sun between sunspot maximum and minimum. Surveys of the calcium lines in the spectra of cool dwarf stars have revealed several examples in which the strengths of the emission peaks have slow, smooth variations. Slowly rotating stars with weak emission lines are most inclined to show periodic changes like the solar cycle, while some of the rapidly rotating stars with strong emission peaks vary either irregularly or with periods generally much shorter than 11 years.

Records of the total visual light of some solar-like stars also show long-term trends. Some stellar variations in *U*, *B*, *V* colours have been found by examining old photographic plates. They are strongest for the RS CVn- and BY Draconis-type binaries, with amplitudes of up to about 0.6 magnitudes in *B* and periods of tens of years. They are hardly present at all for single cool dwarfs except for rapidly rotating stars with strong chromospheric indicators. Maunder minimum-like periods of inactivity may occur for some stars, such as the RS CVn-type star II Pegasi, which showed a lull for some 40 years before resuming cyclical variations in *B* magnitude.

Individual spots or spot groups ('starspots') and active regions are identifiable on some active-chromosphere stars. Thus, short-term variations in calcium H and K and ultraviolet emission line strengths have been attributed to persistent chromospheric active regions rotating on and off the visible surface of the star. Similarly, periodic modulations in *UBV* magnitudes, particularly in RS CVn and BY Draconis binaries, are probably due to starspots, occurring at times of enhanced chromospheric and transition-region emission. The visible light has been observed to dim by as much as about 40% when a large starspot appears. This very large amount must mean that starspots can cover several tenths of the star's surface area – far greater than the amount of solar surface covered by sunspots. Starspots are, like sunspots, cooler by several hundred degrees Kelvin than the surrounding photosphere.

The light modulations due to starspots allow determinations of stellar rotation periods. These could in principle be found from broadening of spectral lines, but in practice the equatorial rotation speeds are usually too small (less than 10 km/s) to measure. Over the past few years, a technique, due to Steven Vogt and colleagues at Lick Observatory, University of California, has been developed for synthesizing images of starspots. It relies on the fact that a spotted (dark) region at some location on the star's surface produces a dip in the profile of an absorption line; this dip is displaced from the centre of the line by a Doppler shift corresponding to the motion of the spot's location towards or away from the observer due to the star's rotation. A large enough spot rotating across a star's surface causes the dip to move from the violet to the red side of a line profile since its velocity is first approaching, then receding. An image of the star's surface features can be synthesized so that the observed profile of its absorption lines is matched. Figure 8.15 shows how starspot groups on one of the components of the RS CVn binary called HR 1099 have been 'Doppler imaged' from repeated observations of line profiles.

Spectroscopic evidence exists for magnetic fields on dMe stars that may be related to those present in starspots. Measurement of fields is not possible using the polarization of the Zeeman σ components of a line, as with solar fields (see §§3.8, 10.1), since without being able to separate areas of each polarity on a star the polarization cancels. However, by a careful comparison of magnetically sensitive and insensitive lines in stellar spectra, average field strengths can be deduced. Infrared lines are better for this than visible-wavelength lines because the Zeeman splitting is larger. Using this technique, sunspot-like fields (about 0.4 T) have been measured for the star AD Leonis covering three-quarters of its visible surface.

Photospheric granulation on other stars might also have been spectroscopically detected. With solar granulation, hot ascending currents form the granules themselves, and cooler descending currents form the intergranular lanes, so that, in integrated sunlight, spectral lines have a larger contribution from the brighter granules, and so show a very slight asymmetry. Such asymmetries exist for the solar-like star α Centauri A, suggesting the presence of such granulation. Using quite different means, direct resolution of stellar convection has recently been achieved with an interferometric technique using a mask with holes in it in front of a large telescope. A bright feature seen on the supergiant Betelgeuse is thought to be the top of a single convection cell in the star's atmosphere, proportionately much larger than cells in the solar convection zone because of a more gradual change of pressure with depth in the star's extended envelope.

The stellar counterpart of solar prominences seems to have been observed on some rapidly rotating stars, in particular a K dwarf star called AB Doradus. Andrew Collier Cameron at Sussex University and colleagues have made repeated observations of the Hα emission-line profile of this star and found evidence for complex clouds of material rotating with the star. The star's

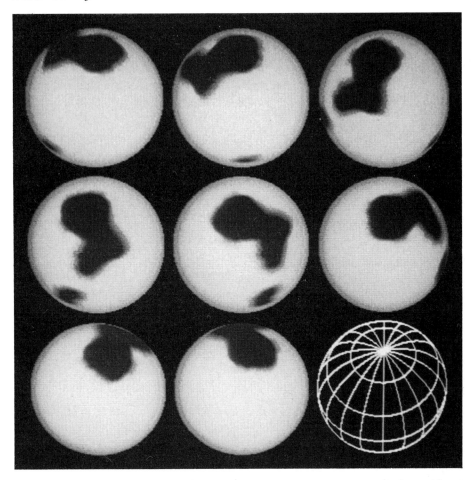

Fig. 8.15 Doppler images of the RS CVn star HR 1099 in 1985, showing the extent of 'starspots', and the orientation of the rotation axis, deduced from spectral line profiles. (Courtesy S. Vogt and A. Hatzer, Lick Observatory)

rotation period is 12 hours, as deduced from starspot modulations of its light, and the Hα profile is broadened by this rotation. Periodically, absorption dips in the profile are observed, going from the violet side of the profile to the red, in a period of from 20 minutes to an hour (Fig. 8.16a). The dips are thought to be due to prominence-like clouds which rotate with the star, located several stellar radii above the star's photosphere, which have line-of-sight radial velocities varying as they transit the disc of the star (Fig. 8.16b). The existence of a whole system of clouds has been deduced, with distances of up to nine stellar radii from the star's centre.

Stellar flares have been observed for many years. Flare-like brightenings were reported on the star DH Carinae by Hertzsprung as early as 1924, and

(a)

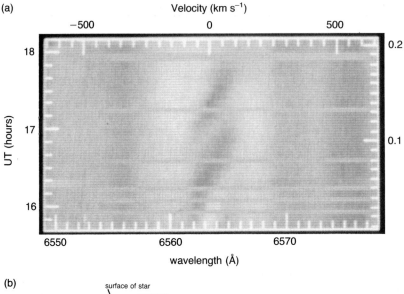

(b)

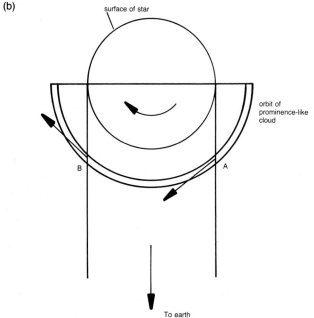

Fig. 8.16 (a) Time series (on 21 January 1986) of the Hα emission-line profile of the rapidly rotating K star AB Doradus, showing absorption features which are believed to be due to prominence-like clouds. (Courtesy A. Collier Cameron) (b) Deduced geometry of clouds rotating round with star. The emission line is due to the star's surface, but the prominence gives rise to an absorption feature when in front of the star; at position A, the prominence is approaching the observer, so the absorption feature is shifted to the violet, but when rotated to position B the prominence is receding, so the absorption feature is shifted to the red.

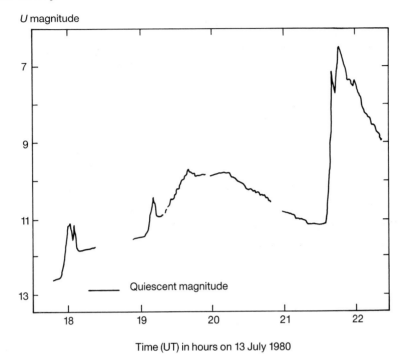

Fig. 8.17 A giant optical flare (*U* magnitude) seen on the star EV Lacertae on 13 July 1980. (Courtesy G. Sh. Roĭzman and V. S. Shevchenko (1982))

on the dMe star L726-8 (a companion of the more famous star UV Ceti) in 1948. By the late 1950s, several dMe stars were known to flare. It was then that Sir Bernard Lovell made a pioneering search for flares at radio wavelengths with the newly completed Jodrell Bank radio telescope, and was rewarded with the first observation of a flare – on UV Ceti – in 1958. This added greatly to their significance, though for some years they were still not widely studied. Their similarity to solar flares became much more obvious when enhancements in X-ray and ultraviolet emission were observed, and more detailed studies were made at visible and radio wavelengths. A great deal has been learned from co-ordinated studies of particular dMe and other active stars in different wavelength regions, using ground-based observatories and spacecraft.

Flares have been most frequently seen on K and M dwarfs with chromospheres, typified by the prototype dMe star UV Ceti. Some eighty of this type are known. The enhancement in visible wavelengths can be enormous, as shown by Fig. 8.17, a light-curve in *U* magnitude for a flare on the dMe star EV Lacertae; the amplitude in this case is 6 magnitudes, i.e. an intensity increase of 250-fold. There is a marked distinction from solar flares in that, while white-light solar flares are rare, stellar flare emission is always 'white-light', i.e. the continuum intensity increases very strongly. It does so very rapidly, often in a few seconds, and falls more slowly. The spectral intensity generally increases towards the blue. Spectral lines such as those in the Balmer series, which are in emission in the non-flaring state, increase their intensities strongly during flares. They start to increase before the continuum does, but they peak later and remain enhanced after the continuum has died away. The Balmer lines have larger enhancements than the calcium H and K lines, and at flare peak may be in emission up to the H15 line (due to electron transitions from the $n=15$ to $n=2$ orbits). Neutral and ionized helium emission lines sometimes occur, as well as those due to metals.

Optical flares are very conspicuous events in cool dwarf stars as they have bright, bluish emission superimposed on the dim red photospheric radiation of the star. The fact that flares seem less common for hotter stars, for which the photosphere is brighter and bluer, may thus be simply a selection effect: flares may be occurring as frequently (this is certainly so for the sun, a G star), but are less noticeable. In fact, flares have been observed in main-sequence stars of all spectral types. The interacting binaries of the RS CVn and W UMa type (§8.4) also produce large flares. Flaring also occurs in young stars, such as the pre-main-sequence T Tauri objects and some rapidly rotating stars in the Pleiades. 'Pseudo-flaring' has also been reported for some giants and supergiants, enhancements of days to several months being recorded in visible and radio (but not ultraviolet or X-ray) wavelengths.

All X-ray observations of flares on cool dwarf stars have hitherto been made in the soft X-ray range, and show the same characteristics as solar flares in the same range, viz. a smooth rise and more gradual decline, with durations of tens of minutes or longer for larger flares. No high-resolution spectra yet exist, but 'broad-band' spectra indicate temperatures of 10×10^6 K to 50×10^6 K for flares on dMe stars and 70×10^6 K or more for those on RS CVn stars. In the ultraviolet, the chromospheric and transition-region emission lines present in the non-flaring state are all enhanced, those emitted at higher temperatures more so than those at lower. There is often a continuum with intensity apparently related to the line intensity.

Early metre-wave radio observations of UV Ceti-type flare stars at Jodrell Bank showed peak radio emission generally following that in visible light by tens of minutes, rather as type IV bursts follow optical emission during solar flares. These and later observations indicate intensities that are sometimes thousands of times greater than for solar flares. At higher frequencies, the light-curves are sometimes extremely complex and show rapid changes as well as more gradual changes. The latter are due to synchrotron emission, while a 'maser' process may give rise to the rapid fluctuations.

The most outstanding feature of stellar flares is the large amount of energy in the most intense of them. Though there are no reliable estimates for flare energies in all wavelengths, some idea of the total energies can be obtained from the X-ray energies. The largest flares observed on dMe stars have X-ray energies of up to 10^{27} J, while even larger X-ray flares (up to 10^{29} J) have been seen on the interacting X-ray binary Algol and the RS CVn binary σ^2 Coronae Borealis. These enormous energies compare with 2.5×10^{24} J for the largest X-ray solar flare. The ultraviolet emission for the largest stellar flares also is much larger than for any known solar flare, with the RS CVn stars having the most energetic and longest-lasting of all flares. The average flare energy can amount to 0.3% of the total luminosity of a flare-productive star, compared with only 0.001% for the sun.

The striking parallels between the various forms of activity on solar-type stars and the sun itself give us reason to believe that, as with the sun, magnetic fields are the controlling influence of stellar activity. The enormous starspots and their observed fields show that they are most likely large flux tubes protruding through the photospheres of stars, as with sunspots. The enhanced calcium K line emission of stellar active regions perhaps arises from magnetic heating in regions of complex fields, though as with the general chromospheres heating by X-ray-emitting coronal loops may play an important rôle. The sudden onset of stellar flares strongly suggests magnetic reconnection as the process for energy release. This may occur by processes similar to the sun for single active stars (e.g. emergence of new flux regions). However, for interacting binaries, there is the added possibility that reconnection occurs in magnetic flux tubes of one star 'colliding' with those of the companion star if both rotate rapidly; this perhaps accounts for the fact that flares on RS CVn binaries are particularly energetic.

Stellar flaring may, incidentally, have galaxy-wide implications. The number of flares occurring on dMe stars, and the fact that dMe stars are extremely numerous in the galaxy (several thousand million), means that the continual flaring on these stars contributes to a background X-ray emission which is concentrated in the Milky Way plane.

8.6   Asteroseismology

The possibility of finding periodic oscillations similar to the solar global oscillations (see §2.3) has been seriously pursued in the past few years. A few positive claims for certain A-type stars have already been made, but most attempts so far have not been conclusive for cool stars like the sun. As the velocity amplitudes of the solar global oscillations are only of the order of 20 cm/s, to observe such low-amplitude stellar oscillations would require special techniques not yet developed. It is, however, possible that the next few years will see some new developments that will result in successful observations.

Chapter 9

Solar energy

Practically all the energy the earth receives is that which the sun radiates. When this energy is received by the earth, it gets converted to numerous forms such as heating of the earth's surface, motion and heating of the earth's atmosphere and oceans. In this chapter, we will deal with how much solar energy there is, how it is measured, its variations and its harnessing, both by natural processes and by human society.

9.1 Amount of solar energy

The solar irradiance is the total amount of solar energy received at the earth outside its atmosphere per unit area per unit time (SI units: W/m^2). Numerous measurements of it from the ground exist over the first half of this century, the most extensive set being that of the Smithsonian Institution programme (§1.10). Their results indicate an irradiance of about $1353\ W/m^2$, constant to within plus or minus 1%. One of the chief uncertainties in these and all ground-based measurements is the correction that must be made for absorption by the earth's atmosphere, in particular the gases ozone, water vapour and carbon dioxide; the correction is large and variable in time, even at the high altitude of the sites from where the Smithsonian measurements were made. This difficulty has been overcome with more recent measurements made with instruments on board rockets and satellites flying above virtually all the atmosphere. Consequently, the extent to which the solar irradiance varies can now be much more definitely stated.

The recent instruments use some of the design features of the 'pyrheliometers' used by Ångström and then by Abbot in the Smithsonian programme (§1.10). The basic operating principle is that the heating of an absorbing cavity produced by sunlight is compared with that produced by a

314

known electric current. The Earth Radiation Budget (ERB) detector on the *Nimbus 7* satellite, launched in November 1978, has a cavity in the form of an inverted cone within a cylinder, all with blackened surfaces. When the instrument is directed at the sun, radiation passing into the cavity suffers almost total absorption and causes the cavity to heat up. The instrument is electrically calibrated by directing it away from the sun and varying a current flowing in a wire wound on the cavity such that the same amount of heating is produced.

Active cavity radiometers, developed by Richard C. Willson, uses the same principle but have an electronic servo-system. Early versions flew on balloons and rockets, and one of more recent design (known as the Active Cavity Radiometer Irradiance Monitor, ACRIM) was carried on the *Solar Maximum Mission* satellite, which operated between 1980 and 1989. Three similar detecting systems were included in ACRIM, with one viewing the sun and the other two periodically compared with the first to check for performance variations. Each system consisted of two identical blackened conical cavities, a primary cavity which was directed at the sun and a secondary or reference cavity that was directed away from the sun and towards a surrounding heat sink; each had an electric heating system. The electronic servo-system kept the primary cavity at a very slightly higher temperature than the reference cavity at all times by varying the power in its electrical heater. Sunlight was alternately admitted to or blocked from the primary cavity by a movable shutter; the difference in electrical power supplied between the open and closed positions gave an estimate of the heating due to the sun and thus the solar irradiance. In an hour's integration of its observations, the ACRIM instrument could detect variations in the solar irradiance down to plus or minus 0.001–0.002%, while the absolute value of the solar irradiance could be determined to better than 0.05%.

Observations with ACRIM and the *Nimbus 7* ERB instruments showed many fluctuations in the solar irradiance of about 0.01% lasting a day or so, with dips of up to 0.25% lasting for about ten days at a time. The ACRIM and ERB measurements are in excellent agreement on the timing of these dips, confirming their reality. The more major dips were found to be clearly associated with the passage of large sunspot groups across the sun's visible hemisphere (Fig. 9.1). When the sun had few spots, the irradiance was relatively high.

The combined observations of the ERB and ACRIM instruments over the 1978–89 period show a clear trend correlating with the solar cycle. Figure 9.2 shows the ACRIM measurements over the lifetime of the *SMM* satellite. At solar maximum, the irradiance fluctuates from high values (few spots) to low (large spots present), while at minimum it remains at a relatively low value; the total range is from 1365 to 1369 W/m^2, or 0.3%. Without the spot-associated dips, the irradiance is not far from a sine wave, greatest at solar maximum and

least at solar minimum; the range of this sine wave is from 1367 to
1369 W/m², or 0.15%.

The relation of the irradiance dips with large sunspots can be made more
exact. By taking the total areas of spots on the sun's visible hemisphere and an
intensity ratio for a spotted to unspotted area on the surface, with an
allowance for limb darkening, one may arrive at an index of 'irradiance deficit'
due to sunspots. This index has been found to correlate generally quite well
with the measured irradiance, but the correlation is less good when the sun is
very spotted. This is probably explained by the presence of the bright faculae,
which are associated with large spot groups, as they appear to make a positive
contribution to the irradiance (see below).

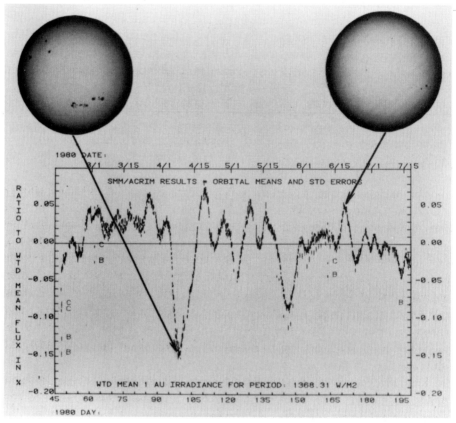

Fig. 9.1 The solar irradiance as measured by the ACRIM instrument on *Solar
Maximum Mission* for a 150-day interval (February–July) in 1980. The
horizontal line indicates the weighted mean irradiance for this period
(1368.31 W/m²), and the values along the vertical axis indicate percentage
changes from this mean value. Large dips are correlated in time with the
passage of large sunspots, peaks with times of few spots, as shown by the
examples of the images of the sun at such times. (Courtesy R. C. Willson
and H. S. Hudson)

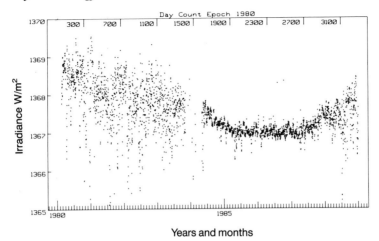

Fig. 9.2 The solar irradiance measured by ACRIM over the period 1980–88, showing sunspot-associated dips at times of large spot activity (1980–84 and 1988 onwards) and the prolonged minimum (1985–87) when solar activity was low. The observational errors are larger for the period between the end of 1980 and the beginning of 1984 because of a lower rate of data acquisition and the coarse pointing of the *SMM* spacecraft in this interval. (Courtesy R. C. Willson and H. S. Hudson)

The implication of the irradiance dips is that sunspots block a small amount of the sun's total energy. But the energy generation at the sun's core is almost certainly quite independent of activity on the sun's surface, and it is not possible that it could vary in periods as short as the few days a large spot may last – it takes radiation several million years to pass from the sun's core to its surface. Any short-term changes in the energy generation at the core, then, would be completely smoothed out by the time radiation appeared at the photosphere. In other words, the energy blocked by a sunspot cannot be truly 'missing', but must reappear at some later time. However, it does not reappear at, for instance, solar minimum, since the ACRIM results show that the irradiance is least then.

The most likely explanation is that the energy is stored over periods of a few days, then reradiated. It could be stored, for example, in the magnetic field associated with the sunspot. A large amount of energy is involved – 10^{29} J for the largest spot-associated dips – and to store this energy, spots with their associated fields must penetrate to very large depths. Alternatively, the energy might temporarily go into the form of Alfvén wave motion, reappearing as radiation as the waves are dissipated. Finally, energy might temporarily be stored in the convective zone around the spot, slowly re-emerging by turbulent eddies after the spot has decayed. Whatever the energy storage mechanism is, it is apparent that faculae near spot groups are areas where the missing energy blocked by spots eventually reappears. They apparently 'over-compensate' for the dimming due to a spot as their combined effect is to produce an irradiance slightly higher at solar maximum, apart from the dips, than at minimum.

Note that the calculation in §3.4 to find solar luminosity from irradiance was based on the assumption that the sun radiates uniformly in all directions. The ACRIM and ERB results indicate that this is not quite so, for it depends on the disposition of large spots on the visible hemisphere. A mean irradiance was used for this calculation, viz. $1368\,\text{W/m}^2$, based on the ACRIM measurements.

9.2   Solar insolation

The amount of solar energy received at a location on the earth's surface averaged over a period of time – the solar *insolation* – is much less than that corresponding to the mean solar irradiance. During the day, the insolation varies according to the sun's altitude both because of the varying angle between the normal to the earth's surface and the sun's direction and because of the varying amount of the earth's atmosphere that the sunlight has to shine through. At night the insolation is zero. There are annual variations too, depending on the latitude of the location and any seasonal variations in cloudiness. There is, in addition, a small annual variation (with a range of 7%, greatest in January) due to the varying distance between the earth and sun.

When the sun is overhead at a location, the amount of atmosphere it shines through is least and there is therefore least attenuation by the atmosphere. The total energy received at ground level under such circumstances is redu-ced from the solar irradiance value, $1368\,\text{W/m}^2$, to about $1000\,\text{W/m}^2$. The attenuation is much greater at certain wavelengths than at others, and the resulting spectrum is known as an AM1 ('air mass of one unit') spectrum. More typically the sun is not overhead and the attenuation by the atmosphere is correspondingly greater. Thus, when at an altitude of about $30°$ (or 'zenith angle' of $60°$), an AM2 ('air mass of two units') spectrum results, the air mass being approximately proportional to the secant of the zenith angle.

Figure 9.3 compares the solar spectrum outside the earth's atmosphere ('AM0', or zero air mass) with the spectrum of AM1 and AM2 radiation (note that the AM0 spectrum is identical to that of Fig. 3.5). The AM1 and AM2 spectra are reduced at particular spectral regions owing to absorption by molecules in the earth's atmosphere (notably water, carbon dioxide, oxygen and ozone), and there is a general reduction at shorter wavelengths owing to absorption by ozone (wavelengths less than 320 nm: §7.5) and to scattering by molecules, aerosols and small particles in the atmosphere.

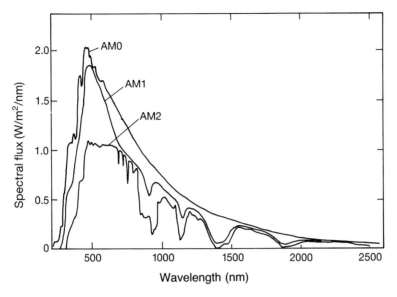

Fig. 9.3 Spectral irradiance of the sun for AM0 sunlight (that received outside the
earth's atmosphere) compared with that for air masses of one (AM1) and
two (AM2).

The solar insolation at a location averaged over a year, taking account of
cloud cover and diurnal and annual variations, is 20% or less of the solar
irradiance; desert areas like the Sahara have the greatest insolation, with
annual averages of about $300 \, W/m^2$, and polar regions the least, with average
values of less than $100 \, W/m^2$. Figure 9.4 shows the geographical distribution
of this averaged insolation.

Insolation is not synonymous with 'sunniness', as cloudy regions of the
earth, such as north-western Europe including Britain, receive as much as
60% of their insolation as 'diffuse' radiation from all over the sky, and only
40% as 'direct' sunlight. This diffuse radiation is much bluer than direct
sunlight. Tropical locations have a much greater proportion (up to 90%) of
their annual insolation as direct sunlight.

9.3   Harnessing solar energy: photosynthesis

Organic food is required by all living beings as building blocks for more
complex material and as their source of energy. The energy is released by
breaking down the molecules of this organic material, mainly by combination
with the oxygen of air (i.e. respiration), with carbon dioxide and water as
waste products. Living beings destroy this material at such a rate that life on
the earth would disappear in only a few years were it not for a single process

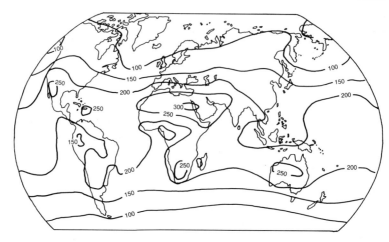

Fig. 9.4 The global distribution of annual mean insolation (W/m²), or solar energy
received on a horizontal surface averaged over a year. (Courtesy UK
Section of the International Solar Energy Society)

that Nature provides for counteracting this destruction – that of photosynthesis in green plant life. By photosynthesis, plants build carbohydrate molecules (with release of oxygen) from carbon dioxide and water, using light
energy from the sun: chemical energy is formed from the sun's radiant
energy. The rate of photosynthesis for green plants is far greater than the
reverse process of respiration, even though photosynthesis can only occur in
daylight while respiration occurs all the time. A plant as a result forms excess
amounts of carbohydrate, which it transforms into fats, proteins and other
material which it then stores. These stored substances become the organic
foodstuffs that the animal kingdom uses for its energy supply.

Photosynthesis is almost entirely accomplished by molecules of the green
pigment chlorophyll which gives most plant life its characteristic colour.
Absorption of sunlight by the chlorophyll molecules, assisted by other molecules such as the carotenoids, allows photosynthesis to occur. An example is
the production of the carbohydrate glucose ($C_6H_{12}O_6$) as follows:

$$Sunlight + 6H_2O + 6CO_2 \rightarrow C_6H_{12}O_6 + 6O_2.$$

Carbon dioxide is 'reduced' by the hydrogen of water to form the glucose
molecule with the release of an oxygen molecule.

The details of how the chlorophyll operates in a plant are complex. The
molecules of chlorophyll and other pigments are arranged in protein structures which are held together in organs called chloroplasts. Several hundred
chlorophyll molecules act together to form what is called a photosynthetic unit
(PSU). A photon of solar radiation incident on a plant is absorbed first by a set
of 'antenna' chlorophyll molecules in the PSU, with the photon's energy

transferred through the antenna by successive excitations and de-excitations of molecules. It is finally trapped by the excitation of another type of chlorophyll contained in the 'photo-reaction centre' of the PSU. The energy is then supplied to an electron that is extracted from a water molecule. This electron, now with a raised energy, then makes possible the chemical reduction of carbon dioxide.

Photosynthesis is not a particularly efficient process, with only about 2% of solar radiant energy incident on a plant being transformed into chemical energy. However, much of this inefficiency is due to reflection or other processes that reduce the amount of light incident on the chlorophyll and other pigments. The amount of energy that is actually absorbed by them is transformed to chemical energy at 35% efficiency.

The operation of photosynthesis over the ages by plants, including aquatic organisms, has tended to reduce the amount of carbon dioxide and increase the amount of oxygen in the atmosphere, leading to the present proportions of these gases in the atmosphere. The remnants of this ancient plant life now appear as coal, oil and natural gas – the 'fossil fuels'. They are in fact concentrates of solar energy. The extensive burning of them, especially in recent times, to release this energy in a form more usable to human society is returning greatly increased amounts of carbon dioxide to the atmosphere (see §7.7), so reversing the trend that Nature has effected for so long.

9.4   Harnessing solar energy: thermal systems

The burning of fossil fuels provides an overwhelming proportion of the energy used by industrialized countries at present. In 1986, oil accounted for 34% of the UK's energy needs, coal 34%, natural gas 25%, while nuclear, hydro-electric and all other power sources combined accounted for only about 7%. However, fossil-fuel burning is generally inefficient (more of the energy extracted is actually rejected than is converted to a usable form) and gives rise to much pollution. At the present rates of consumption, global resources of fossil fuel will run out in less than 150 years, and local resources (such as in Europe) within the next few decades.

The limited lifetime of fossil fuels has led to searches for alternative energy sources, these being greatly spurred on by the rise of oil prices in the mid-1970s and concerns for the environment. Serious consideration is now being given to projects involving tidal, wind, wave, geothermal, biomass and nuclear fusion sources of energy. Equally, energy direct from the sun has received much attention, the attractions being that its harnessing is pollution-free and it is the energy form closest to being 'renewable' in the sense that the sun is for all practical purposes a limitless source.

Solar energy can be extracted in several ways, and we will consider those that show most promise of wide applicability. In this section, we discuss solar energy for heating purposes (photothermal systems).

Over the past decade or so, there have been numerous architectural projects in various parts of the world in which specially designed, 'passive solar' buildings conserve solar heat gained during the day for use at night. Thus, walls and roofs can be arranged so that exposure to the sun is maximized, with improved insulation and the use of building features such as 'water walls' and materials such as rock or concrete to store heat. There are generally no mechanical or electrical components.

The basis for 'active' solar heating systems is that sunlight falling on an absorbing surface causes the surface to heat up; solar energy is converted to heat energy, which may then be extracted by mechanical or electrical means for useful, in particular domestic, purposes.

We can find the temperature which a perfect (black body) absorbing surface would attain if placed perpendicularly to the sun's rays, rather along the lines of the calculation in §3.4 to find the sun's effective temperature. We assume that AM2 sunlight, amounting to about $800\,W/m^2$, is incident on, and is perfectly absorbed by, the surface. It is heated to a temperature T_A, say, and radiates away an energy σT_A^4 (σ is the Stefan–Boltzmann constant) per second per square metre equal to that incident. Thus, with $\sigma=5.67\times10^{-8}$ SI units, T_A is $(800/5.67\times10^{-8})^{1/4}=345\,K$, or $72\,°C$. Under AM1 sunlight ($1000\,W/m^2$), T_A is $364\,K$ ($91\,°C$). In practice much higher temperatures can be attained.

A photothermal system might consist of an absorber in the form of a flat plate, either stationary in a position most exposed to sunlight (e.g. south-facing for a northern-hemisphere location) or driven to follow the sun across the sky. The flat plate must be well insulated from its surroundings to reduce loss of heat by conduction. When the surface is heated, it radiates in a wavelength region that (by Wien's law) depends on its temperature. At around $344\,K$, the radiation is largely in the infrared. To minimize this infrared emission, the surface can be made to be 'selectively' absorbing, i.e. absorbing as much visible-wavelength sunlight as possible but radiating little in the infrared. Absorbers with 'black chrome' coatings, these having the selectively absorbing properties desired, have been widely used. The addition of a glass cover over the absorbing surface reduces heat losses due to convection by air near the surface; evacuation of air between the cover and surface reduces these losses still further. Eventually absorbing surface temperatures of around $700\,K$ ($430\,°C$) can be achieved.

Several methods of extracting energy from flat-plate collectors have been successfully tried, generally involving the circulation of a liquid passing through the collector and transferring the heat to a domestic water supply by means of a heat exchanger. Figure 9.5 schematically illustrates a system in which the liquid – for example, a glycol solution, to prevent freezing during

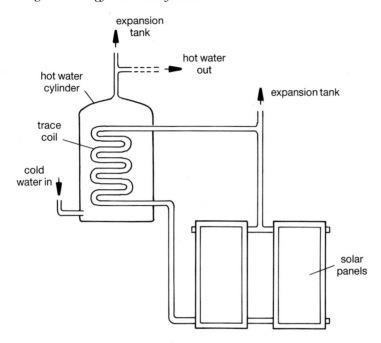

Fig. 9.5 Arrangement for domestic heating system using solar panels and
thermosyphon circulation with trace coil attached to hot water cylinder.

cold weather – circulates by natural convection (a 'thermosyphon' system).

Flat-plate collectors have been successfully used in the UK and other places with a relatively cloudy climate, where the amount of direct sunlight received is less than the diffuse light. Figure 9.6 shows a school in Netley, Hampshire, with space and water heating provided by flat-plate collectors in the roof of the building. For places with a high proportion of direct sunlight, systems which concentrate solar radiation with mirrors can be used, raising the temperature of the absorber to very high values. Mirrors in the form of parabolic dishes are a possibility, with the absorber at the focus, but they must be driven to follow the sun. Non-imaging 'Winston' devices are currently being used, in which a parabolic mirror directs sunlight to a hollow silver cone, with shape so designed that any ray undergoes no more than one reflection before it reaches an output hole; by such means incident solar energy can be concentrated to levels of $44 \times 10^6 \, \text{W/m}^2$, many thousands of times the solar irradiance value. Another design is to use a large array of driven flat mirrors ('heliostats') that simulate a large-aperture concave mirror, focusing sunlight on to an absorber at the top of a tower. Figure 9.7 shows the solar 'Power Tower' project near Albuquerque, New Mexico. Some 222 heliostats direct up to 6 MW of solar energy at the top of a tall concrete tower, where a 'receiver' in the form of metal tubes containing fluid such as water or molten salt is positioned. The fluid when heated is used to boil water in a steam generator, and the steam drives a turbine to produce electricity.

Large-area ponds as collectors of solar energy have been used for various purposes. Shallow ponds in the roofs of houses, for example, help to reduce high interior temperatures during the day; the energy retained by the pond is then made available for heating the house at night. Solar ponds have been used as thermal power stations, e.g. in the Dead Sea in Israel. Normally, a body of water heated by exposure to sunlight maintains a fairly uniform temperature owing to convection currents that are set up; deeper water (especially if the pond bottom is black) is heated more, rises to the surface and so displaces water at the top which is cooled by evaporation. Solar ponds are designed to reduce this convection. They consist of a salt solution which has increasing concentration (and is therefore denser) towards the pond bottom. The heated lower layers remain denser than the cooler, top layers, so a temperature gradient builds up. In the Dead Sea project, the temperature difference is used to generate electricity, but the heat can also be directly extracted by removing the deepest layer of fluid, passing it to a heat exchanger, then returning it to the bottom of the pond.

Fig. 9.6 Solar-heated building in a school at Netley, Hampshire, UK. (Courtesy Energy Technology Support Unit, Harwell Laboratory)

Fig. 9.7 The Solar Thermal Test Facility ('Power Tower') near Albuquerque, New
Mexico. Some 222 heliostat mirrors track the sun, reflecting sunlight on to
a 'receiver' at the top of a tower, where fluid contained in metal tubes is
heated and used to boil water to steam which drives a turbine to generate
electricity. (Courtesy Sandia National Laboratories, New Mexico, USA)

9.5   Harnessing solar energy: the solar cell

The conversion of solar energy to electricity was first accomplished with a
'solar cell' of semiconductor material in 1954 at the Bell Laboratories in the
USA. Although of low efficiency (an output equal to 6% of the incident solar
energy), the cell's applicability was demonstrated a year later with an array
generating 9 W of power which was used to operate a telephone repeater.
This success led to an extensive application of solar cells to spacecraft,
starting with the NASA satellite *Vanguard I* in 1958. But only in recent times
has a much wider use been found for solar cells. Megawatt solar-cell power
stations are already in use and larger stations are planned or under construction. The prospects for providing a substantial fraction of the world's future
energy needs by such means seem very promising.

A solar cell generates electricity by the photovoltaic principle which can be
explained as follows. The element silicon has a nuclear charge of 14 and so
normally has 14 electrons. Ten of them are in orbits that are, by quantum-
mechanical laws, filled, while the remaining four electrons are in an outer or
'valence' orbit. In a crystalline state, silicon has its atoms arranged in regular
patterns forming a lattice structure; each atom retains its four valence
electrons, but they are all shared with neighbouring silicon atoms. Alternatively, the atoms can be regarded as being held together by chemical
'bonds', each bond consisting of a pair of shared electrons.

The valence electrons are susceptible to being removed from their parent atoms. Their removal is not, however, quite the same as ionization. On becoming detached, the electrons are free in the sense that they can travel along the crystal lattice but not removed altogether. In energy-level terms, they are said to be raised from a 'valence' band of energies to a 'conduction' band. An electric field applied to the solid causes detached electrons to move in a particular direction, setting up a current.

Increasing the temperature of a crystal can take electrons out of the valence band to the conduction band. The crystal is thus a 'semiconductor', i.e. does not conduct electricity at low temperature but does so at higher temperatures. Electrons may also be raised to the conduction band by the addition of energy in the form of a photon. For crystalline silicon, the photon must have an energy of at least 1.8×10^{-19} J for this to occur, corresponding to a wavelength of less than 1.1 μm. Photons of visible and near-infrared light, then, are able to do this. When, by increase of temperature or exposure to light, electrons are moved to the conduction band, vacancies or 'holes' are created in the valence orbits; these holes act as positive charges, and move through the lattice in a direction opposite to the conduction-band electrons when an electric field is applied.

Adding small numbers of impurity atoms in the lattice structure, known as doping, gives rise to useful properties. Thus, doping crystalline silicon with atoms having five instead of four electrons in the valence orbit, for example, means that extra electrons are added to the conduction band, i.e. it assists the conductivity of the crystal. Correspondingly, doping with atoms having three electrons adds holes or positive charges to the material. The conductivity is called *n*-type in the first case, *p*-type in the second. A junction can be formed by fusing a piece of *n*-type silicon with a piece of *p*-type silicon. The excess electrons in the *n*-type silicon will then diffuse to the *p*-type, the electron depletion causing a net negative charge, and the excess holes in the *p*-type silicon will diffuse to the *n*-type, the depletion of holes causing a net positive charge. An electric field is therefore set up across the boundary.

A solar cell using crystalline silicon uses such a '*p–n*' junction, with the two types of crystal in the form of thin wafers mounted on a glass or plastic substrate (Fig. 9.8). There are metal contacts on either side of the device. The cell is designed so that a photon of sunlight incident through the *n*-type silicon layer is absorbed as closely as possible to the electric field region of the junction, the absorption giving rise to an electron–hole pair. When the two metal contacts are connected by an external circuit, the electrons and holes flow round in opposite directions to give a current, which flows for as long as sunlight is incident on the cell. Solar energy is thereby converted to electricity. In their manufacture, cells are wired together to make modules, and panels or arrays are built consisting of several such modules.

Several improvements have been made to this basic design. Alternative

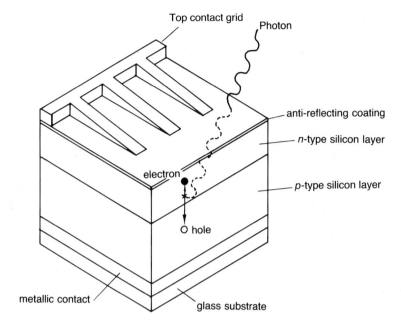

Fig. 9.8 A silicon solar cell. A photon from sunlight incident from the top is
absorbed in the electric field region at the junction of the *n*-type and *p*-type
silicon layers. An electron–hole pair is formed, with the electron attracted
upwards, the hole downwards, so that when the metal contacts of the cell
are connected by an external circuit a current flows.

semiconductor materials are now being used. Thus, the use of amorphous
instead of crystalline silicon allows the wafers to be much thinner (1 μm
instead of 50 μm), so reducing manufacture costs. Also, anti-reflection coating on the *n*-type silicon wafer allows more solar photons through to the
electric-field region. Concentrating sunlight on the cells with mirrors
increases their performance still further.

These and other sophistications have resulted in a steady increase of cell
efficiency and decrease in cost per peak watt output (defined to be the
maximum power output from a cell in an ideal day). Efficiencies approaching
30% (35% for certain types of cell) are now being realized. These are rather
lower than the efficiency in burning coal or oil to drive steam turbines
generating electricity. However, such a comparison cannot be considered
valid: inefficient use of coal and oil, which are expensive fuels, costs money,
but inefficient use of solar energy costs nothing since solar energy itself is
free.

Apart from their inclusion in spacecraft and certain consumer devices such
as pocket calculators and wrist-watches using small amounts of power, solar
cells are now being used in water pumps, boats and machinery demanding

much more power. The most serious application to date has been the building, in the USA and Japan, of electricity-generating stations. The largest, at Carissa Plains in California, consists of solar cell arrays covering 0.6 square kilometres, and has a rating of 7 MW. This is small compared with some power stations running on fossil or nuclear fuels, but there are plans for much larger installations.

In a recent development, semiconductor material has been used to generate power by a photoelectrochemical principle rather than the photovoltaic principle. By this, a semiconductor is placed in contact with a water-based electrolyte solution. On exposure to sunlight, the electrons released in the semiconductor pass to the electrolyte rather than another piece of semiconductor. Otherwise the cell has a similar design to the photovoltaic cell. Corrosion of the semiconductor can be avoided using a very thin coating of osmium. Efficiencies as high as 15% have been achieved.

9.6   Harnessing indirect forms of solar energy

Solar energy occurs on the earth in many indirect forms – ocean waves, winds and in plant life – and efforts have been made to harness this energy. It may not be immediately obvious that some of these energy forms are indeed solar energy in a derived sense, as for example hydro-electric power. In this case, it is solar energy that results in the evaporation of sea water, which later falls as rain, which feeds rivers whose energy, on flowing to the sea once more, is used to drive turbines in a hydro-electric plant. This is already an established way of harnessing energy, and accounts for a sizeable fraction of energy needs of some countries. The potential for its further development is very large, though there are concerns about the large amount of land needed for dammed water and the possibility of a catastrophic failure of a dam once built.

The heating of the earth's surface and lower atmosphere by the sun results in fluctuations of air pressure and therefore in winds. The harnessing of wind power has been used for centuries to propel ships and grind corn. But there has been renewed interest in it in recent times and, after hydro-power, it has become the form of solar energy with the greatest economic potential. The oil price rises of the 1970s produced much development of wind turbines to generate electricity from wind, especially in the US, but also in European countries such as Denmark. In southern California, at the Altamont Pass, there are several thousand wind turbines producing in excess of 1 GW of electric power which is supplied to a local electricity company. In the UK, several major wind farms are planned or being built, with individual wind turbines of up to 350 kW power rating. The largest single machine operating at present in the UK is a 3-MW, 60-m diameter, turbine at Burgar Hill in the Orkneys. A number of research programmes, funded by government and

Fig. 9.9 The Wind Turbine Test Site of the Energy Research Unit at the
Rutherford Appleton Laboratory, Oxfordshire. The largest turbine (far
left), built by Windharvester Ltd, has a 17-m diameter rotor and a power
rating of 45 kW. Among the others are a Northwind 16-kW turbine and a
5-kW vertical-axis turbine. Meteorological instruments on towers enable
the wind environment to be constantly monitored. (Courtesy Rutherford
Appleton Laboratory)

industry, are being carried out to develop and evaluate machines and to assess
the resource. The Energy Research Unit of the Rutherford Appleton
Laboratory in Oxfordshire is running several wind-related experimental projects in collaboration with universities (Fig. 9.9), and wind data bases and
computer facilities are being used to select possible future sites for wind
farms.

The principles involved in harnessing wind power are fundamentally the
same as those used for the past 50 years, though the technology is much
changed. Nowadays light-weight materials for the rotating blades such as
fibreglass have replaced steel (this also having the benefit of not causing
interference to television reception) while the speed of the blades can be
electronically controlled to maximize the energy taken from the wind. For
small power systems exposed to high wind, penetration problems caused by

the short-term variability of wind speed can be compensated for by using energy stored in flywheels. Both vertical-axis and horizontal-axis turbines are being developed, although most machines are of the latter type.

The very considerable advantages of wind power are now being recognized. Wind turbines can produce electricity cost-effectively in a relatively environmentally friendly way, with no release of pollutants or, most importantly, carbon dioxide.

Related to wind energy is energy from ocean waves. Although extensive harnessing of wave energy is not considered to be a likely possibility in the near future, several designs have been proposed for water turbines floating on the sea surface to produce sizeable amounts of electric power.

Solar energy used in the photosynthesis process can be extracted from living green plant matter, or biomass. Unlike the fossil fuels coal, oil and natural gas, biomass is available over most regions of the earth. In certain forest-related enterprises, electricity is at present generated from the burning of wood to drive steam turbines. The burning of biomass to release energy has the advantage over the burning of coal in that very little ash is formed and no sulphur dioxide. Carbon dioxide is, however, released, but the amount exactly equals that extracted by the plant matter during photosynthesis. It is thought that there is considerable potential for the driving of gas turbines by biomass, particularly sugar-cane waste, which would first be gasified using high-pressure air and steam.

Apart from the generation of electricity, fuels have been successively synthesized using solar energy. Ethanol has been produced from maize in the US and, combined with petrol, is sold in small amounts as 'gasohol'. There is considered to be potential in the production of hydrogen gas as a fuel via solar energy. It is envisaged that electricity produced by the sun could be used to decompose water into its constituent hydrogen and oxygen by electrolysis, i.e. passing a current through water. Hydrogen is environmentally ideal as a fuel since on burning it produces virtually no other product than water.

There is little doubt that solar-derived energy will feature more prominently in the future as fossil-fuel supplies become depleted and concerns about a clean environment increase. The potential for developing at least some of the technologies mentioned in this chapter seems to be very great, though at present there still needs to be recognition of their value by government and other authorities.

Chapter 10

Observing the sun

Our understanding of the processes occurring in the sun and its atmosphere, described in previous chapters, is based on careful and painstaking observations made in the visible part of the spectrum using conventional telescopes and, over the last 50 years or so, in the radio, ultraviolet, X-ray and other parts of the electromagnetic spectrum. In this chapter we will discuss instrumentation used to observe the sun, including some projects planned for the future. Generally, the instrumentation is distinct from that used, e.g., for observing stars, galaxies or other faint astronomical objects as the sun's radiation is so much more intense. This applies most obviously to visible-light (or 'optical') telescopes but it also applies to telescopes and detectors operating in other parts of the spectrum. There are, then, many ground-based and spacecraft 'solar observatories' in which the instruments are devoted to looking at the sun. We shall be describing some examples of these as used by solar astronomers (§§10.1–3). The sun can also be observed easily using very small and inexpensive telescopes, these being within the range of amateurs, and in 10.4 we shall see how this can be done safely and properly.

10.1 Optical instruments and observatories

The sun can be observed with telescopes that are either refractors or reflectors, i.e. that use a lens or concave mirror respectively to focus the light. As there is plenty of light available, the lens or mirror can have a long focal length, so that a large image can be produced. This is illustrated in Fig. 10.1 for a refractor (the same principles apply to a reflector). Rays of light from the sun are practically parallel as the sun can be considered infinitely distant. Those from, say, the north (N) limb are brought to a focus at N', those from the south (S) limb to a focus S'. In the figure, f is the lens or mirror focal length, equal to the distance of the solar image from the lens or mirror, and *d* the diameter of the solar image. The sun's angular diameter is about 32 arc minutes or 0.0093 radians (varying from 31′ 28″ arc to 32′ 32″ arc over a year), so *d* equals 0.0093*f*. Thus, the large reflecting telescope at Sacramento Peak Observatory, focal length 54.9 m, produces a solar image with diameter 51 cm. Ideally, a telescope's resolution – the ability to separate objects close together in the sky – is determined by the process of light diffraction, the tendency for light rays to depart from straight-line propagation and to bend round an obstruction. Because of this, a point source of light viewed through a telescope appears as a tiny light disc surrounded by alternately dark and light rings. Two closely spaced stars are just resolved if their separation is 0.14/*D* arc seconds, where *D* is the aperture (diameter of the telescope's main lens or mirror) expressed in metres (this assumes that the stars' light is 'white', with an effective wavelength of 560 nm). Telescopes with larger apertures should therefore have better resolution. However, the resolution of a ground-based solar telescope is generally worse than the diffraction limit, and is determined by turbulent air currents in the light path. These occur in the immediate environment of the telescope, caused by the heating of the ground, and within the optical system itself as a result of heating of the telescope parts.

George Ellery Hale's experiments at building solar telescopes around the turn of the century produced some discoveries which have had lasting influence on the design of large modern instruments. One was that atmospheric turbulence markedly decreases with height above the ground and that vertical light-paths are better than horizontal ones (see §1.6). He found what seemed to be an ideal site for observing at Mount Wilson, where the Snow and tower telescopes were built. Like Hale's telescopes, many modern large solar telescopes are situated on mountain tops with sunlight constantly fed

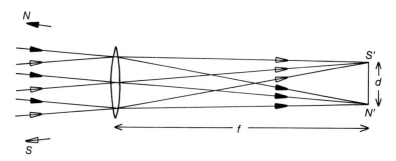

Fig. 10.1 Image of sun produced by the objective of a refractor. The direction of the north limb of the sun is *N* (imaged at *N*′), that of the south limb *S* (imaged at *S*′). The image diameter *d* is proportional to the objective's focal length – *d*=0.0093*f*, where *f* is the focal length (*d* and *f* are in the same units).

Fig. 10.2 The McMath Solar Telescope at the Kitt Peak station of the National
Solar Observatory, near Tucson, Arizona. (Courtesy National Solar
Observatory/Kitt Peak)

into them by rotating mirrors ('heliostats') high above the ground. The large
vacuum tower telescope at the National Solar Observatory (Sacramento
Peak), New Mexico (altitude 30 200 m), for example, has this design, with its
1.63-m-diameter mirror at the foot of the tower, some 55 m below ground
level. By evacuating the air in the tower, convection currents within the
telescope are almost entirely eliminated.

The McMath solar telescope at the National Solar Observatory (Kitt
Peak), Arizona, the world's largest, is a striking piece of architecture as well as
a major scientific instrument (Figs. 10.2, 10.3). The main telescope optics are
contained in a diagonal shaft aligned with the earth's rotation axis. A heliostat
reflects sunlight down its 152-m length to an off-axis paraboloid mirror,
1.6 m in diameter, 52 m below ground level (the view in Fig. 10.3 is down
towards the main mirror). Rays reflected from this mirror pass upwards to a
flat mirror at ground level and the solar image is formed in an observation
room containing spectrographs. The extremely long focal length of the telescope gives a uniquely large diameter for the solar image – 85 cm.

The mountain-top locations of these and other solar observatories do not
entirely avoid the problems of poor image quality through air turbulence
induced by ground heating and that of cloud cover. The Big Bear Solar
Observatory near Pasadena, California, is located in the middle of a high-

altitude lake, giving it the advantage of cooler surroundings and a laminar air flow over the lake surface. Excellent, diffraction-limited solar imaging has been achieved at the Swedish Solar Observatory on La Palma in the Canaries; the site there – a mountain top surrounded by sea – seems to be close to ideal for much of the year. This is a vacuum tower telescope rather like Sacramento Peak, but with the imaging done by a relatively modest (0.5 m) aperture achromatic lens mounted on the entrance window at the top of the tower.

The success of the Swedish Solar Observatory's telescope has led to designs of telescopes with even better resolution. The Large Earth-based Solar Telescope (LEST) is currently being planned by a consortium of European countries, the USA, Australia and China. It will consist of a very compact reflecting telescope with a 2.4-m-diameter mirror (larger than the McMath telescope) and a focal length of only 5.5 m. An additional mirror will re-image the Gregorian focus, directing the light down a 25-m-long tower on which the compact telescope will be mounted. Spectrographs and other equipment will be located at the foot of the tower. It is expected that a resolution of 0.1 arc sec. (70 km on the sun) will be achieved.

Fig. 10.3 Light path in the McMath Solar Telescope, looking down towards the main mirror, viewed from a point approximately at ground level. (Courtesy National Solar Observatory/Kitt Peak)

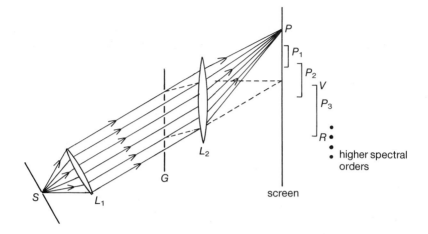

Fig. 10.4 Optical scheme for a grating spectrograph. Light is incident on a slit *S*, and a lens *L*$_1$ collimates the light which is diffracted by the transmission grating *G*. A second lens *L*$_2$ produces a focused spectrum on the screen: *P* is the zeroth order (with no dispersion), and *P*$_1$, *P*$_2$, *P*$_3$ the successive spectral orders having violet (*V*) light nearest *P*, red (*R*) furthest.

After a solar image has been formed with a telescope, the chief means of examining solar radiation is the spectrograph. Diffraction gratings are generally used to disperse the radiation (i.e. analyse by wavelength), and these consist of flat or curved surfaces across which are ruled numerous, equally spaced and parallel grooves. Light is either transmitted by or reflected from the ungrooved parts of the grating, but scattered by the grooved parts and effectively lost. Figure 10.4 shows the principle of the instrument. Sunlight is incident on a transmission grating via a narrow slit *S* and a lens (*L*$_1$) that 'collimates', i.e. makes parallel, the light from the slit. A central bright image of the slit is formed at point *P* on the screen or photographic plate, brought to a focus by a second lens *L*$_2$; light is not dispersed for this image. Because of diffraction, light is allowed through the unruled parts of the grating *G* which act as narrow slits; wavelets of light emanate from each of the slits as if they were themselves sources of light. These interfere with one another, producing a series of spectra, or 'orders', along the screen (*P*$_1$, *P*$_2$, *P*$_3$, etc.), with decreasing intensity but increasing spread of wavelengths, and with shortest wavelengths (violet, *V*) closest to the central image.

Astronomers use a wide variety of grating spectrographs, but many are variations on this basic design. One commonly in use at solar observatories is the Littrow spectrograph (Fig. 10.5), favoured because of its compactness. This has a single 'autocollimating' lens, which serves as a collimator for light incident on the reflection grating and as a camera lens for photographing the resulting spectrum.

A spectroheliograph is a type of spectrograph that allows the solar disc to be observed at a particular wavelength. This is accomplished as indicated in Fig. 10.6. To the usual spectrograph arrangement is added a second, or 'exit', slit where the spectrum is formed and at the wavelength of a prominent Fraunhofer (e.g. Hα) line; the spectral resolving power is usually good enough to permit location of the exit slit at a particular point within the line profile, e.g. its central part (core) or in the wings. The whole instrument is moved such that the first, or 'entrance', slit viewing the sun passes over the entire disc. The photographic plate, exposed while the instrument is in motion, thus records an image of the sun taken at this wavelength.

A later evolution of the spectroheliograph has the exit slit at positions on either side of the centre of an absorption line, in the red (R) and violet (V) wings (Fig. 10.7), to detect velocity patterns on the sun. A pair of spectroheliograms is simultaneously formed at these positions. An area of the sun's surface that is rising (i.e. approaching an observer) causes the line to be violet-shifted, and one sinking (receding from an observer) causes the line to be red-shifted. One of the spectroheliograms so formed thus has rising areas appearing light, sinking areas dark, while the other has rising areas dark, sinking areas light. The two spectroheliograms can be 'photographically cancelled', i.e. the photographic negative of one is placed on the positive print of the other; on the resulting 'Dopplergram', dark areas are rising, light areas sinking (or vice versa) and grey areas those with no appreciable velocity shift.

Further information about photospheric motions may be acquired from the photographic cancelling of successive Dopplergrams, taken a few minutes apart. A 'Doppler difference' image obtained by this technique, with no change of the exit slits (R and V), reveals the presence of any accelerated motions. A 'Doppler sum' image, obtained when the second Dopplergram has the exit slits R and V reversed compared with the first, enhances the Doppler signal of any material due to a moving part of the sun's surface if its motion persists between taking the two Dopplergrams. It was from Doppler sum images that the discovery of the supergranulation and five-minute oscillations was made (§§3.1, 2.3). The careful matching of prints and negatives to

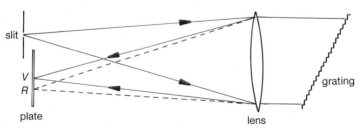

Fig. 10.5 The Littrow spectrograph. A single lens acts as collimator and camera lens for photographing the spectrum.

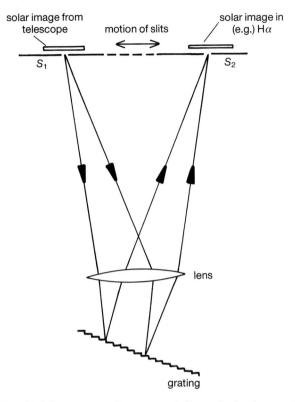

Fig. 10.6 Arrangement for a spectroheliograph. A telescope (not shown) produces a
solar image at slit S_1, and light passes to a grating via a lens, which also
produces a focused image at a particular wavelength (e.g. Hα), part of
which is sampled by an exit slit S_2. A whole-disc solar image in Hα is
formed on a photographic plate behind S_2 when the slits S_1 and S_2 are
moved across the two images.

achieve photographic differencing which was necessary for the successful
application of these procedures can now be done much more quickly and
conveniently by electronic techniques.

Broadly the same methods have been used in the construction of
magnetographs, or instruments to measure solar magnetic fields (§3.8). Large
sunspot fields (about 0.4 T) can be measured directly from the Zeeman
splitting of Fraunhofer lines into the undisplaced π component and the two
displaced σ components. The much weaker fields away from sunspots give
rise to only a broadening of lines, but can be measured with the Babcock
magnetograph using the fact that the π component is plane polarized, and the
two σ components circularly polarized. Generally, the circular polarization of
the σ components is easier to measure, giving the longitudinal component of
the magnetic field. We recall that circularly polarized light can be resolved
into two beams of plane-polarized light, one retarded with respect to the other

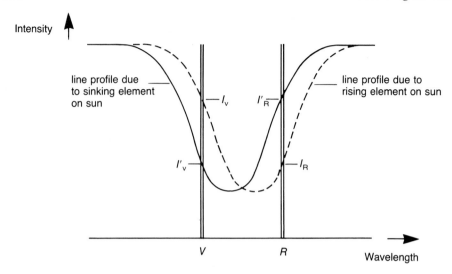

Fig. 10.7 Principle behind the Dopplergram. The exit slit of a spectroheliograph is placed at two locations within the profile of a single Fraunhofer line in the solar spectrum, on the red and violet wings. Motions of particular solar features can be detected by the fact that the line is Doppler-shifted, producing different intensities at positions R and V. Two spectroheliograms at these positions can be photographically cancelled to give an image that shows rising or sinking features on the solar disc as dark or light.

by a quarter of a wavelength (Fig. 3.12c). When incident on an electro-optic crystal, one of the beams can be advanced or retarded by a quarter of a wavelength, leaving the other unaffected. The circular polarization can consequently be transformed into plane polarization, with planes that are 90° to each other for right-handed and left-handed circular polarization. The plane-polarized light may then be analysed with a Polaroid filter or equivalent.

The Babcock magnetograph has the same optical arrangement as a spectroheliograph, with the exit slit positioned in the spectrum in one of the wings of a line sensitive to the Zeeman effect. The electro-optic crystal and Polaroid analyser are placed in front of the entrance slit. An oscillating voltage is applied to the electro-optic crystal causing the σ components of the line to appear and disappear at the same frequency as its light is transmitted or blocked by the Polaroid. The light signal at the exit slit, which may be photo-electrically detected, thus consists of a steady (or DC) component plus an oscillating (or AC) component, the latter being translatable into solar magnetic field strength. The original Babcock system detected magnetic fields in a single area on the sun, but by using an array of tiny probes along the exit slit, connected by optical fibres to small light detectors (either miniature photo-multiplier tubes or charge-coupled devices, CCDs), the magnetic field can be measured at many points along a straight line. Complete magnetic field information may then be obtained by moving the entrance slit across the solar image, in the manner of a spectroheliograph. The magnetogram of Fig. 3.14, produced by an instrument at the National Solar Observatory (Kitt Peak), was formed by field measurements at some 600 000 area elements over the solar disc.

The total strength and direction of the field can be obtained by *vector* magnetographs. They operate using special polarization analysers that can detect both circular and plane polarization, so that the plane polarization of the π component as well as the circular polarization of the σ components can be measured. Owing to the small amount of polarization in the π component, more precautions must be taken than with a Babcock-type magnetograph, e.g. the use of telescopes directly pointed at the sun rather than fixed telescopes fed with light from a heliostat since mirrors introduce plane polarization.

Observation of the corona outside the rare times of total eclipses is possible from high-altitude observatories with exceptionally clear atmosphere using coronagraphs. Light from the inner corona is so faint (surface brightness a million times less than the photosphere) that the slightest amount of stray photospheric light easily overwhelms it. Great precautions are therefore necessary in coronagraphs to remove this stray light. Lyot's instrument, made in 1930, was the first to be used for successful photography of the corona, and its basic design is still used (Fig. 10.8). An objective lens (L_1), highly polished and as blemish-free as possible, forms an image of the sun at the focus F_1, where an occulting disc blocks out the intense photospheric light. Any stray light produced by the first lens is eliminated as far as possible; thus, light diffracted at the edge of the lens is removed with a field lens (L_2) that forms an image of lens L_1 on a diaphragm D, while a small screen at its centre (S) blocks light formed by reflection at the surfaces of L_1. A third lens (L_3), just behind D and S, forms an image of the solar corona (I). Any heating of the instrument by direct sunlight is minimized to prevent image distortions. The coronal radiation has better contrast when observed in a wavelength range narrowed around the FeXIV 'green' or FeX 'red' coronal spectral lines. However, a wide-band coronagraph has been developed at the Mauna Loa (Hawaii) observing station of the High Altitude Observatory, detecting coronal emission from its polarization. Strictly, the Mauna Loa instrument is a coronameter, since the radiation is detected photo-electrically, with diode arrays. The various spacecraft instruments mentioned in Chapters 5 and 6 also had the same basic design of a Lyot coronagraph, with various narrow-band filters and Polaroids allowing the corona to be observed with particular wavelength regions and in different directions of plane polarization.

10.2 Solar radio telescopes

As radio waves are simply electromagnetic radiation with a much longer wavelength than visible light, the same optical principles still apply, though the reception of radio signals from the sun (or any other source) requires the use of aerials instead of cameras or other light detectors. In its simplest form an aerial is a dipole, consisting of a metal rod or length of wire, having a length equal to half the wavelength of the radiation being detected; the dipole is divided into two half-lengths, the inner ends of which are connected by cables to a receiver. A radio wave incident on the aerial has an associated oscillating electric field which induces a small alternating current in the aerial halves. This current passes to a receiver where it is electronically amplified. The sensitivity of such a system is poor, as is its ability to define the direction of a source in the sky such as the sun (the 'directionality'). But improvements can be made. Thus, for wavelengths of several centimetres up to about 3 m (frequencies about 10 GHz to 100 MHz), with the dipole in the form of a rod, the addition of a reflector rod or mesh on the side opposite to the source gives enhanced sensitivity, and director rods spaced at certain intervals in front of the dipole narrow the directionality. The result is a Yagi aerial, much favoured by amateur astronomers and in general domestic use for the reception of television and FM radio signals. It is suitable for wavelengths less than 3 m (frequencies more than 100 MHz). For much shorter radio wavelengths, e.g. around 1 cm, paraboloid dishes are suitable; the radiation is reflected and brought to a focus as in an optical telescope. Such dishes may be fully steerable if not too large, and objects like the sun can be resolved by scanning the dish across them. Large dishes that are fixed can also resolve objects as they pass over the sky with the earth's rotation.

   The resolution of a single radio telescope when determined by diffraction is much worse than that of an optical telescope with the same aperture. Fortunately, the problem can be circumvented using *interferometers*. These consist of two or more radio telescopes that are separated by large distances

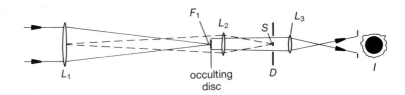

Fig. 10.8 Optical scheme for the Lyot coronagraph. An objective L_1 focuses sunlight on to an occulting disc at focus (F_1). A field lens (L_2), diaphragm (D) and small screen (S) remove light produced by diffraction and reflection from lens L_1. A third lens L_3 forms an image (I) of the corona.

(a) Two-element interferometer

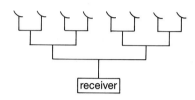

(b) Multi-element interferometer

Fig. 10.9 Two-element (a) and multi-element (b) radio interferometers. The polar
diagram shows the directionality of the two-element array.

but linked together by cables to a common receiver, where the signals are
mixed. Figure 10.9a shows the simplest form, one consisting of two radio
telescopes or 'elements'. If the two telescopes are directed parallel to each
other and are scanned across a point source in the sky, the two received
signals go successively in and out of phase with each other depending on the
difference of the path lengths from the source to each telescope. The signal
produces a pattern or 'polar diagram' like that in Fig. 10.9a, with several
lobes, the strongest one being in the direction of the source. The same
pattern is produced if the telescopes are stationary and the source passes
across the sky as the earth rotates: this arrangement is called a 'drift' interferometer. The angular widths or 'fatness' of the lobes determines the resolution
of an interferometer. This can be improved to a very narrow beam plus small
side-lobes by increasing the number of telescopes to form a 'multi-element'
interferometer (Fig. 10.9b). The telescopes may be arranged in a straight line
to give high resolution in one dimension on the sky, or in a T-shape pattern
for two dimensions.

Several arrays are, or have been, used in the metre-wavelength regions for
observing the sun and they have greatly added to our knowledge of how bursts
propagate into the outer corona (§6.3). The Culgoora radioheliograph in
Australia, which produced the images of the type IV burst of Fig. 6.21, was
one of the largest arrays until its dismantling in 1984; it consisted of 96 dishes
arranged in a circle 3 km in diameter. Another important array, also no longer
in operation, was the Clark Lake Radio Observatory in southern California,
owned by the University of Maryland. It was a T-shaped array with 720

aerials, having an east–west baseline 3 km long and a north–south one 1.8 km long, achieving a resoluton of a few arc minutes at 2 m wavelength. The Nançay Radioheliograph in France, recently modified, operates in the decimetre to metre range (150–450 MHz) with 18 dishes along an east–west baseline 3 km long and 24 dishes along a north–south baseline 1 km long. A radioheliograph under construction at the Nobeyama Radio Observatory in Japan, with 84 elements also in a T-shaped array and working at 375 MHz (80 cm wavelength) is expected to have a resolution of 7 arc seconds and very high time resolution.

A number of arrays have been used for observation at centimetre and millimetre wavelengths. The Very Large Array (VLA) in New Mexico has been used for observing the sun at 2, 6 and 20 cm wavelength (15, 5 and 1.5 GHz). It is an array of 27 dishes, each 25 m in diameter (Fig. 10.10). The dishes can be moved to form a Y-shaped pattern with each arm up to 27 km long, the resolution then being equivalent to a telescope aperture of tens of kilometres, and comparable to that of the largest optical telescopes. For solar work, however, the dishes are arranged in compact arrays. At 2 and 6 cm, solar features as small as a few arc seconds have been resolved. Other high-frequency arrays include that at the Owens Valley Radio Observatory and the RATAN-600 interferometer in the Crimea. A new array called BIMA, operated by US universities (Berkeley, Illinois and Maryland), will eventually have nine elements and will observe particularly solar flares at wavelengths of 2 cm, 1 cm and 3 mm with high sensitivity and 1 arc second resolution.

Single large, steerable dishes can be used at very short wavelengths, down

Fig. 10.10 The Very Large Array in New Mexico. The 27 dishes are generally arranged in a compact Y-shaped configuration for solar work. (Courtesy National Radio Astronomy Observatory and AURA)

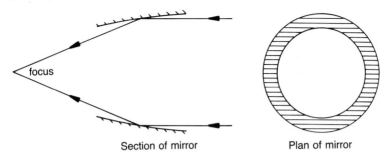

Section of mirror   Plan of mirror

Fig. 10.11 Section and plan views of an X-ray telescope. Rays can be reflected from the mirror surface only if they are at grazing incidence.

to the submillimetre region. Successful observations have been made with the UK/Netherlands James Clerk Maxwell Telescope, a 15-m steerable dish located at 4200 m altitude on the volcanic mountain of Mauna Kea in Hawaii. The site is renowned for low atmospheric water content, allowing transmission down to wavelengths of 0.3 mm. Eventually it is hoped that solar observation at these wavelengths will give us information about the chromosphere from where much of this radiation is emitted.

10.3 Solar spacecraft observatories

Observation of the sun in regions outside the visible and radio parts of the spectrum is only possible above the earth's atmosphere, so requiring the use of rockets, satellites or (for some purposes) high-altitude balloons. The spectral regions concerned include ultraviolet, X-ray and gamma-ray radiation. For the ultraviolet region (we arbitrarily define this to be the wavelength range 100–350 nm), telescopes and spectrographs are basically the same as visible-wavelength instruments, though any lenses used must be made of quartz or equivalent material since glass absorbs ultraviolet radiation. In the extreme ultraviolet and soft X-ray regions (say, 0.1 nm to 100 nm), considerable modifications must be made. The reflectivity of such radiation off a silvered surface becomes very small unless the radiation is incident at a grazing angle, typically 1°–2°. This means that paraboloid reflecting telescopes can only be used away from the 'nose' or vertex of the paraboloid, as indicated in Fig. 10.11. The collecting area of the mirror, as viewed from the source, is very small, being in the form of a narrow ring, but this is usually not a great restriction for observing the sun which is still an intense source at these wavelengths. The radiation must also be used at grazing incidence on spectrograph reflection gratings. A recent development is the use of special layered surfaces that enable radiation at these wavelengths to be reflected off surfaces at 'near-normal' (i.e. almost 90°) incidence angles: Fig. 5.11, a very high-resolution photograph of the soft X-ray sun, was taken with such optics. For wavelengths shorter than about 1 nm, mirrors are usually dispensed with in favour of fine collimators to direct the radiation. These consist of several grids, each containing two sets of extremely fine wires, perpendicular to one another. Radiation is blocked by the wires of successive grids unless it is incident at precisely 90° to the plane of the grids, so an X-ray detector behind the collimator 'sees' X-rays from only this direction. An X-ray image of the sun can be built up by moving the instruments in a sequence of positions over the sun's disc, monitoring the amount of X-ray emission at each point.

Ultraviolet detectors such as channel electron multipliers rely on the fact that the photons are energetic enough to liberate electrons from particular surfaces, which can be 'multiplied' to give a measurable voltage pulse. X-ray photons, being even more energetic, can cause the ionization of the atoms of a gas, a fact used in X-ray detectors consisting of gas-filled chambers such as the Geiger–Müller counter. Hard X-rays and gamma rays are detected using blocks of scintillation crystal such as sodium iodide; X-rays passing through them produce light pulses that are observed and counted with photomultipliers. Techniques can be used to eliminate what would be a sizeable 'background' emission due to the passage of cosmic rays through the detector and producing the same kind of signals as X-rays.

Spectral observations of X-rays can be made with gas-filled counters or scintillation detectors since it can be arranged that the size of the voltage pulse in the former and the light pulse in the latter produced by an X-ray photon depend on its energy, i.e. wavelength. The resolution is, however, quite crude – typically 30% of the photon energy observed – but the cryogenic cooling of crystal detectors has improved this considerably. In the soft X-ray region (photon energies less than 10 keV, or wavelengths larger than 0.1 nm), crystals of various materials can be used to diffract X-rays of a particular wavelength. By rotating the crystal, i.e. varying the angle of incidence that the solar X-rays make with the crystal face, different wavelengths are diffracted, and so a spectrum can be formed (this was how Fig. 5.7 was obtained).

Several X-ray and ultraviolet imaging instruments (telescopes and scanning collimated detectors) and spectrometers (i.e. 'photoelectric' spectrographs) were flown on rockets and satellites in the 1950s and 1960s (1.12). A significant step forward occurred with the NASA spacecraft *Skylab* (Fig. 10.12) which was launched in May 1973. It was the first manned observatory, with crews of astronauts being ferried up to the spacecraft for periods of several weeks on three occasions during the nine-month-long mission. The spacecraft, which had been planned and developed for more than ten years before launch, weighed 100 tons and was placed in an orbit 435 km above the earth. This altitude gradually decreased and *Skylab* eventually re-entered the earth's atmosphere in 1979. An accident shortly after launch resulted in the removal

Fig. 10.12 The *Skylab* spacecraft in orbit. The solar instruments were housed in the Apollo Telescope Mount, furthest from the camera in this view. (Courtesy NASA)

of a solar panel and part of the protective panelling, and the first astronaut rendezvous had to be postponed until solutions to these problems could be devised.

Skylab carried an array of instruments called the Apollo Telescope Mount (ATM) designed to study the sun in various wavelengths. The instruments, all built by US research groups, included a white-light coronagraph, a soft X-ray telescope, a soft X-ray spectroheliograph, an ultraviolet spectrograph and extreme ultraviolet spectroheliograph, and an extreme ultraviolet spectrometer/spectroheliometer. Apart from the last instrument, the data were obtained in mostly photographic form. Astronauts were able to direct instruments at particular features on the sun by using images from two on-board Hα telescopes, and were able to respond quickly – often within minutes – to solar phenomena that had been reported by ground-based observers.

Although it is almost 20 years since *Skylab* flew, the data from the ATM instruments are still being analysed. Much emphasis was placed on co-operative studies of solar phenomena, such as flares, active regions and coronal holes, comparing data sets from several instruments as well as from ground-based solar observatories. The great interest in solar flares, particularly since *Skylab*, has been further encouraged by the results from several

large unmanned spacecraft as well as the *Spacelab 2* mission on the US Space
Shuttle. The NASA spacecraft *Solar Maximum Mission* (Fig. 10.13), launched
in February 1980, was intended to study solar activity in solar cycle 21 which
had peaked in late 1979. It carried a group of seven instruments primarily
designed to observe flares or associated phenomena in a wavelength range
extending from white-light to gamma rays. These included a white-light
coronagraph, ultraviolet and soft X-ray spectrometers, as well as some instruments not included in the *Skylab* complement – a hard X-ray spectrometer,
an imaging spectrometer sensitive to soft and intermediate-energy X-rays and
a gamma-ray spectrometer. It also carried the solar irradiance monitor,

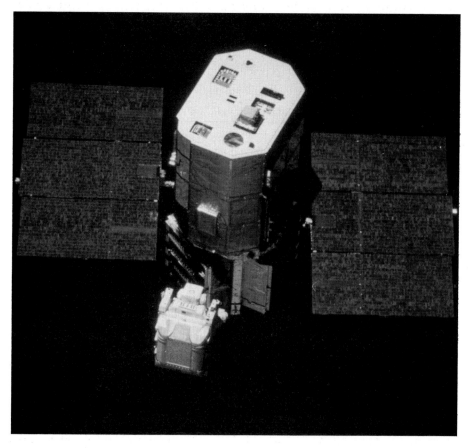

Fig. 10.13 The *Solar Maximum Mission* in orbit at the moment when an astronaut
(George Nelson) from the nearby *Challenger* Shuttle attempted to de-
spin *SMM* with his one-man vehicle, the 'Manned Manœuvrable Unit'.
This attempt failed, but *SMM* was later successfully captured by the
Shuttle's Remote Manipulator Arm and brought into the Shuttle bay for
repairs to its failed attitude (pointing) unit and some of the instruments.
In this view, the panel of *SMM* facing the sun contains the entrance
windows of the instruments and star-trackers. (Courtesy NASA)

ACRIM (§9.1). As the spacecraft was unmanned, the data had to be recorded electronically and telemetered down to ground stations. Like *Skylab*, the spacecraft had an eventful history, suffering an early failure in the attitude control unit, responsible for keeping the spacecraft accurately pointed at the sun. *SMM* had been built to a modular design so that repairs could be more easily made, either in orbit or on the ground following a retrieval by the Space Shuttle. It was thus decided to dedicate a Shuttle mission to repair *SMM* by removing the module (one of three) containing the attitude control unit and replacing it with a new one. The Shuttle Flight eventually occurred in April 1984. The capture of *SMM* was accomplished with the Shuttle's remote manipulator arm. Repairs on *SMM*, now loaded in the Shuttle bay, were done by two astronauts who had been previously trained for the operations. After the week-long Shuttle mission, *SMM* was returned to orbit and continued operating until re-entry into the earth's atmosphere, in December 1989.

Other satellites operating over the period of cycles 21 and 22 included the US satellite *P78-1* and the Japanese *Hinotori* satellite. The *P78-1* spacecraft was launched in 1979 and carried the SOLWIND coronagraph (§6.4) and soft X-ray spectrometers, which obtained the first observations of the broadening and shifts of soft X-ray lines at the solar flare impulsive stage (§6.3). It operated successfully until destroyed in an anti-satellite missile test in 1985. *Hinotori* carried a soft X-ray spectrometer, obtaining the highest-resolution spectra to date, and an imaging hard X-ray instrument.

Spacelab 2 was a week-long mission on the *Challenger* Space Shuttle in the summer of 1985. A large number of scientific instruments – biological, astronomical and solar – filled the bay of the Shuttle, and were operated by astronauts. The four solar experiments – designed to study ultraviolet and visible radiation from the sun – were mounted on a German-built pointing system (Fig. 10.14) which, after initial problems, was able to acquire features on the sun with an accuracy of 1 arc second. The instruments included an extreme ultraviolet spectrometer (CHASE) to measure the coronal helium abundance, an ultraviolet high-resolution telescope and spectrograph (HRTS), and a high-resolution optical imaging polarimeter (SOUP). The success of the mission promoted the possibility of a second flight of the solar instruments, but the disastrous *Challenger* accident in 1986 led to an indefinite postponement of this.

The Japanese *Yōkō* spacecraft (Fig. 10.15), launched in August 1991, is studying the high-energy aspects of flares with hard and soft X-ray imaging instruments, as well as a sensitive X-ray crystal spectrometer observing lines of very hot ions. The Soft X-ray Telescope, built by the Lockheed Palo Alto Research Laboratory, uses a charge-coupled device to detect the soft X-ray image at the focus of a grazing-incidence reflector. Figure 10.16 is a full-sun image obtained during the first month of the spacecraft's operation, and shows active-region and other loop structures with a resolution (2 arc seconds) that is better than either *Skylab* or *SMM*.

Although not specifically a solar mission, the NASA satellite *Gamma-Ray Observatory* (*GRO*), also launched in 1991, has on board some high-energy instruments with large fields of veiw that enable the sun to be observed. Hard X-ray solar flares seen by one of them are being catalogued and analysed.

Several spacecraft have explored the terrestrial and other planetary

Fig. 10.14 The solar instruments on *Spacelab 2* were mounted on an instrument pointing system, and can be seen in this view looking towards the aft section of the *Challenger* Shuttle. (Courtesy NASA)

Fig. 10.15 The Japanese *Yōkō* spacecraft in an artist's impression. High-energy
instruments including soft and hard X-ray telescopes on board view the
sun. (Courtesy Institute of Space and Astronautical Science, Japan)

magnetospheres and the interplanetary medium, not only near the sun but at
the edge of the solar system, notably *Pioneers 10* and *11* and *Voyagers 1* and *2*
(§7.2). Our knowledge of the nature of interplanetary space has largely come
from instruments on these and other spacecraft which have measured
energies and composition of the charged particles making up the solar wind as
well as magnetic and electric fields. Much information has come from instruments imaging the terrestrial auroral zone, like the *Dynamic Explorers* (see
Figs. 7.17, 7.18 and 7.20) and an experiment on the Swedish *Viking* spacecraft, all launched in the 1980s. The spacecraft *Ulysses*, a joint European
Space Agency (ESA) and NASA mission, was launched in 1990 from the
Space Shuttle into a complex orbit that, in 1994–95, will take it over the polar
regions of the sun, enabling it to explore in detail for the first time the high-
speed solar wind thought to originate from polar coronal holes. This will be
accomplished by a 'gravity assist' from Jupiter, to which *Ulysses* is at first
travelling, allowing the spacecraft to leave the ecliptic plane and so, on returning to the sun's vicinity, orbit over both solar poles.

The increasing sophistication and consequent rising costs of space
instrumentation in recent years has put solar physics and astronomy into the

realm of 'big science', and many years of preparation and planning go into spacecraft before they are launched. There is now a general reluctance on the part of many governments to supply large funds for such research, especially as it is frequently viewed as having no immediate commercial application. Nevertheless, the 1990s and beyond should see several solar spacecraft, with not only the USA (NASA) and the CIS but the Europeans (national governments and ESA), the Japanese and others playing major rôles in their development. In addition, some high-energy instruments are expected to fly on long-duration balloon missions.

Two of the largest missions that will fly in the mid-1990s are joint ESA–NASA ventures. The *Solar and Heliospheric Observatory (SOHO)*, due for launch in 1996, will orbit round an equilibrium position between the earth and sun known as the inner Lagrangian point, some 1.5×10^6 km sunward of the earth (Fig. 10.17). Here solar observing will enjoy full-time coverage, without interruption (unlike most previous solar spacecraft) by earth eclipses.

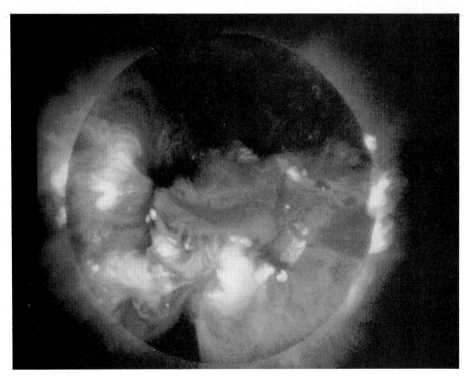

Fig. 10.16 One of the first images acquired by the Soft X-ray Telescope on the *Yōkō* spacecraft of the Japanese Institute of Space and Astronautical Science. (The Soft X-ray Telescope experiment is a Japan/US collaboration involving the National Astronomical Observatory of Japan, the University of Tokyo and the Lockheed Palo Alto Research Laboratory. The US work is supported by the National Aeronautics and Space Administration.)

Fig. 10.17 Solar spacecraft for the 1990s (artist's impression). The *Solar and Heliospheric Observatory* (*SOHO*, lower left) will observe the sun from an equilibrium point between the sun and earth, while the *CLUSTER* group of four satellites will investigate the three-dimensional structure of the earth's magnetosphere and surrounding solar wind. Each project is being jointly undertaken by NASA and the European Space Agency. (Photo ESA)

This will be ideal for a French-built helioseismology instrument on board, since solar oscillations with very long periods can be studied. It will also carry spectrometers and particle detectors for investigating the corona and solar wind. The *CLUSTER* project, to be launched soon after, will be a three-dimensional investigation of the solar wind and earth's magnetosphere using four satellites in orbits that will take them in and out of the magnetosphere. A CIS multi-spacecraft mission due to be launched at about the same time is expected to enhance the scientific capabilities of *CLUSTER*. Other future spacecraft investigating the solar wind and terrestrial magnetosphere include the NASA *WIND* and Japanese *Geotail* missions. The CIS *Koronas* spacecraft, due for launch into high-inclination orbits in the early to mid-1990s, will have X-ray, white-light and very-low-frequency radio wave detectors on board to study solar flare activity. High-resolution soft X-ray imaging instruments are planned for future satellites in the NOAA *GOES* series.

A number of other spacecraft are currently being defined or planned, and may become reality by the end of the century. These include the NASA *Orbiting Solar Laboratory*, which will have imaging and spectral instruments covering the range from X-rays to white-light, and the ambitious *Solar Probe* which will measure small solar structures from an orbit with perihelion of only three solar radii $(2.1\times10^6\,\text{km})$ from the sun's surface; it will have a conical shield to protect it from the sun's heat, with instruments mounted in its shadow. ESA hope to provide a solar physics element for the *Columbus* Space Station, which may consist of a set of instruments, including an imaging high-resolution spectrometer in the extreme ultraviolet spectrum to study the chromosphere, transition region and corona.

10.4 Observing the sun for amateurs

Practically all solar observation was, until the turn of the century, carried out by amateur astronomers, and up till quite recent times amateur observations were proving very useful in following sunspot and other activity, e.g. during the International Geophysical Year period of 1957–58. The sun is nowadays observed with highly sophisticated instrumentation, both ground-based and space-borne, and amateur observations no longer have the major value that they used to have. Nevertheless, a very large number of amateurs do faithfully follow solar activity with telescopes and other equipment, and doing so derive a great deal of enjoyment and interest. In this section we describe some of the observing methods within reach of most amateurs.

We must start with a most important warning. The abundance of light and heat emitted by the sun means that *directly observing the sun with just a small telescope or pair of binoculars is extremely dangerous and is likely to result in permanent blindness*. This is the case even when the sun is dimmed by mist or

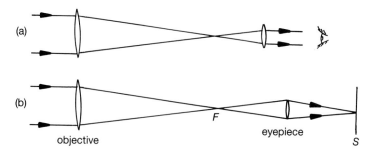

Fig. 10.18 (a) Ray diagram for telescope when used for directly viewing stars or planets. (b) Ray diagram for telescope when projecting image of sun on to a screen. The eyepiece is moved outwards from its usual position.

cloud. The use of darkened filters or 'sun-caps' is also not recommended, as there is still an unsafe amount of heat radiation transmitted and in addition there is a risk of their splintering under intense heat. Solar 'diagonals', eyepieces transmitting a small fraction of sunlight by means of double or multiple reflections off prism surfaces, are possibly fairly harmless for direct viewing, but a perfectly safe method of observing, and the only one recommended here, is by projection of the solar image on to a screen beyond the telescope eyepiece.

When a telescope – refractor or reflector – is used for observing stars, the planets or the moon, the eyepiece, considered as a single lens (though usually it consists of several lenses), is placed such that parallel light emerges (Fig. 10.18a): the focal point of the eyepiece coincides with that of the objective lens for a refractor (as in the figure) or mirror for a reflector. These parallel light rays are concentrated by the eye to form a sharp image. To observe the sun with such a telescope, a projected image can be formed by moving the eyepiece out from this position, the light rays from the eyepiece converging to a point on the screen S in Fig. 10.18b.

For solar observation, refractors are generally considered better than reflectors, though either can be used. To avoid overheating of the eyepiece, a reflector should have the main mirror unsilvered. Small refractors – up to 12 cm aperture – are ideal for observing; larger telescopes should be stopped down, again to avoid overheating the eyepiece.

The projection screen may be formed from white card glued to a wooden frame and attached to the telescope with an extendable mounting arm. A card attached to the telescope at the objective end can be used to throw a shadow over the screen, giving the projected image better light contrast. Alternatively, a wooden box may be used (Fig. 10.19) that is open at the screen end on two sides to allow sunspots and faculae on the solar disc to be drawn on tracing paper attached to the screen. A graticule of squares drawn on the screen assists in recording the positions of spots. Accurate positions of spots on a

drawing are much more easily obtained if the telescope is equatorially mounted and clock-driven to follow the diurnal motion of the sun, i.e. its motion
across the sky due to the earth's rotation. Moving the telescope's eyepiece in
and out and varying the screen distance alters the size of the projected image
of the solar disc, and in principle any size may be chosen, though too large an
image will be very dim and too small very bright. In practice, a projected
image diameter of about 10 cm is most suitable unless details of spot groups
are to be recorded, when a larger size should be used.

The orientation of the projected solar disc – i.e. which point is north, which
east, etc. – may be determined from the diurnal motion of the image. The
square graticule on the screen can be orientated such that sunspots drift along
one set of lines with the sun's diurnal motion: this then defines the east–west

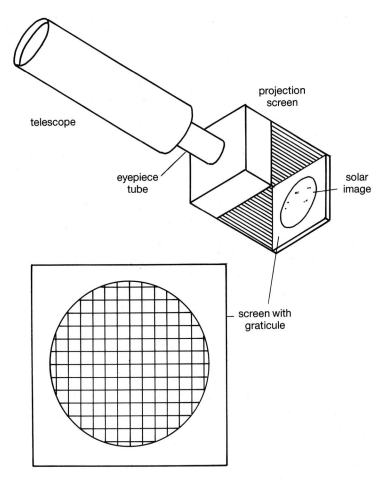

Fig. 10.19 Projection screen attached to the eyepiece of a telescope for safe viewing
of the sun. The graticule of squares on the screen allows the positions of
sunspots to be more easily recorded.

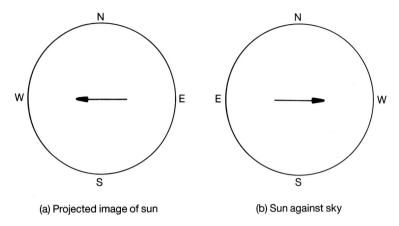

(a) Projected image of sun   (b) Sun against sky

Fig. 10.20 Orientation of the solar disc when projected on to a screen and on the sky. The arrows indicate the apparent motion of the sun because of the earth's rotation.

direction of the image. The leading edge of the image in the drift direction is the *western* edge (i.e. nearest to the western horizon). If the telescope is an astronomical one as opposed to a terrestrial one, i.e. has an eyepiece consisting of one lens instead of two, the projected image has the orientation as shown in Fig. 10.20a. Note that this orientation has east and west transposed (is 'laterally' inverted) compared with the sun's disc on the sky background (Fig. 10.20b, drawn for a northern-hemisphere observer).

Following the progress of developing sunspot groups, especially if very complex, is often of great interest. This can be done by a series of drawings done daily at high magnification as the group crosses the solar disc by solar rotation. The formation of light-bridges and the proper motion of small spots (see §6.1) can be easily followed with small telescopes.

Sunspot numbers can also be estimated very easily, and if done consistently over several years, the eleven-year number cycle can be traced out. Spots can be counted using the Wolf formula, $10g+f$, where g is the number of separate spot groups and f the total number of spots, or simply the number of spot groups.

The latitudes and longitudes – the 'heliographic' co-ordinates – of spots can also be found. To show how, we first describe the co-ordinate system of the sun and its appearance from the earth. The sun, which can be regarded as a perfect sphere, rotates on an axis that is inclined by about 7° to the perpendicular to the earth's orbital plane. The earth's rotation axis is inclined by about $23\frac{1}{2}°$ to its orbital plane. The directions of the sun's and earth's rotation axes and the earth's orbital plane are all effectively fixed in space. Referred to the centre of the sun, the northern end of the earth's rotation axis is along the direction OE_N, that of the sun's rotation axis along OS_N, and the perpendicular to the earth's orbital plane is along OP_N. The angle between the directions OE_N and OS_N is about 26°.

We now consider how the sun appears from the earth (Fig. 10.21). The north end of the rotation axis is inclined to the north pole of the disc by an angle P which varies over the course of a year by up to 26° west and east, the angle 26° being that between the rotation axes of the sun and earth. During the year, successively the sun's northern pole, then its southern pole, are revealed to the earth. The angle of inclination, or what is the same thing the latitude of a point at the disc's centre, is often denoted by the symbol B_0. It, too, varies over the year, with maximum 7°, equal to the angle between the sun's rotation axis and perpendicular to the earth's orbital plane. Precise values for the angles P and B_0 can be obtained from *The Astronomical Almanac* or the *Handbook* of the British Astronomical Association. Figure 10.21 illustrates the orientation of the sun's rotation axis and equator (and correspondingly the apparent paths of sunspots across the solar disc) at four times during the year.

The solar longitude system needs some comment. Sunspot motions provide proof that the sun rotates, but there is no recognizable permanent feature to which longitudes can be referred, i.e. there is no equivalent to the Greenwich meridian. For this reason, an arbitrary system is used, introduced

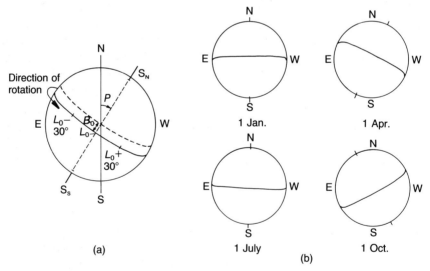

Fig. 10.21 (a) Definition of the P and B_0 angles and the solar longitude system. The north, south, etc., points of the solar disc as seen in the sky are indicated by N, S, etc. The sun rotates from the east (E) to west (W) limbs. The rotation axis of the sun is the line joining S_N and S_S, also called the central meridian. The longitude of the central meridian is L_0, and *decreases* with time. (b) The changing orientation of the solar axis (due to varying P and B_0 angles) is illustrated by the solar discs for four times during the course of the year.

by Carrington; it is defined on the basis of a sidereal rotation period of 25.38 days (synodic period 27.2753 days), with zero solar longitude passing the central meridian on 9 November 1853 (synodic rotations are sometimes given sequence numbers, starting with this date). Solar longitudes are reckoned in the direction of rotation, i.e. towards the west (see Fig. 10.21a), so that, seen from the earth, the longitude of the central meridian, denoted L_0, *decreases* with time. It does so by about 13.2° per day. Values of L_0 can be obtained, like P and B_0, from *The Astronomical Almanac* or *Handbook* of the British Astronomical Association.

The heliographic co-ordinates of sunspots and faculae observed on a particular day can be found by two methods. The first and easiest involves comparing a drawing with an appropriate grid of latitudes and longitudes on transparent sheets. Eight such grids, drawn for values of B_0 from 0° to 7° in 1° steps, allows positions to be found for any time in the year – when B_0 is negative (from $-1°$ to $-7°$), the grid with appropriate positive value is simply turned upside down. The sunspot drawing on tracing paper should have north, east, south and west indicated on it, the orientation having been determined by sunspot drift as described earlier. The solar rotation axis should then be drawn across according to the value of P for the day of observation: the convention for the sign of P is that the north end of the rotation is east of the north point of the image if P is positive, west if P is negative. The drawing is now placed under the grid with the B_0 value for the day of observation, and the grid orientated such that the rotation axis on the drawing coincides with that indicated on the grid. Latitudes and longitudes of spots and faculae may then be estimated from comparison with the grid. Longitudes can be referred to either the central meridian or to the Carrington system with L_0 for the day of observation. Figure 10.22 shows a typical sunspot drawing with latitude/longitude grid superimposed, illustrating how spot co-ordinates can be read off. It is generally possible by this method to estimate heliographic co-ordinates with an accuracy of about 1° for features near the centre of the sun.

The second method involves measuring the 'polar' co-ordinates of a solar feature, i.e. its distance from the sun's centre, in units of the radius of the solar image, and the position angle θ of the line joining the feature to sun centre with respect to the north–south axis of the drawing. The convention is that θ is 0° for the north direction, 90° for east, 180° for south and 270° for west. The measured r and θ can be converted to angular distance of the feature from the sun's centre, and, with values of P, B_0 and L_0 for the time of observation, to latitude and longitude. Formulae for this are given in Appendix 2 which also contains a computer program in BASIC language suitable for small personal computers which does the calculation.

Observation of spot positions continued for a few days can reveal any proper motions, and checks can be made to see whether spots with different latitudes reflect the differential rotation rates illustrated in Table 3.1 (§3.2). Over a period of several years, during which the sunspot cycle will have noticeably developed, spot latitude measurements should show a gradual decrease of average spot latitudes. A 'butterfly diagram' like Fig. 1.6 can be constructed from many measurements. It is always exciting to see the first small spots at high latitudes (above 30° north or south) of the new cycle when they appear.

Observing the sun in the light of Hα used to be strictly in the professional province, but nowadays the cost of Hα filters is within the range of amateurs. Using them, prominences on the limb can be observed and their changes recorded.

Photography of the sun has been very successfully tried by some amateurs, though it is by no means an easy pursuit because of the sun's great intensity. Cameras with very short (less than 1/100 second) exposure times and slow (fine-grain) film must generally be used. Whole-sun photographs showing

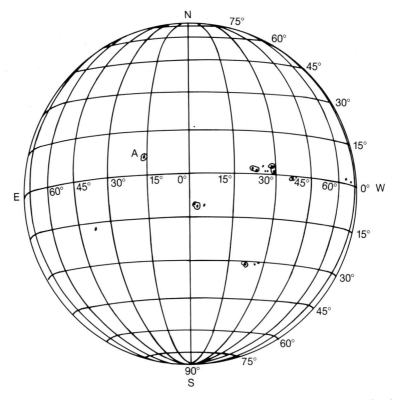

Fig. 10.22 A sunspot drawing with latitude and longitude grid appropriate for the value of B_0, $-5°$, on the day in question. Spot co-ordinates can be directly read off from the grid. Thus, spot A is approximately at latitude 6°N, longitude 17°E. If the value of L_0 were 150° for the time of observation, the Carrington longitude of spot A would be 133°.

large sunspots can be achieved using only a fine pinhole in a screen and a lens-less camera to record the image, but naturally the use of a telescope gives improved results. Projected solar images can be photographed using the set-up described earlier. In this, the projection screen used for visual observation of spots is replaced by a camera with its lens removed. To reduce the solar intensity, the objective can be stopped down and filters inserted. A camera with 35-mm film can be used to photograph the whole sun, but only with some difficulty: as only very small image sizes (up to 25 mm in diameter, the width of 35 mm film) can be obtained, exposure times must be very carefully judged to show sunspots. Details of spots can be more successfully photographed if the solar image size is much larger than 25 mm. The use of larger films or plates (e.g. film size 60×90 mm for, say, a 50-mm-diameter solar image) gives much improved results. Photographs of improved quality result when the solar image is directly photographed. The size of the image is increased using a negative or 'Barlow' lens in the optical path between the objective (lens or mirror) and focal point.

Amateur radio observations of the sun have been made using a variety of equipment, particularly Yagi aerials for low frequencies (metre wavelengths) but also parabolic reflectors for higher frequencies. The time variations of radio emission at a particular frequency can be plotted out with a pen-recorder connected to the receiver. Thus, type I noise storms and bursts from strong flares can be observed. Although radio spectral observations are not within the reach of most amateurs, the slow drift of type II bursts could conceivably be observed with two receivers operating at slightly different frequencies.

Epilogue: future directions

The rapid advances over the past 40 years or so, with space instruments, sophisticated ground-based telescopes and improved theoretical understanding, have still not enabled us to answer many fundamental questions about the sun. The following is a selection of questions that remain open but are probably near to some sort of resolution within the next decade or two. A future edition of this book may well give the answers.

(a) What the detailed structure of the sun's interior is, including the nature of large-scale convection flows (e.g. the giant cells) and whether the sun is rotating faster or more slowly with increasing depth.

(b) Whether there is a solar solution to the neutrino problem: if so, whether our conception of nuclear fusion as the source of the sun's energy needs to be modified.

(c) The nature, or even existence, of the solar dynamo to generate the magnetic fields that give rise to solar activity.

(d) The nature of photospheric fine structure (e.g. the exploding granules) and chromospheric fine structure (e.g. the cool carbon monoxide clouds and bright cell points) and the probably related problem of heating the chromosphere.

(e) Whether the corona is heated (and the solar-wind flow is driven) by dissipation of MHD waves, steady magnetic reconnection, nano-flaring or some other process.

(f) The rôle and manner of magnetic reconnection to provide the energy of flares.

The next few years should see more insight into the solar–stellar connection, i.e. how solar-like are the chromospheres, coronae, spots and flares on

360

especially the dMe stars. There could be increased understanding of the stellar phenomena from what we know about the sun, and a better appreciation of the context of solar phenomena.

Finally, the sun is, for all human purposes, a limitless source of energy, and in the next decade or two there is bound to be a much more widespread use of solar energy. There should be considerable advances in solar-cell technology, both of the photovoltaic and photoelectrochemical types, the use of thermal systems for heating buildings and the application of wind power. Undoubtedly governments and industry still do not appreciate how limited the resources of fossil fuels are becoming or what the impact of their continued consumption on the environment really will be, but it is clear that there will have to be major changes to alternative forms of energy very soon.

Appendix 1

Physical and astronomical constants

Physical and astronomical constants referred to in the text are listed here with their values in SI units. Other constants are given in the Preface, and statistics of the sun are listed in Table 8.1 (§8.1).

Physical constants

Velocity of light $(c) = 2.998 \times 10^8$ m/s
Mass of electron $= 9.109 \times 10^{-31}$ kg
Mass of proton $= 1.673 \times 10^{-27}$ kg
Radius of first Bohr orbit of hydrogen atom $= 5.29 \times 10^{-11}$ m
Gravitational constant $= 6.67 \times 10^{-11}$ N m^2/kg^2
Stefan–Boltzmann constant $(\sigma) = 5.67 \times 10^{-8}$ W/m^2/K^4

Astronomical constants

Astronomical unit, AU (mean earth–sun distance) $= 149\,597\,870$ km $= 215$ solar radii
Light-time for 1 AU (time for light to travel 1 AU) $= 499.0$ s
Parsec $= 3.08 \times 10^{13}$ km
Light-year $= 9.46 \times 10^{12}$ km
Earth's mass $= 5.976 \times 10^{24}$ kg
Earth's mean radius $= 6371.0$ km (equatorial radius is 6378.2 km)
Jupiter's equatorial radius $= 71\,300$ km (11.2 times the earth's)
Diameter of galaxy $= 25\,000$ parsecs
Mass of galaxy $= 1.4 \times 10^{11}$ times mass of sun
Sun's distance from centre of galaxy $= 10\,000$ parsecs

Appendix 2

Finding the heliographic co-ordinates of a sunspot

The polar co-ordinates of a sunspot, measured from a drawing, may be converted to heliographic co-ordinates as follows (see §10.4 for details). The sunspots are assumed to have been drawn on a disc with north point determined as described in §10.4. The values of P, B_0 and L_0 for the time of observation are then found from tables in *The Astronomical Almanac* or the *Handbook* of the British Astronomical Association. Figure A.1 illustrates the notation used in the following. The radius of the drawing, r_0, is measured (e.g. in mm). Using the north point of the drawing as the reference direction, the polar co-ordinates of a sunspot X are determined: these are the radial distance r of the spot from the centre of the disc O measured in the same units as r_0, and the position angle θ, i.e. the angle between the reference direction and the direction OX, measured from $0°$ to $360°$ in the sense north–east–south–west–north. With S the sun's angular radius (or semi-diameter), also obtained for the day of observation from *The Astronomical Almanac* or BAA *Handbook*, the angle ϱ is found from

$$\varrho = \arcsin(r/r_0) - Sr/r_0$$

The heliographic latitude B of the spot is then found from

$$\sin B = \cos\varrho \sin B_0 + \sin\varrho \cos B_0 \cos(P-\theta)$$

and the longitude difference ΔL between the spot and the central meridian from

$$\sin\Delta L = \sin(P-\theta)\sin\varrho / \cos B$$

The sign of ΔL is positive if the spot is west of the central meridian, negative if east. The Carrington longitude L of the spot can be found from

$$L = L_0 + \Delta L$$

The following computer program, written in BASIC language, will enable the values of B, ΔL and L to be found for a spot, as well as its angular distance from the sun's centre, given its polar co-ordinates r and θ, the measured radius r_0 and the values of P, B_0, L_0 and S.

363

PROGRAM COORDS

```
10   DEF FNASN(A) = ATN(A / SQR(1.0 − A * A))
20   INPUT "Type in value of P ", P
30   INPUT "Type in value of B0 ", B0
40   INPUT "Type in value of L0 ", L0
50   INPUT "Type in sun's semi-diam. (arcmins) ", S
60   INPUT "Type in radius (mm) of drawing ", R0
70   INPUT "Type in r (mm) of spot ", R
80   INPUT "Type in theta (degrees) of spot ", THETA
90   PMINTHR = (P − THETA) / 57.3
100  B0R = B0 / 57.3
110  RHO = FNASN(R / R0) − (R / R0) * S * 60.0 / 206265
120  SINB = COS(RHO) * SIN(B0R) + SIN(RHO) * COS(B0R) *
     COS(PMINTHR)
130  BR = FNASN(SINB)
140  SINDEL = SIN(PMINTHR) * SIN(RHO) / COS(BR)
150  DEL = 57.3 * FNASN(SINDEL)
160  L = L0 + DEL
170  IF L>360.0 THEN L = L − 360.0
180  B = 57.3 * BR
```

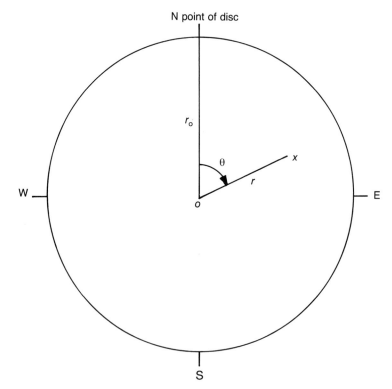

A.1 Diagram of a sunspot X on the sun to illustrate the notation used in text.

190   RHOD = RHO * 57.3
200   PRINT "Values of B, DEL, L for spot are ";B;DEL;L
210   PRINT "Values of rho (degrees) ";RHOD
220   INPUT "Calculate co-ords. of another spot on same drawing?
       (y/n) ";ANS$
230   IF ANS$="Y" OR ANS$="y" THEN GOTO 70 ELSE PRINT
       "End of calculation"

Glossary of selected terms

active region Region of the solar atmosphere, from the photosphere to the corona, associated with the emergence of a sub-photospheric magnetic field; sizes range from small, short-lived regions to large areas of weak field. Active regions usually comprise enhanced chromospheric and coronal emission, sometimes sunspots and filaments (prominences on limb), and are the site of flares.

Alfvén wave A wave motion occurring in a magnetized plasma in which the magnetic field lines oscillate in a transverse motion without a change in magnetic field strength. The tension in the field lines acts as a restoring force.

asteroseismology Study of global oscillations (helioseismology) as applied to stars.

aurora A display of light, usually seen from polar regions of the earth, characterized by red, green and other colours and rapidly changing forms, and associated with geomagnetic activity, initiated by particle emission from the sun.

auroral ovals Zones, in the shape of oval belts in near-polar latitudes, of maximum occurrence of aurorae.

bipolar region Region of the photosphere in which the magnetic field shows distinct leading and following magnetic polarities; includes sunspot regions, active regions without sunspots and small, ephemeral regions.

chromosphere Region of the solar atmosphere above the photosphere and temperature minimum characterized by a rise in temperature with height and much fine structure, mostly in the form of a network of bright patches.

366

convection zone Region comprising the outer one-third of the solar interior where energy generated at the core is transported by convection.

corona High-temperature region of the solar atmosphere above the chromosphere consisting of almost fully ionized plasma contained in closed magnetic field loops or of plasma expanding outwards along open magnetic field lines.

coronagraph Instrument for observing the corona in white light, with photospheric light blocked out by an occulting disc and with precautions for removing all traces of stray light.

coronal hole Region of the corona where magnetic field lines are open and there is very low X-ray and white-light emission arising from a low density of plasma. Coronal holes are the source of the high-speed streams of the solar wind.

coronal mass ejection Outward-moving cloud of coronal material often associated with erupting prominences, sometimes with flares (which occur after the start of the ejection).

dynamo A means by which the kinetic energy of an electrically conducting body is converted to magnetic energy.

effective temperature The sun's effective temperature is the temperature of a black body emitting the same amount of radiation as that emitted by the sun, and is equal to $5778\,K$.

Evershed flow A flow observed near sunspots, outwards at the level of the photosphere, but inwards and downwards at higher levels.

excitation Atomic process in which an atom or ion is raised to a higher-energy state by taking one of its electrons from one orbit to another further out.

facula Light region of the photosphere, associated with sunspots, seen near the limb; resolvable into facular bright points.

fibrils Dark elongated features seen in $H\alpha$ spectroheliograms in and around active regions forming a pattern, thought to delineate the chromospheric magnetic field.

filaments Long, dark features occurring near magnetic inversion lines which appear as (bright) prominences on the limb; they are masses of relatively cool and dense material suspended in the corona. They may be divided into quiescent and active-region (sunspot) types, with an active third group (including sprays, surges and loop prominences) associated with flares.

filigree   Fine structure visible in the photosphere, collectively forming the photospheric network.

flare   Explosive release of energy within an active region resulting in a wide range of phenomena in electromagnetic radiation, ejections of particles and mass, wave motions and shock waves, with durations of from minutes to a day.

flash spectrum   The emission-line spectrum due to the chromosphere seen at the beginning or end of totality during a solar eclipse.

granulation   A cellular pattern seen in white light in the photosphere. Individual granules are bright, often polygon-shaped regions, with average size about 1100 km and lifetime 18 minutes, separated by darker inter-granular lanes.

heliographic co-ordinates   Latitude and longitude of a feature on the sun.

helioseismology   Study of sound waves that propagate through the solar interior and manifest themselves by oscillations at the solar surface.

heliosphere   Region surrounding the sun where the solar wind flows, terminated by the heliopause (perhaps 50–100 AU from the sun) where the wind pressure balances that in interstellar space.

high-speed stream   Stream within the general solar wind having speed of up to 800 km/s, thought to originate from coronal holes.

hydrostatic equilibrium   In an atmosphere or stellar interior, a state of equilibrium in which the inward gravitational force of overlying material is balanced by the outward force of gas and radiation pressure.

insolation   The amount of solar energy received at a particular location on the earth's surface averaged over a period of time.

ion   An atom in which one or more electrons is removed.

ionization   Atomic process in which an electron is removed from an atom or ion.

ionosphere   Region of the earth's atmosphere which is partly ionized by the sun's ultraviolet and X-ray emission.

irradiance   Solar irradiance is the amount of solar energy received at the earth outside its atmosphere per unit area per unit time (SI units: W/m^2).

limb   The visible edge of a celestial object, e.g. the sun.

local thermodynamic equilibrium (LTE)   An assumption made in the study of stellar and solar atmospheres that the excitation and ionization of atoms and the radiation emitted by those atoms at some level of an atmosphere are determined by local values of temperature and density.

luminosity   The amount of energy radiated by a body per unit time (SI units: W).

magnetic inversion line   The line where the observed photospheric longitudinal (line-of-sight) magnetic field is zero (sometimes, but inaccurately, known as the neutral line).

magnetic storm and substorm   A magnetic storm is a disturbance in the earth's magnetic field observed all over the earth due to the passage of a high-speed stream in the solar wind. A substorm is a magnetic disturbance observed in polar regions only, and is associated with changes in direction of the interplanetary magnetic field as it encounters the earth.

magnetograph   Instrument for measuring the sun's (usually photospheric) magnetic field using the Zeeman effect. Normally, the longitudinal (line-of-sight) magnetic field is measured, but the transverse component is measured in addition with a vector magnetograph. A *magnetogram* is a map of measured magnetic field.

magnetohydrodynamic (MHD) wave   Wave which has the properties of sound (compressive) wave and non-compressive, transverse Alfvén wave.

magnetosphere   The magnetic cavity surrounding the earth or other planets with a magnetic field, formed by the interaction of the planetary magnetic field with solar wind as it flows past.

Maunder minimum   A period, lasting from 1645 to 1715, when there was an apparent dearth of sunspots and other indicators of solar activity.

network   Chromospheric and photospheric features arranged in a cellular structure, associated with supergranular cells and magnetic field.

neutrinos   Particles which have extremely small or zero mass (in which case they travel at the speed of light), emitted by certain nuclear reactions (especially radio-active decay).

penumbra   Outer part of a fully developed sunspot, consisting (in white light) of fine filaments which may be regular and radial or irregular and consist of small grains.

photosphere   In white light, the visible surface layers of the sun from where most of the sun's energy escapes to space.

plage   Bright emission seen especially in Hα and calcium H and K lines in active regions.

plasma   An ionized gas (i.e. composed largely of electrons and positive ions) which is electrically neutral over a sufficiently large volume.

pore   Small, short-lived dark area in the photosphere out of which a sunspot may develop.

prominence   See **filament**.

pyrheliometer   Instrument for measuring solar irradiance by heating effects.

radiative zone   The inner two-thirds of the solar interior in which the transport of energy generated at the sun's core is by radiative processes.

recombination   Atomic process in which an electron is joined to an ion or atom.

SNU   Solar neutrino unit, equal to 10^{-36} neutrino reactions per target atom per second.

solar cycle   A cyclical variation in solar activity, with period of about 11 years, measured by, e.g., the number of sunspots. There is a magnetic cycle of 22 years.

solar wind   A flow of particles – mainly electrons and protons – from the sun into interplanetary space, carrying with it the interplanetary magnetic field.

source function   Ratio of the emission coefficient (amount of radiation emitted per unit time per unit volume of a gas) to the absorption coefficient (fraction of radiation subtracted from a beam per unit distance travelled through the absorbing material).

spectrograph   Instrument for producing a photograph of the spectrum of an object. A *spectrometer* is the equivalent instrument in which a photoelectric device measures the spectrum, and with a *spectroscope* the spectrum is viewed visually.

spectroheliograph   Instrument for taking photographs (spectroheliograms) of the sun at specific wavelengths, particularly of strong Fraunhofer lines (in which case the chromosphere is imaged).

spectrohelioscope   Similar to a spectroheliograph except that the image is viewed by the eye.

spicule   Spike-like jet seen in the upper part of the solar chromosphere, either at the limb or in the spectroheliograms taken especially in the red wing of the Hα line.

spray   See **filament**.

standard solar model   Mathematical description of the solar interior, giving the variation with radius of, e.g., temperature and pressure, based on observations and physical laws.

sudden ionospheric disturbance   Disturbance in the earth's ionosphere as a result of a burst of X-rays or ultraviolet radiation during flares which produces extra ionization.

sunspot   Dark area in the photosphere with temperatures cooler than the general photosphere, associated with strong magnetic fields. Spots appear in pairs and groups with a general bipolar pattern. They appear as dark structures in Hα light and have depressed amounts of coronal X-ray emission above them.

supergranulation   A convection pattern consisting of large (30 000 km diameter) cells in which flow is mostly horizontal and outward from cell centres but which has weak upward flow at cell centres and downward flow at cell boundaries.

surge   See **filament**.

torsional oscillations   Zones of alternating fast and slow rotation appearing in the photosphere, moving from poles to equator.

transition region   Region of the solar atmosphere between the chromosphere and the corona characterized by a large rise of temperature from about 20 000 K to 2 000 000 K.

umbra   The inner region of a fully developed sunspot; sunspots without penumbra consist of only umbra. Characterized by an effective temperature of 4160 K and strong (0.4 T) magnetic field.

unipolar region   Large area, often to the poleward side of the sunspot belt, showing weak magnetic field of a single polarity.

Wilson effect   The effect by which the umbra of a regular sunspot appears displaced towards the sun's centre as the spot approaches the limb; supposedly due to a depression in the umbral level.

X-ray bright point   Small X-ray-emitting region in the corona associated with a bipolar magnetic region.

zodiacal light   Faint cone of light seen in the evening or morning sky, especially from low latitudes, which is an extension of the F (dust) component of the solar corona.

Bibliography

PREFACE

ALLEN, C. W. *Astrophysical Quantities* (3rd edition). London: The Athlone Press, 1973.

SI *The International System of Units* (4th edition). London: HMSO, 1982.

CHAPTER 1

ABETTI, G. *The Sun* (2nd edition). London: Faber & Faber, 1963.

BRAY, R. J. AND LOUGHHEAD, R. E. *Sunspots*. New York: Dover Publications, 1964.

CHAPMAN, S. *Solar Plasma, Geomagnetism and Aurora*. New York: Gordon & Breach, 1964.

EATHER, R. H. *Majestic Lights: The Aurora in Science, History, and the Arts*. Washington, DC: American Geophysical Union, 1980.

EDDY, J. A. The Maunder minimum. Article in *Science*, vol. 192, p. 1189, 1976.

HINE, A. *Magnetic Compasses and Magnetometers*. Toronto: University of Toronto Press/Adam Hilger, 1968.

MEADOWS, A. J. *Early Solar Physics*. Oxford: Pergamon Press, 1970.

STEPHENSON, F. R. Historical eclipses. Article in *Scientific American*, vol. 247, October 1982.

THACKERAY, A. D. *Astronomical Spectroscopy*. London: Eyre & Spottiswoode, 1961.

YAU, K. K. C. AND STEPHENSON, F. R. A Revised Catalogue of Far Eastern Observations of Sunspots. Article in *Quarterly Journal of the Royal Astronomical Society*, vol. 29, p. 175, 1988.

CHAPTER 2

BAHCALL, J. N. *Neutrino Astrophysics*. Cambridge University Press, 1989.

EDDY, J. A. (ed.). *The New Solar Physics*. AAAS Selected Symposia series. Boulder, Colo.: Westview Press, 1978. Chapters by Raymond Davies and John C. Evans (p. 35) and Henry A. Hill (p. 135).

LEIBACHER, J. W., NOYES, R. W., TOOMRE, J. AND ULRICH, R. K. Helioseismology. Article in *Scientific American*, vol. 253, September 1985.

PARKER, E. N. Magnetic fields in the cosmos. Article in *Scientific American*, vol. 249, August 1983.

SCHWARZSCHILD, M. *Structure and Evolution of the Stars*. New York: Dover Publications, Inc., 1958.

CHAPTER 3

ALLER, L. H. *Atoms, Stars, and Nebulae* (3rd edition). Cambridge University Press, 1991.

BRAY, R. J., LOUGHHEAD, R. E. AND DURRANT, C. J. *The Solar Granulation* (2nd edition). Cambridge University Press, 1984.

GIBSON, E. G. *The Quiet Sun*. NASA Special Publication SP-303, 1973.

ZIRIN, H. *Astrophysics of the Sun*. Cambridge University Press, 1988.

CHAPTER 4

BRAY, R. J. AND LOUGHHEAD, R. E. *The Solar Chromosphere*. Cambridge University Press, 1984.

GIBSON, E.G. *The Quiet Sun*. NASA Special Publication SP-303, 1973.

ZIRIN, H. *Astrophysics of the Sun*. Cambridge University Press, 1988.

CHAPTER 5

GIBSON, E. G. *The Quiet Sun*. NASA Special Publication SP-303, 1973.

KUNDU, M. R. *Solar Radio Astronomy*. New York: John Wiley & Sons, 1965.

SHKLOVSKII, I. S. *Physics of the Solar Corona* (2nd edition). Translated by L. A. Fenn. Oxford: Pergamon Press, 1965.

ZIRIN, H. *Astrophysics of the Sun*. Cambridge University Press, 1988.

ZIRKER, J. B. (ed.). *Coronal Holes and High-Speed Wind Streams: A Monograph from* Skylab *Solar Workshop I*. Colorado Associated University Press, 1977.

CHAPTER 6

ABETTI, G. *The Sun*. London: Faber & Faber, 1962.

BRAY, R. J. AND LOUGHHEAD, R. E. *Sunspots*. New York: Dover Publications, 1964.

KUNDU, M. R. *Solar Radio Astronomy*. New York: Interscience Publishers, John Wiley & Sons, 1965.

NOYES, R. W. *The Sun, our Star*. Cambridge, Mass.: Harvard University Press, 1982.

ORRALL, F. Q. (ed.). *Solar Active Regions: A Monograph from* Skylab *Solar Workshop III*. Colorado Associated University Press, 1981.

STURROCK, P. A. (ed.). *Solar Flares: A Monograph from* Skylab *Solar Workshop II*. Colorado Associated University Press, 1980.

TANDBERG-HANSSEN, E. *Solar Prominences*. Dordrecht, Holland: D. Reidel Publishing Co., 1974.

ZIRIN, H. *Astrophysics of the Sun*. Cambridge University Press, 1988.

Highlights of the *Solar Maximum Mission* results are given by K. J. H. Phillips, *The Solar Maximum Mission, 1980–89*, published in *Astronomy Now*, vol. 4, no. 7, July 1990. Technical results are summarized in *Energetic Phenomena on the Sun – The* Solar Maximum Mission *Flare Workshop Proceedings* (M. Kundu and B. Woodgate, eds.). NASA Conference Publication no. 2439, 1986.

CHAPTER 7

BARRY, R. G. AND CHORLEY, R. J. *Atmosphere, Weather and Climate*. London and New York: Methuen, 1982.

BRANDT, J. C. AND CHAPMAN, R. D. *Introduction to Comets*. Cambridge University Press, 1981.

EDDY, J. A. (ed.). *The New Solar Physics* (American Association for the Advancement of Science Selected Symposia Series). Boulder, Colo.: Westview Press, 1978. Article by A. J. Hundhausen, p. 59.

HUNDHAUSEN, A. J. *Coronal Expansion and Solar Wind*. Berlin: Springer-Verlag, 1972.

RATCLIFFE, J. A. *An Introduction to the Ionosphere and Magnetosphere*. Cambridge University Press, 1972.

STOLARSKI, R. S. The Antarctic ozone hole. Article in *Scientific American*, vol. 258, January 1988.

CHAPTER 8

ALLER, L. H. *Atoms, Stars, and Nebulae* (3rd edition). Cambridge University Press, 1991.

BOHM-VITENSE, E. *Introduction to Stellar Astrophysics*. Cambridge University Press, 1989. Volume 1: *Basic Stellar Observations and Data*. Volume 2: *Stellar Atmospheres*.

EDDINGTON, A. S. *Internal Constitution of the Stars*. Cambridge University Press, 1988.

KAPLAN, S. A. *The Physics of the Stars* (translated by R. Feldman). Chichester, UK: John Wiley & Sons, 1982.

MEADOWS, A. J. *Stellar Evolution* (2nd edition). Oxford: Pergamon Press, 1978.

SHKLOVSKII, I. S. *Stars: Their Birth, Life, and Death*. San Francisco: W. H. Freeman & Co., 1978.

CHAPTER 9

FAHRENBRUCH, A. L. AND BUBE, R. H. *Fundamentals of Solar Cells*. New York: Academic Press, 1983.

HAMIKAWA, Y. Photovoltaic power. Article in *Scientific American*, vol. 256, April 1987.

WEINBERG, C. J. AND WILLIAMS, R. H. Energy from the sun. Article in *Scientific American*, vol. 263, September 1990.

WILSON, J. I. B. *Solar Energy*. London: Wykeham Publications, 1979.

CHAPTER 10

BAXTER, W. M. *The Sun and the Amateur Astronomer*. Newton Abbot, Devon: David & Charles, 1973.

HEYWOOD, J. *Radio Astronomy Simplified*. London: Arco Publications, 1963.

KRISCIUNAS, K. *Astronomical Centers of the World*. Cambridge University Press, 1988.

LEARNER, R. *Astronomy through the Telescope*. New York: Van Nostrand Reinhold, 1981.

RONAN, C. A. *Invisible Astronomy*. London: Eyre & Spottiswoode, 1969.

ROTH, G. D. (ed.). *Astronomy: a Handbook* (translated by A. Beer). New York: Springer-Verlag, 1975.

TAYLOR, P. O. *Observing the Sun*. Cambridge University Press, 1991.

ZIRIN, H. *Astrophysics of the Sun*. Cambridge University Press, 1988.

The Astronomical Almanac. London: HMSO and Washington, DC: US Government Printing Office. Published annually.

Handbook of the British Astronomical Association. Published annually by the British Astronomical Association, available from the BAA, Burlington House, Piccadilly, London W1V 9AG.

Advice on solar observing for amateurs may be obtained from the Solar Section of the BAA at the above address.

References

The following are full references to published works mentioned in figure captions and tables.

ALLEN, C. W. 1973, *Astrophysical Quantities*, 3rd edn, The Athlone Press, London.

ARNAUD, M. AND ROTHENFLUG, R. 1985, *Astron. and Astrophys. Suppl. Ser.*, **60**, 425.

ATHAY, R. G. 1981, in *The Sun as a Star* (ed. Jordan), CNRS, Paris, and NASA, Washington, DC, Ch. 4.

BAHCALL, J. N. 1989, *Neutrino Astrophysics*, Cambridge University Press.

BALTHAZAR, H. *et al.* 1986, *Astron. and Astrophys.*, **155**, 87.

BAME, S. J. *et al.* 1976, *Astrophys. J.*, **207**, 977.

BIEBER, J. W., SECKEL, D., STANEV, T. AND STEIGMAN, G. 1990, *Nature*, **348**, 407.

BIRD, G. A. 1964, *Astrophys. J.*, **140**, 288.

BLACKWELL, D. E., DEWHIRST, D. H. AND INGHAM, M. F. 1967, *Adv. Astron. Astrophys.*, **5**, 1.

BROMAGE, G. E., PHILLIPS, K. J. H., DUFTON, P. L. AND KINGSTON, A. E. 1968, *Mon. Not. R. Astron. Soc.*, **220**, 1021.

CANFIELD, R. C., DE LA BEAUJARDIÈRE, J.-F. AND LEKA, K. D. 1991, *Phil. Trans. Roy. Soc., Lond. A*, **336**, 381.

CHAPMAN, S. AND BARTELS, J. 1962, *Geomagnetism*, Oxford University Press.

CHENG, C.-C. *et al.* 1988, *Astrophys. J.*, **330**, 480.

CHUPP, E. L. 1984, *Ann. Rev. Astr. Astrophys.*, **22**, 359.

CHUPP, E. L. *et al.* 1982, *Astrophys. J. (Letters)*, **263**, L95.

DENNIS, B. R. 1985, *Solar Phys.*, **100**, 465.

DENNIS, B. R. AND SCHWARTZ, R. A. 1989, *Solar Phys.*, **121**, 75.

DERE, K. P., BARTOE, J.-D. AND BRUECKNER, G. E. 1989, *Solar Phys.*, **123**, 41.

378

DUVALL, T. L. AND HARVEY, J. W. 1983, *Nature*, **302**, 24.

EDDY, J. A. 1978, in *The New Solar Physics* (ed. Eddy), AAAS Selected Symposium no. 17, Westview Press, Boulder, Colo., Ch. 2.

FARMAN, J. C., GARDINER, B. G. AND SHANKLIN, J. D. 1985, *Nature*, **315**, 207.

FRANK, L. A. AND CRAVEN, J. D. 1988, *Rev. Geophys.*, **26**, 249.

FURTH, H. P., KILLEEN, P. H. AND ROSENBLUTH, M. N. 1963, *Phys. Fluids*, **6**, 459.

GABRIEL, A. H. 1976, *Phil. Trans. Roy. Soc., Lond. A*, **281**, 339.

GABRIEL, A. H. *et al.* 1971, *Astrophys. J.*, **169**, 595.

GOSLING, J. T. AND MCCOMAS, D. J. 1987, *Geophys. Res. Letts*, **14**, 355.

GREVESSE, N. 1984, *Physica Scripta*, **T8**, 49.

HERTZBERG, L. 1965, in *Physics of the Earth's Upper Atmosphere* (eds. Hines, Paghis, Hartz and Fejer), Prentice-Hall, London.

HEYVAERTS, J., PRIEST, E. R. AND RUST, D. M. 1977, *Astrophys. J.*, **216**, 123.

HOUGHTON, J. T. 1986, *The Physics of Atmospheres*, Cambridge University Press.

HOUGHTON, R. A. AND WOODWELL, G. M. 1989, *Scientific American*, April issue.

HOWARD, R., TANENBAUM, A. D. AND WILCOX, J. M. 1968, *Solar Physics*, **4**, 286.

HOWARD, R. A., MICHELS, D. J., SHEELEY, N. R., JR, AND KOOMEN, M. J. 1982, *Astrophys. J.*, **263**, L101.

HUNDHAUSEN, A. J. 1987, in *Proceedings of the Sixth International Solar Wind Conference*, Estes Park, 1987 (eds. Pizzo, Holzer and Sime), NCAR Technical Note no. TN-306, Boulder, Colo., p. 181.

ICHIMOTO, K. AND KUROKAWA, H. 1984, *Solar Phys.*, **93**, 105.

JORDAN, C. AND WILSON, R. 1971, in *Physics of the Solar Corona*, D. Reidel, Dordrecht.

KAPLAN, S. A. 1982, *The Physics of Stars*, John Wiley & Sons, Inc., New York.

KUNDU, M. R. 1965, *Solar Radio Astronomy*, John Wiley & Sons, Inc., New York.

LABS, D. AND NECKEL, H. 1968, *Zeitschrift f. Astrophys.*, **69**, 1.

LEIBACHER, J. W., NOYES, R. W., TOOMRE, J. AND ULRICH, R. K. 1985, *Scientific American*, September issue.

LEMAIRE, P. AND SKUMANICH, A. 1973, *Astron. and Astrophys.*, **22**, 61.

LEMAIRE, P. *et al.* 1978, *Astrophys. J.*, **223**, L55.

LINSKY, J. L. 1985, *Solar Phys.*, **100**, 333.

MCINTOSH, P. S. 1990, *Solar Phys.*, **125**, 251.

MEYER, J.-P. 1985, *Astrophys. J. Suppl. Ser.*, **57**, 151.

NEWKIRK, G. 1971, in IAU Symposium No. 43 on *Solar Magnetic Fields* (ed. Howard), D. Reidel, Dordrecht, p. 547.

NICOLAS, K. R. *et al.* 1981, in *The Physics of Sunspots* (eds. Cram and Thomas), AURA, Sunspot, New Mexico, p. 167.

PARKER, E. N. 1963, *Interplanetary Dynamic Processes*, Interscience Publishers, New York.

PARKINSON, J. H., MORRISON, L. V. AND STEPHENSON, F. R. 1980, *Nature*, **288**, 548.

PHILLIPS, K. J. H. AND KEENAN, F. P. 1990, *Mon. Not. R. Astron. Soc.*, **245**, 4P.

PIDDINGTON, J. H. 1976, in IAU Symposium No. 71 on *Basic Mechanisms of Solar Activity* (eds. Bumba and Kleczek), D. Reidel, Dordrecht, p. 395.

PNEUMAN, G. W. AND KOPP, R. A. 1971, *Solar Phys.*, **18**, 258.

PRIEST, E. R. 1972, *Solar Magnetohydrodynamics*, D. Reidel, Dordrecht.

RIDDLE, A. C. 1970, *Solar Phys.*, **13**, 448.

ROĬZMAN, G. SH. AND SHEVCHENKO, V. S. 1982, *Sov. Astron. Letts*, **8** (2), 85.

SAITO, K. 1970, *Sky and Telesc.*, August issue.

SHEELEY, N. R. 1980, *Solar Phys.*, **66**, 79.

SNODGRASS, H. B. 1984, *Solar Phys.*, **94**, 13.

SPICER, D. S. 1977, *Solar Phys.*,, **53**, 305.

SPRUIT, H. C. 1981, in *The Physics of Sunspots* (eds. Cram and Thomas), AURA, Sunspot, New Mexico, p. 98.

STURROCK, P. A. 1980, in *Solar Flares* (ed. Sturrock), Colorado Associated University Press.

TOKUMARU, M. *et al.* 1991, *J. Geomag. Geoelect.*, **43**, 619.

VAN LOON, H. AND LABITZKE, K. 1983, *New Scientist* (8 September issue).

VANSPEYBROECK, L. P., KRIEGER, A. S. AND VAIANA, G. S. 1970, *Nature*, **227**, 818.

VERNAZZA, J. E., AVRETT, E. H. AND LOESER, R. 1981, *Astrophys. J. Suppl. Ser.*, **45**, 635.

WALTER, F. M. *et al.* 1983, *Activity in Red-Dwarf Stars* (eds. Byrne and Rodonò), D. Reidel, Dordrecht, p. 445.

WHITE, O. R. 1964, *Astrophys. J.*, **140**, 1164.

WHITE, O. R. AND LIVINGSTON, W. C. 1981, *Astrophys. J.*, **249**, 798.

XUEFU, L. AND CHENGZHONG, L. 1983, *Activity in Red-Dwarf Stars* (eds. Byrne and Rodonò), D. Reidel, Dordrecht, p. 485.

Index

381